Cell Signalling

Cell Signalling

SECOND EDITION

John T. Hancock

Reader in Molecular Biology
School of Biosciences
Faculty of Applied Sciences
University of the West of England, Bristol

OXFORD
UNIVERSITY PRESS

OXFORD

UNIVERSITY PRESS

Great Clarendon Street, Oxford OX2 6DP

Oxford University Press is a department of the University of Oxford.
It furthers the University's objective of excellence in research, scholarship,
and education by publishing worldwide in

Oxford New York

Auckland Cape Town Dar es Salaam Hong Kong Karachi
Kuala Lumpur Madrid Melbourne Mexico City Nairobi
New Delhi Shanghai Taipei Toronto

With offices in

Argentina Austria Brazil Chile Czech Republic France Greece
Guatemala Hungary Italy Japan South Korea Poland Portugal
Singapore Switzerland Thailand Turkey Ukraine Vietnam

Oxford is a registered trade mark of Oxford University Press
in the UK and in certain other countries

Published in the United States
by Oxford University Press Inc., New York

A catalogue record for this title is available from the British Library

Library of Congress Cataloging in Publication Data
(Data available)
ISBN 978-0-19-926467-4   24319244

10 9 8 7 6 5 4 3

Typeset by Newgen Imaging Systems (P) Ltd., Chennai, India
Printed in Great Britain
on acid-free paper by
Ashford Colour Press Ltd, Gosport, Hampshire

To James,

who died before we had a chance to get to know him,
and of course to

Annabel and Thomas,

the best kids a dad could ask for.

CONTENTS

This is the second edition of *Cell Signalling*, and it hopefully will be useful for those who wish to find out more about the subject. The original book was written with the intention that it would be useful for degree students in their second and final years of undergraduate studies. It was also hoped that the book would be of interest to those studying for a medical degree, or those embarking on postgraduate studies, where they were perhaps new to this area or needed to refresh earlier studies. It is therefore hoped that for the second edition this vein has continued. However, in rewriting the text, many new diagrams have been added, with colour included to highlight the important aspects of the figures, and the text altered in an attempt to make it more readable and accessible, perhaps at the detriment of added details. Also, many of the original references have been removed from the Further reading sections, but new ones, mostly reviews, added. Therefore, for some of the primary sources used to write the original book, the reader is referred to the first edition.

The field of cell signalling continues to be very large and to expand at what seems like an exponential rate, with a great deal of new literature published in the journals every month. Here the aim was to describe the main components used by cells in their communications, with discussion on the ways in which they interact. This is not an approach taken by many books which cover this area of biochemistry, but it is becoming apparent that pathways are not just single lines of signal transducers leading simply from the original signal to the cellular response, but an intricate interplay of many components, with one pathway having profound effects on another. Therefore the discussion of such components in relative isolation allows the knowledge gained about one particular component to be used when considering many pathways, not just the one in which it was first discovered. The aim was also to have a generic approach, as the ubiquity of some of these systems means that an understanding of the mechanisms is relevant to all organisms throughout nature and not just mammals or indeed the animal kingdom.

Once again, it has proved impossible to do justice to every area in this field and the overall emphasis still reflects my interests, so some may feel that I have overlooked their particular area of research. To those that feel ignored I apologize.

Chapter 1 contains further reading which includes several excellent chapters published in major textbooks covering cell signalling, as well as other books published on this topic, and as such would be a good place to start further reading.

The book starts with a overview chapter, encompassing many of the ideas and discussions which could be applicable to all areas of cell signalling. Chapters 2 and 3 concentrate on the use of extracellular signals and their perception, while Chapters 4–8 concentrate on the intracellular components which are thought to be of most importance. Chapters 9–11 take more specific examples of signalling pathways and try to show how these components might come together to form complete pathways, including a more-extensive coverage of the process of apoptosis in a new chapter (Chapter 11). Another new chapter at the end (Chapter 12) brings some ideas and thoughts together, and summarizes some of the discussion from the rest of the book.

As with the original, the content of the book was greatly shaped by many conversations and e-mails with colleagues. I am indebted to many of my colleagues at the University of the West of England (UWE), Bristol, such as Richard Osborne who read chunks of chapters for me, and collaborators elsewhere. In particular I would like to thank Steve Neill, who many years ago instigated a new course which inspired the original book, and has remained a valuable colleague and friend, and Radhika Desikan, who has encouraged me along the way and dropped many papers on my desk! I would also like to thank those with Bristol connections who have helped over the years: including David Barford, Jeremy Tavaré, and Guy Rutter. Thanks also go to Chris Shorland and his team for ICT support. I must also thank my editor, Jonathan Crowe, who not only persuaded me to do the second edition, but who has made very useful comments along the way. Lastly I would like to again thank my wife Sally-Ann, for her never-ending help and, not least, encouragement.

■ ABBREVIATIONS

AA	arachidonic acid
ABA	abscisic acid
ADP	adenosine diphosphate
AF-1	activation function 1
AF-2	activation function 2
AIF	apoptosis-inducing factor
AP-1	activator protein 1
Apaf-1	apoptotic protease-activating factor 1
βARK	β-adrenergic receptor kinase
AMP	adenosine monophosphate
AMPK	5′-AMP-activated protein kinase
AOS	active oxygen species
ARF	ADP-ribosylation factor
ATP	adenosine triphosphate
Boss	Bride of Sevenless
bp	base pair
cADPr	cyclic ADP-ribose
$[Ca^{2+}]_i$	intracellular calcium ion concentration
cAMP	cyclic adenosine monophosphate (adenosine 3′,5′-cyclic monophosphate)
cAPK	cAMP-dependent protein kinase (PKA)
CAP	Cbl-associated protein (see Chapter 9)
CAP	catabolite gene-activator protein (see Chapter 4 and 5)
CAPP	ceramide-activated protein phosphatase
CARD	caspase-activation and -recruitment domain
caspase	cysteinyl-aspartate-specific protease
CCE	capacitative Ca^{2+} entry
cCMP	cyclic cytidine monophosphate (cytidine 3′,5′-cyclic monophosphate)
cDNA	complementary DNA
CGD	chronic granulomatous disease
cGMP	cyclic guanosine monophosphate (guanosine 3′,5′-cyclic monophosphate)
cGPK	cGMP-dependent protein kinase
cIMP	cyclic inosine monophosphate
CK	cytokinin
CRADD	caspase and RIP adaptor with death domain
CRD	cysteine-rich domain
CRE	cAMP-response element

CREB	CRE-binding protein
CRP	cAMP receptor protein
DAF-2DA	diaminofluorescein diacetate
DAG	diacylglycerol
DBD	DNA-binding domain
DED	dead effector domain
DGK	diacylglycerol kinase
DISC	death-inducing signalling complex
DNA	deoxyribonucleic acid
DR	death receptor
EDRF	endothelium-derived relaxing factor (now known to be NO)
EET	epoxyeicosatrienoic acids
EGF	epidermal growth factor
ER	endoplasmic reticulum
ERK	extracellular signal-regulated kinase
ESI-MS	electrospray ionization mass spectroscopy
FAD	flavin adenine dinucleotide
FADD	Fas-associated protein with death domain
FGF	fibroblast growth factor
FLIM	fluorescence lifetime imaging microscopy
FLIP	Flice-like inhibitory protein
FMN	flavin mononucleotide
FRET	fluorescence resonance energy transfer
FSH	follicle-stimulating hormone
GA	gibberellin
GABA	γ-aminobutyric acid
GAP	GTPase-activating protein
GDP	guanosine diphosphate
GEF	guanine nucleotide-exchange factor
GFP	green fluorescent protein
GM-CSF	granulocyte-macrophage colony-stimulating factor
GMP	guanosine monophosphate
GNRP	guanine nucleotide-releasing protein
GRK	G protein-coupled receptor kinase
GSH	glutathione (reduced)
GSNO	S-nitrosoglutathione
GSSG	glutathione (oxidized)
GTP	guanosine triphosphate
GTPγ	guanosine 5'-[γ-thio] triphosphate
HCR	haem-controlled repressor
HETE	hydroxyeicosatetraenoic acids
HPETE	hydroperoxyeicosatetraenoic acids
HRE	hormone-response element

Hsp	heat-shock protein
IAA	indole-3-acetic acid
IAP	inhibitor of apoptosis protein
IFN	interferon
IGF	insulin-like growth factor
IL	interleukin
$InsP_3$	inositol 1,4,5-triphosphate (IP_3)
$InsP_3R$	inositol 1,4,5-triphosphate receptor
$InsP_4$	inositol 1,3,4,5-tetrakisphosphate (IP_4)
$InsP_6$	inositol hexaphosphate (IP_6)
IP_3	inositol 1,4,5-trisphosphate
IP_4	inositol 1,3,4,5-tetrakisphosphate
IP_6	inositol hexaphosphate
IRS1	insulin receptor substrate 1
ISPK	insulin-sensitive protein kinase or insulin-stimulated protein kinase
JA	jasmonic acid
JAK	Janus kinase
JIPS	jasmonate-induced proteins
LBD	ligand-binding domain
LH	luteinizing hormone
LHC	light-harvesting chorophyll a/b complex
L-NAA	L-N^ω-aminoarginine
L-NMA	L-N^ω-methylarginine
LPA	lysophosphatidic acid
LPC	lysophosphatidylcholine
LPS	lipopolysaccharide
MALDI-TOF	matrix-assisted laser desorption ionization-time-of-flight
MAP kinase	mitogen-activated protein kinase
MAPKK	mitogen-activated protein kinase kinase
MAPKKK	mitogen-activated protein kinase kinase kinase (MEK kinase)
MEK	MAP/ERK kinase
MEKK	MEK kinase
MLCK	myosin light-chain kinase
mRNA	messenger RNA
$NAADP^+$	nicotinate adenine dinucleotide phosphate
NAD^+	nicotinamide adenine dinucleotide (oxidized)
NADH	nicotinamide adenine dinucleotide (reduced)
$NADP^+$	nicotinamide adenine dinucleotide phosphate (oxidized)
NADPH	nicotinamide adenine dinucleotide phosphate (reduced)

NF-κB	nuclear factor κB
NMR	nuclear magnetic resonance
NO$^\bullet\Sigma$	nitric oxide radical
NOS	nitric oxide synthase
PA	phosphatidic acid
PAGE	polyacrylamide gel electrophoresis
PAP	phosphatidate phosphohydrolase
PC	phosphatidylcholine
PCD	programmed cell death
PCR	polymerase chain reaction
PDE	phosphodiesterase
PDGF	platelet-derived growth factor
PDK1	3-phosphoinositide-dependent kinase
PE	phosphatidylethanolamine
PGHS	prostaglandin G/H synthase (cyclooxygenase)
PH domain	Pleckstrin homology domain
P_i	Inorganic phosphate
PI	phosphatidylinositol
PIP_2	phosphatidylinositol 4,5-bisphosphate
PKA	cAMP-dependent protein kinase
PKB	protein kinase B
PKC	protein kinase C
PLA_2	phospholipase A_2
PLC	phospholipase C
PLD	phospholipase D
PMCA	plasma membrane Ca^{2+}-ATPase
PtdIns	phosphatidylinositol (PI)
PtdIns 3-kinase	phosphoinositide 3-kinase
PtdIns 4-kinase	phosphoinositide 4-kinase
PtdIns4P	phosphatidylinositol 4-phosphate
PtdIns P_2	phosphatidylinositol 4,5-bisphosphate (PIP_2)
PtdIns P_3	phosphatidylinositol 3,4,5-trisphosphate
PTP	protein tyrosine phosphatase
RNA	ribonucleic acid
RNAi	RNA interference
RNS	reactive nitrogen species
ROS	reactive oxygen species
RTK	receptor tyrosine kinase
RyR	ryanodine receptor
SDS	sodium dodecyl sulphate
SERCA	smooth endoplasmic reticulum calcium ATPase
SH2 domain	Src homology domain 2
SH3 domain	Src homology domain 3
siRNA	small interfering RNA

SM	sphingomyelin
SOD	superoxide dismutase
Sos	Son of Sevenless
SR	sarcoplasmic reticulum
STAT	*s*ignal *t*ransducers and *a*ctivators of *t*ranscription
TGF	transforming growth factor
TNF	tumour necrosis factor
TNF-R	tumour necrosis factor receptor
Trail	tumour necrosis factor-related apoptosis-inducing ligand
VSP	vegetative storage protein
XOR	xanthine oxidoreductase

1

Aspects of cellular signalling

Cell signalling has become a vital and integral part of modern biology, and has an innate complexity. It controls the inner workings of organisms, allowing them to respond, adapt, and survive. However, the basic workings of cell signalling events are not vastly diverse across different organisms, but rather the regulatory needs of organisms' cells are similar. Principles and mechanisms can be seen to be repeated across the kingdoms of species. The molecules which constitute the signals, the ways that they travel to their targets, and the ways that they are perceived are similar, whether the organism is a plant or an animal. The details may be different, and there are many unique aspects that can be seen to break the rules, but in Chapter 1 some of the more overlying aspects are discussed which should help in the understanding of the chapters to follow.

With similar mechanisms and chemistry being used by a wide variety of organisms, tissues, and cells, it is of no surprise that the techniques used for their study are also similar. A technique used to elucidate the signalling cascades of a mouse liver can often be used to study a plant leaf. Some of the common techniques are discussed here.

A brief history is also included, but unfortunately not all landmarks can be discussed. The study of cell signalling has been continuing for over a century, and still seems to be increasing at an alarming rate, but many of the names and events are discussed briefly here, hopefully giving the reader a taste of the immense amount of work which this book tries to cover.

1.1 Introduction

Cell signalling has arguably become one of the most important aspects of modern biochemistry and cell biology. An understanding of aspects of cell signalling is vital to a wide range of biologists, from those who are investigating the causes of cancer to those who are concerned about the impact of environmental pollutants on the ecosystem.

All cells, whether they live as individuals or in a multicellular organism, are bombarded by signals in many forms in a continual manner. It is the ability of organisms, or individual cells within an organism, to sense and respond to their environment that is crucial to their survival. All cells must have the ability to detect the presence of extracellular molecules and conditions, and must also be able to instigate a range of intracellular responses. Such systems have to be carefully orchestrated and controlled. To enable living organisms to do this a complex and interwoven range of signalling pathways has evolved. Signalling systems of a single-cell organism are complex enough, but a multicellular organism has to coordinate the functioning of cells which may be adjacent or a huge distance apart, possibly in the order of metres. Such cells may also have extremely specific and different functions.

The main principles and components behind the signalling mechanisms are essentially the same across the diverse range of organisms; bacteria, fungi, plants and animals. It is the fine detail of the components that gives the specificity or unique functionality which is required for a particular cell. In today's research, the knowledge and understanding of a signalling system in one tissue or species is often used to speed the discovery of an analogous system in a completely different tissue or species. For example, systems discovered and characterized in mammals are now sought in plants, with the tools used for the original research being adopted for this new use. This can greatly enhance the rate of new discoveries but, to add a note of caution, may also throw up some anomalies.

Cell signalling is not only important for the understanding of the functioning of a normal cell, but is of vital importance to understand the growth and activity of an aberrant cell, or that of a cell that is combating adverse conditions. The discovery of oncogenes, genes which cause the uncontrolled growth of cells which may lead to cancerous growths, was heralded as a major breakthrough in the understanding of cancer in humans while the discovery of cytokines held great hopes to be a cure for a variety of diseases. It is the role and impact of such gene products and blood-borne molecules on the cell signalling pathways that ensued in cells which made them so important, and which highlights the importance of an understanding of cell signalling mechanisms to modern biology.

The literature in the field of cell signalling is still growing exponentially, with several journals dedicated to the publication of research in this field. Other journals specifically highlight papers on cell signalling, either in the journal itself or on web pages and in e-mail alerts. This is because once the functioning of a protein is elucidated, it is often the control of its activity and the control of its synthesis to which the researcher turns to get a better understanding of how the protein fits into the overall biochemistry of the cell. The literature is expanding at such an alarming rate, as more researchers realize the importance of the subject to their own interests, that it is hard for any single text to do justice to the subject. The aims of this book are to discuss the main components found in cell signalling pathways, rather than describe many specific pathways which might contain common elements.

1.2 The main principles of cell signalling

The main underlying event for many cell signalling mechanisms is the arrival at the cell of something that requires the cell to respond. This might be the arrival of light photons, as in the case of cells in the eyes of humans and other animals. Likewise, the arrival may be a chemical released from another organism, or from another cell in the same organism. Most, but not all, signals arrive at, and are perceived at, the outer borders of the cell, commonly the plasma membrane. Therefore, a scenario such as shown in Fig. 1.1 is common.

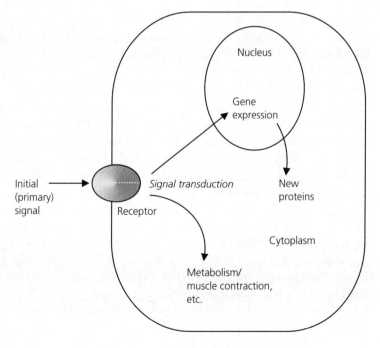

Fig. 1.1 The main event in cell signalling may be the arrival of a signal at the cell surface. A cascade of events commonly follows, carrying the 'message' to its final destination, perhaps the cytoplasm or the nucleus.

Several events follow the arrival of a signal:

- Perception of the signal, usually by dedicated proteins referred to as receptors.
- Transmission of the signal by the receptor into the cell.
- Passing on of the 'message' to a series of cell signalling components, often referred to as a cell signalling cascade, as indicated by 'signal transduction' in Fig. 1.1.
- The arrival of the message at the final destination in the cell.
- A response by the cell, an action carried out so that the outcome is appropriate to the original signal, or stress. In Fig. 1.1, the original signal results in the alteration of activities in the cytoplasm and gene expression in the nucleus.

The response may be anywhere in the cell, and the cascade will be designed to carry the message to a defined place. Some cascades end in the cytoplasm, for example in the control of glycogen metabolism, while others might end in the nucleus, as is seen with the control of gene expression.

Several points should also be borne in mind. A single signal arriving at a cell's surface may lead to more than one outcome, and a single signal may lead to the activation and use of more than one signalling cascade. For example, hormones that signal through phospholipase C initiate both inositol signalling and kinase signalling (see Chapter 6; sections 6.3–6.5). These cascades might share components with other signalling cascades too. Secondly, a extremely small amount of the signal might be perceived by the cell, but a rather large response might be needed. For example, a few molecules of hormone arriving at the cell surface might require a large number of individual enzyme proteins to be activated, and therefore an element of amplification is required by the system. Thirdly, cells need to be selective. Cells do not have the luxury of accepting signals neatly one at a time, but they are likely to be bombarded by them. A myriad of hormones, cytokines, etc. are present in human blood, and a cell may wish to respond to a selection of those simultaneously. However, several may lead to the activation or inactivation of the same signalling component in the cell, and the idea of crosstalk, where one signalling transduction pathway is influenced by another, has been introduced to aid in the elucidation of such complications. It is not to all signals present that a cell responds to, but rather a defined array, tailored for that cell, or type of cell. For example, in a leaf some cells such as the guard cells of the stomata might need to respond to hormones that signal that the plant is short of water, but the rest of the leaf might not need to respond, or some cells of the mammalian immune system might need to respond to a pathogen attack, but not all the cells.

Therefore, in the discussion of cell signalling such principles and factors need to be borne in mind, and hopefully the more in-depth discussion within this book will help to unravel how cell signalling manages both the subtlety and the complexity that is required for cells to survive and thrive. These issues are further discussed in Chapter 12.

1.3 What makes a good signal?

Signals reported in the literature come in a wide variety of shapes and sizes, some of which appear to be almost bizarre, ranging from large proteins to gaseous chemicals, but they generally have a set of common characteristics. Probably the most important characteristic is that the signal must have specificity. A signal must be unique enough to relay a defined signal, and to only be detected by the molecular machinery designed for that detection. If made and perceived incorrectly, the signal will have failed to relay its specific message.

To be an effective signal, a molecule usually is relatively small and therefore able to travel from the site of manufacture to the target site reasonably easily. Extracellular signals can take advantage of the vascular system of the organism to assist in their travels but inside the cell the signal molecule often appears to rely on diffusion, although the cellular cytoskeleton is often thought to play a part. Therefore, intracellular signals need to be small enough to diffuse rapidly and often need to have the capacity to diffuse through membranes, either by the aid of a carrier protein or through their hydrophobic nature. However, as with all rules, there are exceptions, and some signals are very large proteins, or lipids which appear to stay embedded in membranes and not able to travel very far.

A further characteristic which is important is that a signal, whether it is extracellular or intracellular, should be able to be made, mobilized, or altered relatively quickly, allowing the response to be switched on or off rapidly. There are two main ways of creating a signal rapidly. Firstly, an enzyme can be employed to manufacture the signal as and when required, usually from a substrate which is ubiquitous, or at least easily available; perhaps a vital molecule such as adenosine triphosphate (ATP). Secondly, a pre-made signal can be sequestered, and only released, either from a cell or from an intracellular organelle, to be perceived at the required time. Such mechanisms are shown in Fig. 1.2.

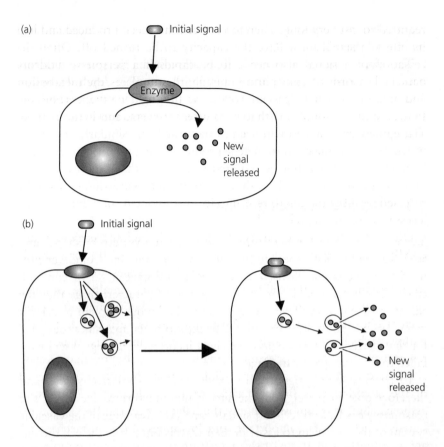

Fig. 1.2 Two common ways in which cells may rapidly muster a signal. (a) They may be made as and when required, the manufacturing need itself being under the control of signals. (b) A signal may be pre-made, and held and sequestered out of the way until needed. Signalling then triggers the release of the signal to allow it to move to its site of recognition.

An example of the first mechanism is exemplified by the production of adenosine 3′, 5′-cyclic monophosphate (cAMP). It is produced enzymatically from ATP inside the cell, by a dedicated enzyme which is under the control of other cell signalling components. Therefore, once initiated by the arrival of a hormone at the cell surface, the signalling cascade can ensure the activation of the enzyme and cAMP is rapidly produced to initiate the next stage of the signalling cascade. The second mechanism is less common in intracellular signalling although it is used for the control of calcium ion concentrations: as calcium ions cannot be created or destroyed, they are sequestered until needed. However, it is a common mechanism for the rapid 'creation' of an extracellular signal. Hormones, such as insulin, are often made and held in intracellular vesicles, and on being triggered the appropriate signalling cascade will induce the movement of the vesicles to the cell surface and the release of the hormones into the extracellular space, perhaps the vascular system of the organism.

Although often the cell has to have a fast response and the signal has to be relayed through the cell as quickly as possible, the responses are rarely required to last very long. Therefore, once it has been produced and had its effect, a signal must have the capacity to be turned off. Often the cessation of a signal also needs to be rapid. If a person is suddenly panicked, a rush of epinephrine (adrenalin) stimulates the metabolism and aids in what has been referred to as the 'fight or flight' response. However, the person will wish to relax after the event, and so the message that epinephrine carried will need to be turned off. Similarly, if a cell is instructed to produce a certain gene product having received an extracellular message, it will not be required to produce that product for ever more, until the cell dies. The stimulation of gene expression needs to be reversed to allow the cell to return to its unstimulated state. How is such a cessation of signals achieved?

Often the signals are physically destroyed, perhaps enzymatically. cAMP is de-cyclized to adenosine monophosphate (AMP) for example (see section 5.6). Alternatively a signal may be re-sequestered, or a receptor internalized into the cell to effectively cease its signalling. Although cells have developed an array of enzymes involved in turning off signals, such as phosphodiesterases and phosphatases, often it is this part of the pathway which is more poorly understood, and in many cases ignored by researchers in the field. For example, many studies will show that a kinase is required to stimulate a certain cellular activity but often the reversal of the stimulation is brushed aside, or the molecular mechanism is assumed. But, it is almost certainly the balance between the on and off signals which is in fact important, and so studies of the off mechanisms are as important as those of the activation processes. For example, the role of phosphatases is as important as the role of kinases in maintaining the level of phosphorylation of proteins in cells (see Chapter 4 for further discussion).

1.4 Different ways in which cells signal to each other

Cells may signal to each other in a variety of different ways. Often the signalling is classified depending on the distance between the signalling cell and the target cell, as indicated in Table 1.1.

If the cells are touching the signalling may simply be through pores in the membranes, such as gap junctions (in animals) or plasmodesmata (in plants), or may be due to a membrane-bound ligand being identified by a receptor in the membrane of a neighbouring cell. If the cells are further apart they may communicate via the release of molecules which are then detected by the target cell or via the transmission of an electrical signal.

> **A ligand** the term ligand is used throughout this book as a generic term for any molecule which binds to specific sites on a protein, such as a hormone binding to its receptor. The word ligand comes from the Latin, *ligare*, which means to bind.

Table 1.1 The types of signalling that could take place in an organism and the distance over which the signalling might have an influence.

Class of signalling	Range of the signalling in an organism	Comment
Electrical	Long range	Propagation of an electrical potential along a cell
Endocrine	Long range	Release and perception of hormones
Paracrine	Short range	Release of perception of extracellular signals
Cell–cell contact	Neighbouring cells	Either by receptor mediated signalling or gap junctions/plasmodesmata
Autocrine	Same cell	Use of diffusible signals

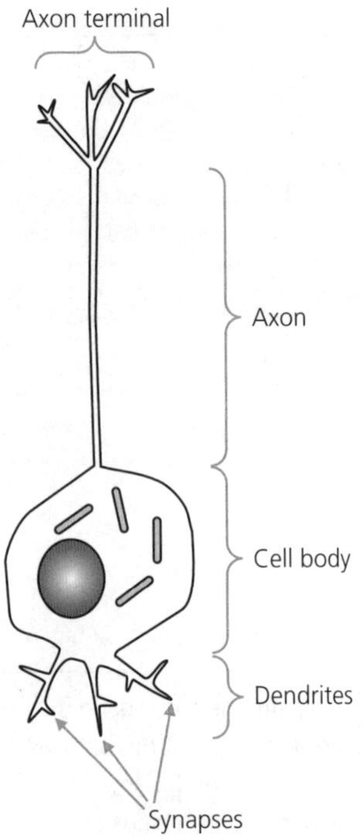

Electrical and synaptic

A fast and efficient method of signalling over long distances is via changes in the electrical potential across the plasma membrane of a cell, such as in the cells known as neurons of the nervous system in animals. Such a potential is propagated along the length of the cell, which is commonly extraordinarily long, and so the signal can therefore travel considerable distances in a body. Neurons usually contain four regions: the axon for conduction of the signal; a cell body containing the nucleus and normal cellular functions; dendrites which receive chemical signals from other neurons or other cells; and axon termini, from where the signal is passed on to the next neuron or target tissue (Fig. 1.3). Neurons communicate through the use of synapses. These are areas of close contact through which the signal can be propagated. Synapses generally fall into two groups, either transmitting the electrical signal directly through gap junctions or chemically by the release of neurotransmitters, which themselves are detected by the target cell.

Although electrical signalling was always thought to be unique to animals, plants also use a primitive type of electrical impulse in response, for example, to wounding. Over relatively long distances the signals are carried by the vascular system, but short signalling distances involving the plasmodesmata have also been seen.

Endocrine

Many signalling systems involve the release of a signalling molecule from one cell, the movement of that molecule to a target cell, and its perception there. Such a scheme is shown in Fig. 1.4a. Endocrine signalling exemplifies this. Here, cells release signalling molecules, such as hormones, which can

Fig. 1.3 A schematic representation of a neuron, showing the four main areas: axon terminal, axon, cell body, and dendrites. The axon section may be extremely long, relatively far longer than shown here, reaching to the outer parts of an organism.

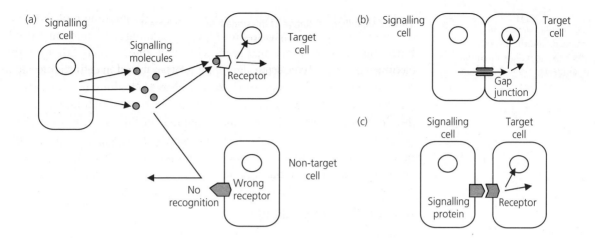

Fig. 1.4 Three of the ways in which cells communicate with each other. (a) By the release of molecules from one cell which are detected by a receptor on another cell. Here, the presence of the wrong receptor on the cell surface will mean that the ligand is not perceived. (b) By the direct transfer of small molecules through a gap junction. (c) By the detection of a membrane protein on the surface of one cell by a receptor on a second cell.

travel vast distances in the organism, usually having an effect in a different tissue. In animals, the bloodstream supplies the route to carry these messengers, but it is relatively slow and unspecific. All the tissues will be bathed in a supply of the signalling molecule and specificity comes from the target cell's ability to detect it. Plants also use endocrine signalling, hormones once again being carried by the vascular system of the organism. Even unicellular organisms have been shown to rely on a type of hormonal signalling; e.g. amoebae relying on diffusion in the surrounding medium to carry the signal to the next organism.

Paracrine

As seen with endocrine signalling a common way for cells to communicate is by the release of molecules that diffuse to and are detected by a second cell. The difference here is that in paracrine signalling the released molecules are detected and have their effect in the local area of the signalling cell. Diffusion of the signalling molecules is very limited because they are either rapidly destroyed by extracellular enzymes or alternatively they are rapidly immobilized on to or taken up by neighbouring cells. An example of such signalling is the release of neurotransmitters from neurons, which have their effect in the neighbouring neuron or in a neighbouring cell such as a muscle cell. A further example is signalling by cells involved in the immune response, where a variety of white blood cells release cytokines and chemokines to modulate the activity of other white blood cells, and perhaps other local cells, in response to an insult to the organism, such as pathogen attack.

Autocrine

In a similar way to paracrine, autocrine describes a local effect of diffusable signalling molecules, but here the term autocrine is used specifically

> **Autocrine** this is signalling to self; that is, a cell signalling to itself.

to describe the mechanism by which the released signalling molecules act on the cell that released them. This type of signalling is common in growth hormones and is sometimes seen with molecules which fall into the eicosanoid group. Autocrine signalling is often found in cells that are in a developmental or differentiating stage and this signalling may be seen as an emphasizing feature. Once a cell has been instructed to follow a certain developmental route, an autocrine signal may ensure that the set route is followed by releasing a signal to emphasize that message. Autocrine signalling is also commonly found in tumour cells where they produce growth hormones stimulating uncontrolled growth and proliferation.

Direct cell–cell signalling

Cells that are in physical contact are often very active in signalling to each other. They can achieve this by either recognition of molecules on each others' surfaces or direct communication through specialized areas of the cell surface.

Receptor–ligand signalling

A common way that cells communicate is through the recognition of surface markers (see Fig. 1.4c). A good illustration of this is the communication of cells in the eyes of the fly *Drosophila* during their development. The fly's compound eyes contain approximately 800 separate photo-responsive units, or ommatidia, which each consist of 22 cells. Eight of these are responsible for photoreception and are called retinula or R cells. During development of ommatidia, the R8 cells express on their surface a protein known as Bride of Sevenless, or Boss. This is effectively a fixed ligand on the surface of these cells. The presence of the Boss protein is detected by another protein, actually a receptor tyrosine kinase (see section 3.2) known as Sevenless, or Sev, which is on the surface of the neighbouring R7 cell. The binding of Sev receptor to the Boss ligand protein triggers the R7 cells to develop and enables the fly to detect ultraviolet light. This system was discovered due to mutations in the Sev protein where no R7 cells were signalled to develop, hence the naming of the protein as Sevenless. Such flies were identified by their inability to detect ultraviolet light and such animals have been extremely useful in the elucidation of signalling pathways.

Other proteins involved in this type of cell-to-cell interaction which have been identified in *Drosophila* include Notch, Armadillo and the product of the *dlg* gene, and such proteins may help to control cell proliferation.

Gap junctions and plasmodesmata

In animal tissues cells are often seen to have an area where the two plasma membranes of adjoining cells are apparently held at a fixed but very small

Son of Sevenless the story of the signalling of the Sevenless mutants continued, as expected, and other downstream components were identified. One of these was a protein which interacts with G proteins, and was called Son of Sevenless, Sos. The Sos protein has now been found in many other signalling pathways, but the name has stuck.

distance apart. These are areas known as gap junctions. The plasma membranes in this area contain proteins which form tubes and when the tubes in the two membranes come into perfect alignment a complete pipe is formed which allows the passage of small molecules directly from one cytoplasm to the other. The passage of molecules of less than 1200 daltons (Da) can therefore take place directly from one cell to the next. For example, signalling molecules such as cAMP or Ca^{2+} ions can move through gap junctions (see Fig. 1.4b).

The tubular structures, known as connexons, in the membranes are composed of a group of six identical proteins called connexins which are arranged in a ring. The connexons protrude slightly from the membrane and therefore the two plasma membranes are held apart by a distance equal to the length of the protrusions from each connexon added together (see Fig. 1.5). Each gap junction of the cells may contain several hundreds of connexons which can be revealed by freeze-fracture techniques.

The protein structure of the connexins has been predicted from their amino acid sequences. At least 11 isoforms coded for by separate genes have been identified and therefore conserved features have been used to determine the likely topology of the polypeptides. It has been predicted that each protein contains four membrane-spanning helices, with a particularly conserved helix sequence being used to line the hole which is made when the six subunits come together.

Dalton a unit of mass used commonly in science. One dalton (1 Da) is equivalent to the mass of a hydrogen atom, and was named after John Dalton (1766–1844). In biological sciences it is often used as the kilodalton (kDa), which is equal to 1000 Da.

Freeze-fracture techniques this method effectively allows the two parts of the bilayer of a membrane to be peeled back to reveal proteins within the membrane. Other techniques with fluorescent tagging will also allow the identification of proteins such as connexons in membranes.

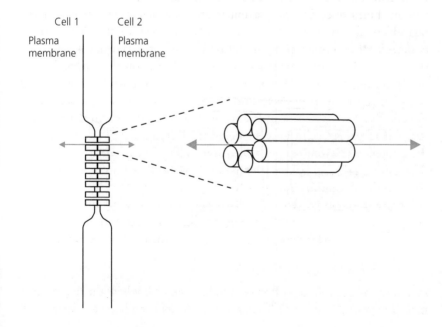

Cell 1   Cell 2

Plasma membrane   Plasma membrane

Fig. 1.5 A schematic representation of a gap junction. This shows the alignment of the connexin proteins forming connexons.

Isoforms isoforms are different forms of a protein. Therefore, such polypeptides would be assumed to have the same function, similar interactions, and similar structures. However, isoforms often differ, sometimes quite subtly, in their kinetics of action, in their control, and in their presence in different tissues. Many proteins in signalling seem to exist in a variety of isoforms, suggesting that the subtlety in their differences is very important.

The expression of the isoforms differs between tissues as does the permeability of the gap junctions, suggesting that the movement between cells is carefully managed. Although a single cell can contain more than one isoform the 12 connexins of a single junction are always of the same type.

As might be expected, the passage through the gap junctions is not a free-for-all but is in fact very well controlled. Gap junctions are rapidly closed if the Ca^{2+} or H^+ concentration rises, for example. The connexins undergo a reversible conformational change, probably working in a similar manner to the aperture on a camera lens. Closure of the gap junctions is of particular importance if the neighbouring cell dies, thus stopping the leakage of material from the still-living cell, or the import of any death signals. The separation of the intracellular part of one cell from the cytosol of another in which there is limited or no control of the input of molecules would be vital to a cell's survival. This may be of particular pertinence during development of tissues, when certain cells undergo cell suicide, while others are required to survive and develop.

Many, but not all, plant cells also have direct cell-to-cell connections but owing to the presence of a cell wall they differ from gap junctions. The connections in plants are called plasmodesmata (Fig. 1.6). The cell walls contain holes lined by the plasma membrane of the cell such that the

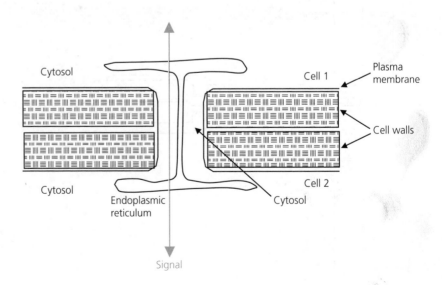

Fig. 1.6 A schematic representation of a plasmodesma. Note that the endoplasmic reticulum and cytosol are continuous across the two cells.

plasma membrane of the cells joined by plasmodesmata can be considered to be continuous. This lined hole is approximately 20–40 nm in diameter and also allows the cytosol to be continuous with the neighbouring cell. Through the hole is a membranous tube, called the desmotubule, which is derived from the endoplasmic reticula of the cells. Therefore a cell's cytosol, plasma membrane and endoplasmic reticulum are shared with the neighbouring cell.

Once again this does not mean the free passage of all molecules between the cells. The cut-off is thought to be approximately 800 Da, allowing the passage of small signalling molecules but not macromolecules such as proteins, although interestingly the transfer of some proteins and even viruses has been reported. Passage through plasmodesmata is also under tight regulation but the processes controlling this are poorly understood.

1.5 Amplification and physical architectures

The mechanisms of signal transduction from one cell to the next and subsequently into the interior of the target cell often need to involve formation of chains of signalling molecules, each passing on the message to the next molecule in the line. An extracellular signalling molecule, or first messenger, perceived by a cell often leads to the production of small and transient signalling molecules on the inside of the cell, often referred to as second messengers (see box). Such intracellular messengers activate or alter the activity of the next component of the transduction pathway, for example, a kinase. Historically, a good example of a second messenger is said to be cAMP, as discussed in Chapter 5.

One very important feature of the production of intracellular messengers and the existence of these cell signalling cascades is the capacity to allow amplification of the original signal, so that the end result is effective in the target cell. Binding of a single hormone molecule to a receptor on the cell surface will not in general cause activation of a single enzyme molecule. Rather, a situation could be envisaged as shown in Fig. 1.7.

Here, binding of ligand will cause the receptor to switch on many molecules of enzyme on the plasma membrane. Such transduction may involve G proteins acting on enzymes such as adenylyl cyclase. Ligand binding to a single receptor will lead to the potential activation of many G proteins, which in turn can potentially activate several adenylyl enzymes each. The activation of each one of these adenylyl cyclase polypeptides will lead to the production of many cAMP molecules causing further amplification of the signal. These cAMP molecules will in turn activate many kinases which will phosphorylate a great many proteins, causing yet

> **Second messengers** the term second messengers was originally coined to describe the signalling molecules which are found to be produced by cells in response to the perception of the first message, or extracellular ligand. However, they are often not the second component in the signalling pathway and the term is now confusing and misleading, and should be avoided.

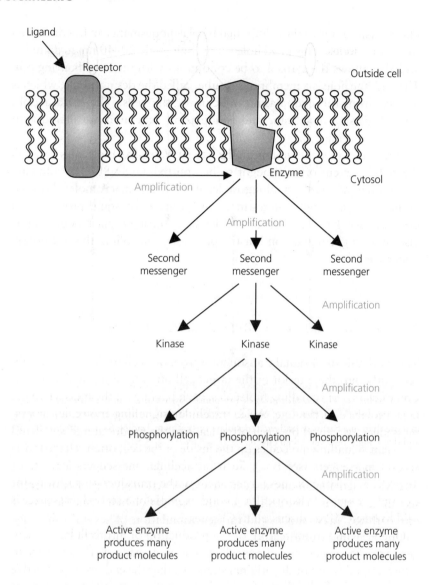

Fig. 1.7 Amplification: an example of how amplification could take place in a signal transduction cascade.

further amplification of the signal. Therefore, it can be seen that the binding of one hormone molecule to the exterior surface of the cell can potentially lead to the activation of hundreds or thousands of enzyme molecules inside the cell, each of which will catalyse many rounds of a reaction. The final signal, causing the desired cell event, will therefore not be reliant on the presence a large concentration of extracellular signals or the vast amount of ligand binding. Relatively small amounts of ligand can have profound intracellular effects if the cellular machinery is in place to allow its perception and this amplification.

However, as discussed above, signal cascades are usually not straightforward pathways with a defined route, going straight from beginning to

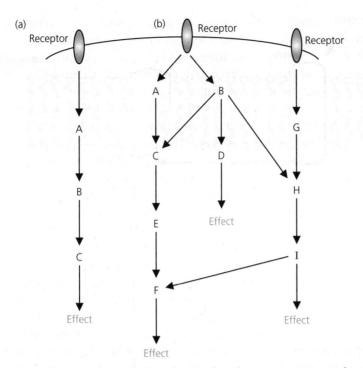

Fig. 1.8 Cell signalling cascades can be viewed as simple pathways, or a more complex web of events. (a) The receptor invokes a response down a single simple pathway leading to an effect. (b) The activation of one receptor may lead to the involvement of components in more than one pathway, and even modulate the pathway leading from another receptor, and there might be several final effects.

end. Most signalling molecules can have an effect on several other different signalling systems in the cell, so that the stimulation of a single element of one pathway may lead to the further stimulation of several other signalling branches (see Fig. 1.8).

1.6 Coordination of signalling

One of the major mysteries in the field of cell signalling is the question of how a cell co-ordinates all the signals that it has to deal with? As discussed above, cells are inundated with a myriad of signals all the time, some of which might require the same response, and others which might require a completely opposite response. Researchers talk about integration of signalling pathways and crosstalk, but there are few examples where there is a clear understanding of how a cell handles the multitude of signals that it has to deal with and how the cell manages different signal pathways simultaneously. It is quite possible, for example, for a cell to detect two or more signals on the outside simultaneously, via its receptors. These receptors will then activate the relevant signalling pathways inside the cell, but quite often these distinct signalling pathways will have common elements. For example, two signals may both use the modulation of intracellular calcium ion concentrations as part of their signalling cascade. How does the cell determine if the alteration of Ca^{2+} concentration is due to turning on of receptor 1 or receptor 2? Or indeed, is interaction of the signalling pathways the key? Does the cell rely on the subtle changes made to one signal by a second signalling pathway and does this give the cell the fine control required? Often, the end result of the activation of receptors may in fact be different or indeed opposite in the cell, and yet both pathways when studied in isolation may again appear to have common components.

An example of such a dilemma is seen when studying the signalling induced in insect gut muscles, where two receptors which apparently have differing effects on the muscle cells both induce the production of diacylglycerol (DAG) and activation of protein kinase C (PKC). How does the cell know what the production of DAG is supposed to mean if it is forming part of two cascades that have differing results? If a common element is used, how is the specificity of any single signal maintained? And yet it appears that it is.

Though we are far from obtaining a full understanding of cell signalling, it is now proposed that there is a greater compartmentalization of cells than classically thought, and this might help to explain why common signals have different effects. In the past, the cytoplasm for example has been thought of as a homogenous organelle, in which one part is the same as any other. How could it not be, as it is just a liquid phase of the cell? However, studies with Ca^{2+} ions and other small molecules has shown that hot-spots exist within the cytoplasm. There are areas of high calcium ion concentrations, and areas with virtually no calcium. Therefore, the calcium signalling can only be perceived by the cell in certain areas of the cytoplasm; that is, where the calcium concentrations are high. How these hot-spots are formed and maintained clearly requires further studies, and will open a way forward for a more full understanding of crosstalk in signalling. However, the idea of hot-spots is not confined to calcium signalling, and the idea has been discussed in other signalling events such as those involving reactive oxygen species and nitric oxide.

To help explain this apparent lack of diffusion across cells which facilitates the specificity of some signalling mechanisms, research has uncovered proteins which have a role as a structural scaffold inside cells, (discussed below), which restrict the movement of signalling components through the cytoplasm. In light of such research a new perception of what a cell actually looks like on the inside is needed before the answers to many of these problems are solved.

Almost opposed to this view of the signalling web of events, along with the proteins of defined signalling function, other polypeptides have also been postulated to be involved in the signalling systems of cells, but not to transfer any message *per se*. Such proteins are thought to have a scaffolding role, holding the relevant components of a cascade in a particular physical environment, allowing efficient transfer of the signal but in doing so restricting their interaction with other structurally and functionally similar components in a related signalling cascade. One such protein is the STE5 polypeptide of yeast, while some mitogen-activated protein kinases (MAP Kinases) in mammals, for example MAP/extracellular signal-regulated Kinase (ERK) Kinase (MEK) Kinase 1 (MEKK1), have been seen to have scaffolding roles too, as well as being active in catalysing phosphorylation. However, such a restriction of movement of components would put a question mark over the potential amplification or divergence of the signal that could take place. If proteins are held and can only further interact with a limited range of other proteins, perhaps only one, then neither amplification nor divergence is possible at this point in the cascade. It is likely that cells have evolved a pay-off whereby amplification is compromised for a vital need to retain specificity of a signal transduction cascade. Examples here would include the MAP kinase pathways, where a message can be relayed through the cell, perhaps with restricted divergence of the signal, but with a defined target, often the expression of genes in the nucleus in this case.

Many signalling cascades do allow for both amplification and divergence, where the arrival of a limited number of signalling molecules at a receptor may invoke a large cellular response, or several responses. For example, a kinase may lead to the phosphorylation of several proteins which have very different effects in the cell, including the activation of other kinases. Likewise, the activation of phospholipase C leads to the release of molecules which ultimately turn on both PKC and calcium signalling pathways, each having their own defined effects in the cell. Conversely, several pathways may lead to the stimulation of the same enzyme, showing convergence of pathways. An example here would be the breakdown of glycogen which can be stimulated by the arrival of a hormone at the cell surface, leading to intracellular signalling through cAMP, or separately, by the release of intracellular stores of calcium.

Therefore, it is misleading to think of cell signalling in simple terms of pathways, each distinct, functioning independently and simultaneously. Rather, signalling is a complex web of events (see Fig. 1.8), with vast amounts of divergence, amplification, and in many cases convergence. A cell will be bombarded by many signals, and will only react to those which it is equipped to recognize. Pathways will need to share common components, and often signals might lead to opposing action on those common components. This complexity has yet to be understood in many cases, but by investigating the potential interactions of all signalling elements in cells, the potential complexity of the signalling web can be determined and in some cases modelled to predict outcomes.

1.7 Domains and modules

Consensus sequence

If the sequences for several related proteins (or deoxyribonucleic acid (DNA)/ribonucleic acid (RNA) sequences) are aligned, and a commonality revealed, then the sequence which represents the common sequence found in these related proteins is referred to as a consensus sequence. The consensus sequence can then be used to hunt for similar regions in newly discovered proteins or proteins of unknown function.

The details of signalling pathways of cells look incredibly complex but quite often protein components contain areas of similarity or homology which are repeated: these are often referred to as domains or modules. Different kinases, for example, might have sequences which can be recognized as catalytically active regions of a polypeptide and although not identical in different proteins a consensus for the sequence in that area can be drawn up. On to this basic structure, other areas may be attached, such as amino acid sequences which are responsible for regulating the activity. Again, this second area may share homology with regions in other proteins which are controlled in the same or similar ways but in proteins which contain a different catalytic region and therefore have a completely different function. Good examples of this are seen with receptors that might have similar transmembrane domains and perhaps even similar ligand-binding domains, but different functional domains which transduce the signal. It can be seen that these polypeptide sequences and regions are like functional modules which can be glued together in an incredibly varied arrangement (Fig. 1.9).

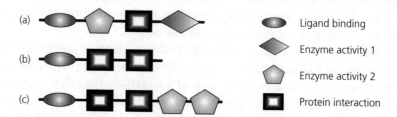

Fig. 1.9 Protein domains. Often proteins can be seen as being made of domains, each with a separate function. During evolution, such domains have been put together in different orders, and different amounts, to create families of related proteins, but each with a specific function and activity.

Evolution appears to have led to a good structure and design within proteins, and then adapted and modified it for other uses, rather than starting the design of a new enzyme or protein from scratch. A good example is found if proteins which contain a redox function are looked at closely. Many cells have several redox pathways, but often they are subtly different, allowing these apparent similar proteins to have a wide range of functions, including proton transfer in energy production, the detoxification of xenobiotics, and of particular relevance here, the production of free radicals such as nitric oxide and superoxide (see Chapter 8).

Redox (reduction/oxidation)

Proteins that have a redox function are involved in the transfer of electrons and as such can exist in both reduced and oxidized states.

Now that such domains are recognized to be similar across a range of proteins, consensus sequences for their primary structure can be determined, and with the use of bioinformatics such domains can be hunted for in new gene sequences or in whole gemones. For example, it is not uncommon to read in the literature that domains such as an EF-hand, which allows finding of calcium ions, have been found in a gene sequence, and the authors will suggest therefore that the resulting protein will be controlled by the presence of calcium ions. However, caution should be exercised here. Just because the domain sequence is present, it does not automatically mean that it is functional, but such bioinformatic data are very constuctive for informing future biochemical studies, where activation profiles and functionallity can be determined experimentally.

As discussed above, some domains are repeated and used in many proteins, and common ones that are seen often in proteins involved in cell signalling are those that are involved in protein–protein interactions. Two very important polypeptide domains found in many proteins which have such a role in cell signalling are the Src homology domains: the so-called SH2 and SH3 domains. Both of these domains are involved in the interactions of proteins, acting like a protein Velcro to hold them together, but the specificity of the two types of sequence differs.

SH2 domains are in general approximately 100 amino acids in length and have a binding affinity for phosphotyrosine amino acids; that is, tyrosine residues of a protein which have been phosphorylated. The domain has the structure of a deep pocket lined with positively charged amino acids which interact with the phosphate group of the phosphotyrosine. Binding is very much reduced if dephosphorylation has occurred. A second region of importance of SH2 domains has been also been used to categorize them into two groups. The first group also contains a second binding pocket which is partly responsible for bestowing the specificity on the binding, as it recognizes a single amino acid of the target polypeptide. The second SH2 group contains a region which is indented and shows binding to a group of mainly hydrophobic amino acids, a little way removed from the target phosphotyrosine residue.

SH3 domains, on the other hand, bind to a sequence on the target polypeptide containing several proline residues, which are often held in a left-handed helical secondary structure, known as a polyproline helix type II.

It is quite common for a single protein to contain both of these domains and even multiple copies of them. They are particularly common where proteins that are not normally associated have to come together leading to activation of one of the proteins. An excellent example of this is provided by adaptor proteins, such as GRB2. GRB2 is a 25 kDa polypeptide that contains one SH2 domain and two SH3 domains and which can be in association with guanine nucleotide-releasing proteins, for example Son of Sevenless (Sos). These proteins are often involved in signalling from receptor tyrosine kinases, such as the epidermal growth factor (EGF) receptor, which are autophosphorylated on binding to their ligand. The formation of the phosphotyrosine residues on the receptor then allows the binding of these adaptor proteins through their SH2 domains. A conformational change will take place, and in so doing cause the activation of associated guanine nucleotide-releasing proteins through the use of the SH3 domains, resulting in the stimulation of a G protein and the activation of the rest of the cascade (see Chapter 5 for details). Other adaptor proteins include Shc, Crk and Nck, although to date these are not as well characterized as GRB2.

Another example of the involvement of SH2 domains is on the signal transducers and activators of transcription (STAT) proteins. STAT proteins are phosphorylated by Janus kinases (JAKs), which become associated with activated cytokine receptors (see Chapter 4). STAT proteins are phosphorylated on tyrosine and because they also contain SH2 domains they dimerize. The dimers so formed then go on to cause the enhancement of transcription. At least three STAT proteins have been identified to date. STAT1α (91 kDa) and STAT1β (84 kDa) arise from the same gene via alternative splicing while STAT2 is slightly larger at 113 kDa.

Other proteins may not contain such domains themselves but on becoming phosphorylated they gain the topology to interact with such domains resident on other proteins. These proteins, often referred to as relay proteins, are also seen in signalling cascades from receptor tyrosine kinases. A good example is insulin receptor substrate 1 (IRS-1). This protein is phosphorylated on multiple tyrosines, so creating sites which can interact with the adaptor protein GRB2 via its SH2 domain. This particular pathway is discussed in more detail in Chapter 9.

> **Topology** the topology of a protein refers to the way the amino acids are arranged in three dimensions, and therefore its tertiary structure.

SH2 and SH3 domains are not the only domains allowing protein–protein interactions. Others include the Pleckstrin homology (PH) domain, which was first identified in the platelet protein pleckstrin. Such domains have now been found in over a hundred human proteins, including Son of Sevenless (Sos), and some calcium-independent PKC isoforms. Within proteins, the PH domain may be anywhere along the sequence. At the sequence level PH domains appear to be a rather diverse group, with limited amino acid homology (10–30% only) but at the structural level they are more conserved. They are usually 100 amino acids long, folded into two anti-parallel β-sheets with an α-helix along the top, the latter being the amino acids of the C-terminal end of the domain. As for their function, they have been associated with interactions of proteins with phosphoinositide lipids, for example as seen with protein kinase B, with interactions of proteins with G proteins and in anchoring proteins to membranes.

There are other protein–protein interaction domains, some of which are discussed in Chapter 11 (section 11.5) and it is very likely that further binding domains will be found as further cloning and sequencing work allows consensus sequences to be deduced.

14-3-3 proteins

A family of polypeptides that have a regulatory role, and act like scaffolds, are the 14-3-3 proteins. Their rather strange name was derived from a survey of brain proteins, but they have now been found to be important in a plethora of pathways, across the kingdoms. They exist in multiple forms in organisms, for example the *Arabidopsis* genome encodes 15 versions of 14-3-3, and are involved in a wide range of signalling pathways. Target proteins for 14-3-3 polypeptides include kinases, phosphatases, ATPase, and nitrate reductase.

14-3-3 proteins act by their ability to bind to phosphorylated proteins. In some instances it is thought that the phosphorylation of a protein may not be sufficient in itself to control that protein. However, interaction of the phosphoprotein with a regulatory peptide, such as 14-3-3, may then enable a full level of control to be reached.

Use of protein crystals

To determine the structure of proteins, often it is attempted to obtain protein crystals, and these are examined using X-ray diffraction.

The sequences of 14-3-3 polypeptides are highly conserved between species, suggesting that the structures are also similar. The central core of the structure in particular is thought to be conserved, while the N- and C-termini seem to be diverged in sequence and also fail to resolve in crystal structures. This suggests that the core has a common function, but that variations in protein interactions and role may be conferred on the protein by the sequences at each end. The 14-3-3 proteins exist as dimers, and in three dimensions the central core forms a double-barrelled W-shaped clamp. This is created by anti-parallel helices of a dimer, where each of the polypeptides forms a channel for interaction with a phosphorylated protein. The sequence to which 14-3-3 proteins binds has also been studied in many proteins, and the consensus R/KXXpS/TXXP (where R is arginine, K is lysine, S is serine, T is threonine, P is proline, PS is phosphoserine, and X is any amino acid) has been derived. Therefore, potentially new interactions can be postulated if such a sequence is found in a protein previously not associated with the 14-3-3 proteins. However, caution has to be exercised, as not all consensus sequences are necessarily used in a protein, and not all interacting proteins will contain the consensus.

As the 14-3-3 proteins are dimers, and each member of the dimer has a protein-interaction site, it is possible that the 14-3-3 acts as a bridge across two phosphoproteins, so bringing those proteins into close proximity. However, 14-3-3 has been shown to bind to non-phosphorylated proteins too, suggesting a wider role for this family of proteins.

1.8 Oncogenes

Oncogenes are genes found in an organism that encode proteins containing a function that leads to uncontrolled growth of cells, and therefore are instrumental in the onset of tumour formation. Therefore discovery of oncogenes and their functions was heralded as a great advance in our fight against cancer and in the understanding of why tumour cells form. The first hint of what was going on was supplied as far back as 1911 by Peyton Rous with his work on chicken tumours and his discovery of tumour viruses. Even today, many of the oncogenes which have been characterized were isolated originally from retroviruses.

However, the discovery of oncogenes really highlights the importance of an understanding of cell signalling as it is now clear that most oncogenes code for proteins which have an influence on cellular signalling mechanisms. They do not all influence a single part of signalling mechanisms but can be found spread through the plethora of signalling components. An oncogene product may well disrupt the way a cell perceives an extracellular signal or indeed the way in which that signal is transmitted through the cell. Therefore, because of their diverse nature, they have been grouped into four classes. Class 1 includes oncogenes which code for growth factors, such as the *sis* gene which codes for platelet-derived growth factor. The oncogenes which encode growth factor receptors make up class 2, and include receptors for EGF, the *erbB* gene, and nerve growth factor gene, *trk*. The oncogenes which code for intracellular signalling components make up the third and fourth classes, the class into which they fit being determined by their intracellular location. Class 3 includes the G protein gene *ras* as well as the kinase genes *src* and *raf*, and all are cytosolic proteins. The nuclear proteins which make up class 4 include transcription factors, *jun* and *fos*, and steroid receptors such as the *erbA* product.

So what is it that makes a gene an oncogene? Often they are versions of normal genes but actually encode defective components of a signalling pathway, which in many instances are in a permanently activated state. For example the *erbB* gene codes for a receptor which has been truncated at both ends and signals to the cell the presence of EGF even in its absence. Likewise, the *ras* oncogene traps the G protein in a guanosine triphosphate (GTP)-bound state, where GTP hydrolysis cannot take place and the G protein remains permanently active, permanently relaying an 'on' signal. Therefore, although many oncogenes are coded for on viral genomes, many are actually coded for by our own genomes too. In the normal state, where the gene products are not defective, the genes are known as proto-oncogenes. The product of the *ras* gene as we shall see later is an instrumental protein involved in critical signalling pathways. It is only when the protein is defective that the gene encoding it becomes classed as an oncogene.

1.9 A brief history

Here, a brief overview of the history of cell signalling will be given (see also Fig. 1.10), but it is impossible to do justice to the subject here, and many researchers will note the absence of some events. Some points of historical interest, not discussed here, are mentioned in later chapters. It is worth pointing the reader to an excellent collection of key papers, edited by Burgoyne and Peterson (see Further reading at the end of the chapter), but of course future work will be undoubtedly lead to further landmarks, and technology has moved on considerably since 1997.

The early days

A quick search using a literature database using any keywords associated with the subject of cell signalling is enough to frighten the most enthusiastic researcher, and it will reveal the vast extent of work in the area and

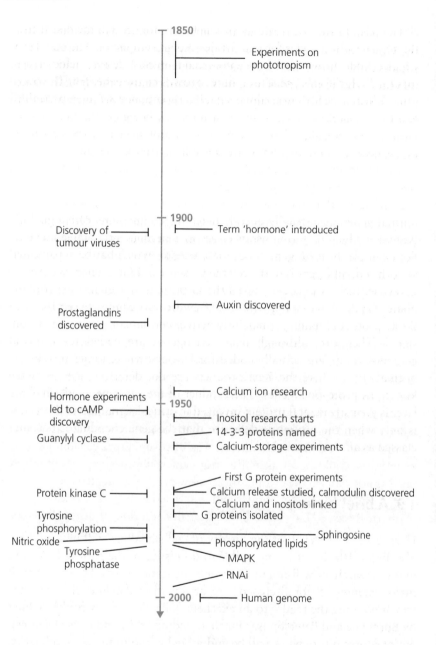

Fig. 1.10 A time line of some of the major findings in cell signalling. RNAi, RNA interference.

the immense interest there is across the whole of biological sciences. However, this interest is not new and has spanned the period of biochemical research. Even in the middle of the nineteenth century, Darwin was experimenting with the phototropism of coleoptiles, and suggested the presence of a substance which was transported around plants, having its effect at a site distant from that of its production and secretion. This substance was discovered by Went in 1928 and was later identified as indoleacetic acid, or, to give it its common name, auxin.

The term hormone, really in an animal context, was introduced from the Greek meaning to excite or arouse, by Starling as long ago as 1905. Lipid-soluble hormones such as prostaglandins however had to wait 30 years to be discovered. This family of compounds was originally named after the organ which was supposed to be their site of production but this has subsequently proved to be misleading. Even today, the number and variety of extracellular signals discovered still seems to be increasing, especially for any researcher in the cytokine/chemokine field.

A growing volume of work in the field

Although extracellular events were beginning to be unravelled in the early part of the twentieth century some of the first clues as to what was happening inside cells came in the late 1940s and 1950s. In 1947 experiments in which small quantities of Ca^{2+} ions were injected directly into the cell showed that an increase in intracellular calcium led to skeletal contractions, but it was not until the early 1960s with the work of the likes of Ebashi and Lipmann that the way calcium ions were stored in cells was beginning to become clear. It took work on another system, that of inositol lipids, and the proposals by the likes of Michell, to elucidate the release of Ca^{2+} from intracellular stores, while proteins which are under the control of Ca^{2+}, such as calmodulin, came to light through the work of Cheung and others in 1970. Complications were introduced to the field of calcium signalling in the 1980s by the work of researchers such as Woods, Cuthbertson, and Cobbold, when it was discovered that calcium ion concentrations could be measured as transient spikes. Other complications, such as hot-spots in the cytoplasm, are now being investigated too.

In the 1950s Earl W. Sunderland and Theodore Rall, working at the Western Reserve University, found that the addition of adrenaline (epinephrine) and glucagon to cells led to the binding of these hormones to cell receptors, and experimentation could show the subsequent production of cAMP. The enzyme responsible for this increase in intracellular cAMP concentration they named adenyl cyclase, but it is now known as adenylyl cyclase or adenylate cyclase. However, it was not until 1989 that Krupinski and colleagues cloned the first mammalian enzyme. Adenosine monophosphate (AMP) and guanosine monophosphate (GMP) were isolated from rat urine in 1963 whereas guanylyl cyclase was identified in the late 1960s. Investigations of the role of adenylyl cyclase and cAMP were further helped by the isolation of adenylyl cyclase-deficient cells, known as cyc$^-$. These cells were first reported in 1975 by Gordon Tomkins, Henry Bourne, and Philip Coffinio working at the University of California.

The involvement of a separate signalling pathway, that of the inositol phosphate pathway, was also starting to unravel in the 1950s. The role of phospholipids, such as phosphoinositides, was first suggested in 1953 when Hokin and Hokin showed that the addition of acetylcholine to cells was found to lead to the incorporation of radioactive phosphate (^{32}P) into what was regarded as a minor lipid, phosphatidylinositol. This pathway was shown to involve the muscarinic receptors. The pathway leading to the biosynthesis of these lipids in the endoplasmic reticulum was reported by Agranoff and Paulus and their respective co-workers in the late 1950s. However, a link was made to another signalling pathway in 1975, when Michell suggested that messengers which stimulated the breakdown of phosphatidylinositol also caused an increase in cytosolic Ca^{2+}. A new pathway involving the membrane inositol lipids was more recently suggested when in 1988 Whitman *et al.* first identified phosphatidylinositol 3-phosphate in transformed lymphocytes. This is now an exciting new area of lipid signalling which is being shown to be implicated in the signal transduction of several systems, including responses to insulin, and the onset of apoptosis. Similarly, evidence in 1986 that sphingosine inhibited PKC has led to elucidation of the sphingomyelin cycle.

Phosphorylation research

Protein phosphorylation is the common end of many signal pathways, and it was over 30 years ago that the regulation of glycogen breakdown by phosphorylation of glycogen phosphorylase was suggested. In this system, the protein kinase was itself regulated by the intracellular cAMP concentration, but 1970 saw the first reports of a kinase which was more sensitive to cGMP than cAMP; that is, cGMP-dependent protein kinase. Another, extremely important kinase, PKC was identified during the 1970s, through the work of Takai, Kishimoto, Nishizuka, and others. However, until 1980, it was thought that only serine and threonine were targets for the phosphorylation of proteins, but it is now clear that one of the major phosphorylation events is the addition of the phosphoryl group to tyrosine residues. It was not until 1988 that Tonks obtained the first partial sequence of a tyrosine phosphatase however, while a year earlier the first MAP kinase was identified through the work of Ray and Sturgill. It has been suggested that the human genome might contain as many as 2000 kinase genes and up to 1000 phosphatase genes, and other genomes are likely to have an equally high number of genes encoding proteins involved in phosphorylation events.

Work on G proteins

The history of the elucidation of the G protein story takes us back to the 1970s. Gilman, working at the University of Virginia, identified G proteins as crucial intermediates in signal transduction in the latter part of the decade and in 1980 Paul Sternweis and John Northup, working in Gilman's laboratory purified the G protein G_s. It was not long afterwards that Lubert Stryer and Mark Bitensky independently discovered the existence of a second G protein, this time, using rod cells from the eye. This G protein was transducin or G_t.

Recent events and the future

Today, it would be folly to claim that all the signalling pathways have been discovered, even if we are not sure of the function of many of them. For example, it was not until 1987 that Moncada, among others, suggested that endothelium-derived relaxing factor was in fact nitric oxide, opening up a whole new field of research, some of the founders of which later received the Nobel Prize for Physiology and Medicine, in 1998. Very recently, the idea of the importance of the reduction and oxidation states of cysteine residues has been discussed, with the term nano-switch, or nano-transducer, being coined. However, with the advances in molecular genetic technology, the rapid rate of cloning of new genes and the human genome project, it will not be long before all potential new members of existing pathways will eventually come to light, but the challenge to understand how all these pathways interact will remain. Perhaps that challenge will be overcome by the use of new technologies, such as RNA interference (RNAi). This relatively new and clever way of reducing the expression of specific genes was discovered in 1998, and has helped to explain the results presented in several previous studies.

1.10 The techniques in the study of cell signalling components

The understanding of the molecular events of cell signalling, like all areas of biochemistry, has been facilitated by the techniques available at the time, and it is probably true that the science of cell signalling has been led by the technology and the popular trends at the time. However, the study of cell signalling requires a holistic approach, where the physiological events which ensue from stimulation still need to be studied, but where the molecular events inside the cell also need to be unravelled.

Early studies were based on the observable events with whole cells, and that sort of study is still instrumental today. If a researcher wishes to know what molecular events make a stomatal aperture on a leaf close, they still need to watch to see if it happens. However, today molecular biology makes a major impact, and even allows us to discover new prospective signalling molecules without even knowing their function, or even if they exist at all!

Labelling techniques

Confocal microscope

A confocal microscope commonly has a scanning laser as a light source, instead of white light, and the focusing nature of the lenses allows the user to optically section a sample and to build up a three-dimensional image. Therefore, it is a powerful technique allowing the researcher to 'look' inside a cell or tissue. Further discussion can be found in section 7.10.

The ability to label proteins and signalling components has been extremely useful. Immunological techniques have been a great benefit to the discovery and the study of the distribution of receptors and, coupled with modern fluorescence techniques, the receptors on a cellular surface may be visualized in a three-dimensional image using confocal microscopic technology (discussed further below and in Chapter 7). Photoreactive radioactive analogues of ligands have also been used in the study, for example, of lysophosphatidic acid (LPA) receptors. Antibodies raised against particular epitopes on a protein have also been invaluable in the elucidation of the interactions between polypeptides, or to knock out a particular active site, so stopping a signal transduction pathway and watching the effect on the cell or cellular function such as transcription. Similarly, if particular polypeptides are thought to undergo an interaction, short peptides can be made and added, so flooding a prospective docking site on one polypeptide and thus potentially upsetting the propagation of the signal or activity.

The use of probes

The combination of fluoresecence and confocal microscopy has also been adopted widely in the absence of antibodies, with some companies able to supply hundreds of probes to investigate the functioning of cell signalling components. Here, a normally non-fluorescent probe, but one that gains fluorescence when in interaction with its target molecule, is added to the cell, and then the subsequent fluorescence seen is proportional for the amount of the signalling molecule present. Such an approach is probably most commonly used to study the levels of calcium in cells, but it can also be used to measure nitric oxide production from cells and a plethora of other events. A further advance in this area is the use of fluorescence resonance energy transfer (FRET) measured with what is referred to as fluorescence lifetime imaging microscopy (FLIM). These technologies allow the tracing of the catalytic activity of fluorescently tagged proteins inside still-living cells, and can also be used to determine the state of activation of proteins in fixed cells, for example in pathological samples.

Pharmacological tools

Pharmacological studies have made a big impression on our understanding of signalling events. Here, the study of signalling pathways involves the addition of stimulators or inhibitors. For example, the question of what happens inside the cell on addition of a particular receptor ligand may be asked. Once a cellular event that is induced has been demonstrated, the method by which it can be inhibited is quite often soon addressed. Other inhibitors, targeted against potential components in a cascade, are also used, and by building up an inhibition profile the possible cascade can be elucidated. This might include kinase or phosphatase inhibitors to modulate particular phosphorylation events, or inhibitors of specific enzymes, such as phopholipases or cyclases.

However, caution is required here. Many of the inhibitors used, when discovered, were thought to be quite specific, but often subsequent studies show that their specificity can be doubted and that they have effects on more than one pathway. Interpretation of the results then becomes more difficult, and the search for better inhibitors continues.

Structure and protein interactions

Once a component of a pathway has been identified, the search for an understanding of its exact mechanism of action usually requires its purification, and often the elucidation of its structure. If the full primary structure of the polypeptide is known the secondary and tertiary structures can either be predicted by computer analysis or, with the pure protein, analysed by time-consuming nuclear magnetic resonance (NMR) or X-ray diffraction studies. Often, once several components of a system have been isolated, they can be recombined in a cell-free reconstitution system where the concentrations and conditions can be carefully controlled and altered. For example, proteins can be added to membrane fractions to get a better understanding of which proteins from the cytoplasm may be activating a membrane-associated activity. Along similar lines, the interaction of proteins and membranes can be studied with surface plasmon resonance, a technique in which the reflected angle of a laser light is dependent on the amount and the mass of an interacting molecule which might interact with an immobilized membrane fraction.

Molecular genetic techniques

The greatest advances made in the finding of new signalling components and in the study of their interactions have come through the use of molecular genetic techniques. One of the most powerful is the use of antisense technology which can be used to specifically knock out the expression of elements in a pathway, and while the use of antisense oligomers has yielded many results, the smart method of choice now is RNA interference (RNAi). Here, specific double-stranded RNA is used. This is introduced into a cell where it is processed into small interfering RNAs (siRNAs), and these subsequently cause the degradation of homologous endogenous mRNA. Therefore, this method can specifically reduce or stop the expression of a particular gene, enabling the researcher to answer the following questions. What would happen if the cell could not express a specific protein, for example a kinase? What is the effect of the lack of that protein on the signalling cascade being studied? Is that protein essential for the signalling pathways used by this particular ligand?

To get DNA clones of signalling components an antibody, or the back translation of an N-terminal protein sequence into a oligonucleotide, can be used to pull a clone from a DNA library. This would prevent the need for much of the protein sequencing undertaken previously. Oligonucleotide sequences based on the consensus sequences of several related polypeptides or complimentary DNA (cDNA) clones can be used to find the genes for more members of a polypeptide family, where the functions have not even been discovered. Such DNA sequences can be ligated into plasmids and overexpressed in cells, so enhancing or reducing the pathway in the cell. For example, the overexpression of a defective kinase (a so-called dominant-negative mutation) in a cell will mean that the defective kinase will preferentially receive the message from the signal pathway, but be unable to undergo any kinase activity, so effectively stopping the pathway at that point. Similarly, the overexpression of a specific phosphatase will mean that a specific kinase activity may be negated.

Microarrays and proteomics

The current technologies which are helping to elucidate the signalling events of cells are those that allow the study of holistic events in cells: microarray analysis and proteomics. Microarray analysis allows the expression profiles of hundreds or thousands of genes to be determined all at the same time, perhaps for all those in a genome. For certain species, particularly microbes, model mammals such as mice, model plants such as *Arabidopsis*, and for humans, the whole genomes have now been sequenced, so that representative example sequences from all the genes can be placed on a microarray and studied simultaneously. Therefore, if a ligand is added to a cell, all the genes which are expressed, or turned off, in response to that ligand can be determined. If signalling has gone awry in cells, such as in cancer, the expression profiles can be compared to those exhibited by normal cells, to unravel the molecular events which might be involved. It is clearly a powerful technique, but like all experiments, the results have to be assessed carefully, and studies to confirm the findings using Northern analysis or real-time polymerase chain reaction (PCR) need to be undertaken, giving a relatively independent and quantifiable measure of gene expression. Secondly, just because a gene is expressed does not mean that the protein will be made and be functional in the cell, and therefore proteomic techniques are required to study the protein profiles of cells. This often involves mass spectrometry and complex bioinformatic analysis, but is again a powerful technology, and one that will be used more commonly in the future.

Computer networks

Signalling pathways are so complex and interwoven that the logic of computer networks is being used to try to understand the complexity of the interaction of the signalling pathways. Much research studies single pathways and events, but it is becoming more important to take a holistic view of the events in cells. It is likely that combinations of certain pathways and their relative contributions to the overall signal will be important for the overall outcome for the cell, and therefore modelling such combinations of pathways and their relative levels of activation will require computers and software development.

1.11 SUMMARY

- The potential for an organism or individual cell to signal to its neighbouring organism or cell and to detect and respond to such signals is crucial for its survival.

- Signals used in biological systems appear to be very diverse, ranging from a simple change in the concentration of an intracellular ions such as Ca^{2+}, to complicated compounds.

- The production of a signal usually needs to be reasonably rapid, particularly for intracellular signals, and it must be able to be conveyed from its site of production to its site of action.

- A signal must be readily reversible.

- Very crucially, a signal has to contain specificity, relaying a defined message that can be translated into the correct cellular response.

- Signals between cells can be via:

  1. electrical potential changes, involving synaptic signalling,

  2. the release of compounds which are detected either over relatively large distances or even by the cell producing the signal; that is, paracrine, endocrine, or autocrine,

  3. the passage of small chemicals through pores in the cell, such as gap junctions and plasmodesmata.

- Signalling pathways can diverge to cause a multitude of cellular changes in response to a single signal, or they can converge, for example, where two or more signals effectively control the same metabolic pathway.

- In nearly all signalling pathways, a great deal of amplification takes place, allowing a small number of signalling molecules to precipitate a large effect.

- Signalling is best viewed as a complex signalling web, with many interactions between pathways and individual signalling components.

- The molecular study of the protein components has often revealed consensus patterns within the polypeptides and many of the proteins can be considered to be made of functional modules.

- Many signalling components have been identified as products of proto-oncogenes, highlighting the importance of these signalling components in the control of cellular proliferation and differentiation, and the study of the functioning of these gene products has helped enormously in the elucidation of many pathways.

- The development of modern molecular biology techniques has allowed the further understanding of how many of these proteins function and may even predict the existence of isoforms of proteins yet to be discovered.

- With the advent of post-genomics technologies, such as microarray analysis and proteomics, cell signalling is entering a new and exciting stage, where holistic studies will inform the future direction of signalling research.

1.12 FURTHER READING

Alberts, B., Johnson, A., Lewis J., Raff, M., Roberts, K., and Walter, P. (2002) *Molecular Biology of the Cell*, 4th edn. Garland Science, New York. Chapter 15 in particular.

Barritt, G.J. (1994) *Communications Within Animal Cells*. Oxford Science Publications, Oxford.

Bowles, D.J. (ed.) (1994) *Molecular Botany: Signals and the Environment*. Portland Press, London.

Bradshaw, R.A. and Dennis, E.A. (eds) (2003) *Handbook of Cell Signalling* (3 volume set) Academic Press. A very specialist book on many aspects of the field.

Cotter, T. (ed.) (2003) *Programmed Cell Death*. Essays in Biochemistry, vol. 39. Portland Press, London.

Hancock, J.T. (1997) *Cell Signalling* 1st edn. Longman, Essex. Contains many primary references removed from this edition.

Heldin, C.-H. and Purton, M. (eds) (1996) *Signal Transduction*. Nelson Thornes, Cheltenham. A collection of essays on topics in the field.

Helmreich, E.J.M. (2001) *The Biochemistry of Cell Signalling*. Oxford University Press, Oxford. A good text but with an over-emphasis on structure and mechanism.

Kornbluth, S. and Pines, J. (2003) Cell division, growth and death. *Current Opinion in Cell Biology* **15**, 645–648 and articles within this issue.

Kumar, S. and Bentley, P.J. (eds) (2003) *On Growth, Form and Computers*. Academic Press. Chapter 3 (pp. 64–81) in particular on the principles of cell signalling.

Lodish, H., Berk, A., Zipursky, S.L., Matsudaira, P., Baltimore, D., and Darnell, J. (1999) *Molecular Cell Biology*, 4th edn. W.H. Freeman, New York. Chapter 20 in particular.

Martinez Arias, A. and Stewart, A. (2002) *Molecular Principles of Animal Development*. Oxford University Press, Oxford. An excellent up-to-date book on development, with discussions on cell signalling.

Plant Cell (May 2002) **14** supplement, S1–S417. An issue dedicated to signalling in plant cells, including aspects of plant development.

Science (May 2002) **296**, 1632–1657. An excellent collection of articles on a range of aspects of cell signalling.

Stryer, L. (1995) *Biochemistry*, 4th edn. W.H. Freeman, New York. An excellent general biochemistry text, but particularly good for the signalling in the eye.

Amplification and physical architectures

Herskowitz, I. (1995) MAP kinase pathways in yeast: for mating and more. *Cell* 80, 187–197. Discussion of scaffolding proteins.

Oparka, K.J. (2004) Getting the message across: how do plant cells exchange macromolecular complexes? *Trends in Plant Science* 9, 33–41.

Šamaj, J., Baluška, F., and Hirt, H. (2004) From signal to cell polarity: mitogen-activated protein kinases as sensors and effectors of cytoskeleton dynamicity. *Journal of Experimental Botany* 55, 189–198. Includes scaffold proteins in discussion.

Coordination of signalling

Dumont, J.E., Pécasse, F., and Maenhaut, C. (2001) Crosstalk and specificity in signalling. Are we crosstalking ourselves into general confusion? *Cellular Signalling* 13, 457–463.

Dumont, J.E., Dremier, S., Pirson, I., and Maenhaut, C. (2002) Cross signaling, cell specificity, and physiology. *American Journal of Physiology Cellular Physiology* 283, C2–C28.

Taylor, J.E. and McAinsh, M.R. (2004) Signalling crosstalk in plants: emerging issues. *Journal of Experimental Botany* 55, 147–149. And articles within this special issue (no. 395) entitled *Crosstalk in Plant Signal Transduction*.

Domains and modules

Alex, L.A. and Simon, M.L. (1994) Protein histidine kinases and signal transduction in prokaryotes and eukaryotes. *Trends in Genetics* 10, 133–138. Illustration of protein modules.

Cohen, G.B., Ren, R., and Baltimore, D. (1995) Modular binding domains in signal transduction proteins. *Cell* 80, 237–248.

Feller, S.M., Ren, R., Hanufusa, H., and Baltimore, D. (1994) SH2 and SH3 domains as molecular adhesives: the interactions of Crk and Abl. *Trends in Biochemical Sciences* 19, 453–458.

Koch, C.A., Anderson, D., Moran, M.F., Ellis, C., Moran, M.F., and Pawson, T. (1991) SH2 and SH3 domains: elements that control interactions of cytoplasmic signalling proteins. *Science* 252, 668–674.

Lemmon, M.A. and Ferguson, K.M. (1998) Pleckstrin homology domains. *Current Topics in Microbiology Immunology* 228, 39–74.

Rebecchi, M.J. and Scarlata, S. (1998) Pleckstrin homology domains: a common fold with diverse functions. *Annual Review of Biophysics Biomolecular Structure* 27, 503–528.

Ren, R., Mayer, B.J., Cichetti, P., and Baltimore, D. (1993) Identification of a 10 amino acid proline rich SH3 binding site. *Science* 259, 1157–1161.

14-3-3 proteins

Fu, H., Subramanian, R.R., and Masters, S.C. (2000) 14-3-3 proteins: structure, function, and regulation. *Annual Review of Pharmacology and Toxicology* 40, 617–647.

Sehnke, P.C., DeLille, J.M., and Feri, R.J. (2002) Consummating signal transduction: the role of 14-3-3 proteins in the completion of signal-induced transitions in protein activity. *The Plant Cell* (suppl.) S339–S354. An excellent review of 14-3-3 proteins, and not just from plants.

Oncogenes

Cantley, L.C., Auger, K.R., Carpernter, C., Duckworth, A., Graziani, A., Kapeller, R., and Soltoff, S. (1991) Oncogenes and signal transduction. *Cell* 64, 281–302.

Rak, J.W. (2003) *Oncogene Therapies (Cancer Drug Discovery and Development)*. Humana Press, Totowa.

Watson, J.D., Gilman, M., Witkowski, J., and Zoller, M. (1992) *Recombinant DNA,* 2nd edn. Scientific Amercian Books, New York. Chapter 18 in particular.

History

Agranoff, B.W., Bradley, R.M., and Brady, R.O. (1958) The enzymatic synthesis of inositol phosphatide. *Journal of Biological Chemistry* 233, 1077–1083.

Ashman, D.F., Lipton, R., Melicow, M.M., and Price, T.D. (1963) Isolation of adenosine 3',5'-monophosphate and guanosine 3',5'-monophosphate from rat urine. *Biochemical and Biophysical Research Communications* 11, 330–334.

Burgoyne, R.D. and Petersen, O.H. (eds) (1997) *Landmarks in Intracellular Signalling*. Portland Press, London.

Cheung, W.Y. (1970) Cyclic 3'-5'-nucleotide phosphodiesterase: demonstration of an activator. *Biochemical and Biophysical Research Communications* 38, 533–538.

Ebashi, S. and Lipmann, F. (1962) Adenosine triphosphate-linked concnetration of calcium ions in aparticulate fraction of rabbit muscle. *Journal of Biological Chemistry* 14, 389–400.

Fire, A., Xu, S.Q., Montgomery, M.K., Kostas, S.A., Driver, S.E., and Mello, C.C. (1998) Potent and specific genetic interference by double-stranded RNA in *Caenorhabditis elegans. Nature* **391**, 806–811. The discovery of RNAi.

Hannun, Y.A., Loomis, C.R., Merrill, A.H., and Bell, R.M. (1986) Shingosine inhibition of protein kinase C and activity of phorbol dibutyrate binding *in vitro* and in human platelets. *Journal of Biological Chemistry* **261**, 12604–12209

Hokin, M.R. and Hokin, L.E. (1953) Enzyme secretion and the incorporation of P^{32} into phospholipids of pancreatic slices. *Journal of Biological Chemistry* **203**, 967–977.

Hunter, T. and Sefton, B.M. (1980) Transforming gene product of Rous sarcoma virus phosphorylates tyrosine. *Proceedings of the National Academy of Science USA* **77**, 1311–1315.

Krupinski, J., Coussen, F., Bakalyar, H.A., Tang, W-.J., Feinstein, P.G., Orth, K., Slaughter, C., Reed, R.R., and Gilman, A.G. (1989) Adenylyl cyclase amino acid sequence: possible channel-like or transporter-like structure. *Science* **244**, 1558–1564.

Kuo, J.F. and Greengard, P. (1970) Cyclic nucleotide-dependent protein kinases. *Journal of Biological Chemistry* **245**, 2493–2498.

Moore, B.W. and Perez, V.J. (1968) Specific acidic proteins of the nervous system. In *Physiological and Biochemical Aspects of Nervous Integration* (F. Carlson, ed.), pp. 343–359. Prentice Hall, Englewood Cliffs, NJ. Early report of 14-3-3 proteins.

Northup, J.K., Sternweis, P.C., Smigel, M.D., Schleifer, L.S., Ross, E.M., and Gilman, A.G. (1980) Purification of the regulatory component of adenylate cyclase. *Proceedings of the National Academy of Science USA* **77**, 6516–6520.

Palmer, R.M.J., Ferrige, A.G., and Moncada, S. (1987) Nitric oxide release accounts for the biological activity of endothelium-derived relaxing factor. *Nature* **327**, 524–526.

Paulus, H. and Kennedy, E.P. (1960) The enzymatic synthesis of inositol monophosphate. *Journal of Biological Chemistry* **235**, 1303–1311.

Rall, T.W. and Sutherland, E.W. (1958) Formation of a cyclic adenine ribonucleotide by tissue particles. *Journal of Biological Chemistry* **232**, 1065–1076.

Rall, T.W., Sutherland, E.W., and Berthet, J. (1957) The relationship of epinephrine and glucagon to liver phosphorylase. *Journal of Biological Chemistry* **224**, 463–475.

Ray, L.B. and Sturgill, T.W. (1987) Rapid stimulation by insulin of a serine/threonine kinase in 3T3-L1 adipocytes that phosphorylates microtubule-associated protein 2 *in vitro. Proceedings of the National Academy of Science USA* **84**, 1502–1506.

Sattin, A., Rall, T.W., and Zanella, J. (1975) Regulation of cyclic adenosine $3',5'$-monophosphate levels in guinea-pig cerebral cortex by interaction of α adrenergic and adenosine receptor activity. *Journal of Pharmacology and Experimental Therapeutics* **192**, 22–32.

Sutherland, E.W. and Rall, T.W. (1957) The properties of an adenine ribonucleotide produced with cellular particles, ATP, Mg^{++}, and epinephrine or glucagon. *Journal of the American Chemical Society* **79**, 3608.

Sutherland, E.W. and Rall, T.W. (1958) Fractionation and characterisation of a cyclic adenine ribonucleotide formed by tissue particles. *Journal of Biological Chemistry* **232**, 1077–1091.

Takai, Y., Kishimoto, A., Kikkawa, U., Mori, T., and Nishizuka, Y. (1979) Unsaturated diacylglycerol as a possible messenger for the activation of calcium-activated, phospholipid-dependent protein kinase system. *Biochemical and Biophysical Research Communications* **91**, 1218–1224.

Tonks, N.K., Diltz, C.D., and Fischer, E.H. (1988) Characterisation of the major protein-tyrosine-phosphatases of human placenta. *Journal of Biology Chemistry* **263**, 6731–6737.

Whitman, M., Downes, C.P. Keeler, M. Keller, T., and Cantley, L. (1988) Type 1 phosphotidylinositol kinase makes a novel inositol phospholipid, phosphatidylinositol-3-phosphate. *Nature* **332**, 644–646.

Woods, N.M., Cuthbertson, K.S.R., and Cobbold, P.H. (1986) Repetitive transient rises in cytoplasmic free calcium in hormone-stimulated hepatocytes. *Nature* **319**, 600–602.

The techniques in the study of cell signalling components.

Alberts, B., Bray, D., Lewis, J., Raff, M., Roberts, K., and Watson, J.D. (1994) *Molecular Biology of the Cell*, 3rd edn. Garland Press, New York. In particular pp.778–782 for discussion on computer networks: removed from the later edition.

Arenz, C. and Schepers, U. (2003) RNA interference: from an ancient mechanism to a state of the art therapeutic application. *Naturwissenschaften* **90**, 345–359. An excellent review of RNAi.

Causton, H.C., Quackenbush, J., and Brazma, A. (2003) *Microarray Gene Expression Data Analysis: a Beginner's Guide*. Blackwell Publishing, Oxford.

Desikan, R., Hagenbeek, D., Neill, S.J., and Rock, C.D. (1999) Flow cytometry and surface plasmon resonance analyses demonstrate that the monoclonal antibody JIM19 interacts with a rice cell surface component involved in abscisic acid signalling in protoplasts. *FEBS Letters* **456**, 257–262. An example of a paper which uses surface plasmon resonance to study cell signalling events.

Desikan, R.A.-H., Mackerness, S., Hancock, J.T., and Neill, S.J. (2001) Regulation of the *Arabidopsis* transcriptome by oxidative stress. *Plant Physiology* **127**, 159–172. An example of a microarray study, looking at cellular responses to a signal.

Hannon, G.J. (ed.) (2003) *RNAi; a Guide to Gene Silencing*. Cold Spring Harbor Press, New York.

Hidaka, H. and Kobayashi, R. (1994) Protein kinase inhibitors. In *Essays in Biochemistry*, vol. 28 (Tipton, K., ed.), pp 73–97. Portland Press, London. Discussion on kinase inhibitors.

Kendall, D.A. and Hill, S.J. (eds) (1995) *Signal Transduction Protocols*. Humana Press, Totowa. A particularly useful collection of protocols for cell signalling research.

Liebler, D.C. (ed.) (2001) *Introduction to Proteomics: Tools for the New Biology*. Humana Press, Totowa.

Matsumoto, B. (ed.) (2003) *Cell Biological Applications of Confocal Microscopy (Methods in Cell Biology)*. Academic Press.

Ng, T., Squire, A., Hansra, G., Bornancin, F., Prevostel, C., Hanby, A., Harris, W., Barnes, D., Schmidt, S., Mellor, H., Bastiaens, P.I., and Parker, P.J. (1999) Imaging protein kinase C-α activation in cells. *Science* **283**, 2085–2089. A study using FRET in conjunction with FLIM.

Palzkill, T. (2002) *Proteomics*. Kluwer Academic Publishers, Dordrecht. ISBN 0792375653.

Polverari, A., Molessini, B., Pezzotti, M. Buonaurio, R., Marte, M., and Delledonne, M. (2003) Nitric oxide-mediated transcriptional changes in *Arabidopsis thaliana*. *Molecular Plant-Microbe Interactions* **12**, 1094–1105. An example of a paper in which the expression of a large number of genes is studied, following the addition of a cellular signal to cells, and it is not surprising that many of the genes encode proteins involved in cell signalling.

Rutter, G.A., White, M.R.H., and Tavare, J.M. (1995) Involvement of MAP kinase in insulin signalling revealed by noninvasive imaging of luciferase gene-expression in single living cells. *Current Biology* **5**, 890–899.

Schena, M. (2002) *Microarray Analysis*. Wiley-Liss, NJ.

Tavaré, J.M., Fletcher, L.M., and Welsh, G.I. (2001) Review: using green fluorescent protein to study intracellular signalling. *Journal of Endocrinology* **170**, 297–306.

Terrian, D.M. (2002) *Cancer Cell Signalling: Methods and Protocols*. Humana Press, Totowa.

1.13 USEFUL WEB PAGES

Although the internet is a vast source of information, much of what is accessible is not well written, not peer reviewed, and potentially misleading. However, several journals and companies have excellent web sites that are valuable sources of information in cell signalling. Among these are:

Nature www.signaling-gateway.org/

Science http://stke.sciencemag.org

Molecular Probes www.probes.com An excellent site from a company that supplies fluorescent probes for molecular biological studies.

Calbiochem www.calbiochem.com An excellent site by a company supplying chemicals for cell signalling research; an internet site full of information.

2 Extracellular signals: hormones, cytokines, and growth factors

One of the basic tenets of cell signalling is that two cells need to communicate with each other. They may be a long way apart, or they may be neighbours, but still they often need to signal to each other. Many types of molecules are used as signals, and often the way they are classified appears to be a little arbitrary, but here the main groups are discussed. Animal hormones are commonly used as intercellular signals, and plants use hormones too.

The group of peptides known as the cytokines is used by animals for short-range signalling, and new members of this group of signals seem to be discovered on a regular basis. Some of this group were thought to be elixirs with which to cure many diseases, but their full potential has yet to be uncovered.

On the other hand, long-range signalling between organisms involves the pheromones, and these too are discussed here.

It is by the uncovering of the multitude of intercellular signals used in nature that an understanding of the messages that are sent between cells can be gained, giving an insight into how cells respond and survive. Often it is the dysfunction of these messages that leads to disease, and a greater understanding may also lead to ways to modulate such messages. Such manipulation may allow organisms to survive in new conditions; for example, perhaps new plants can be developed which require less water, or new drugs can be developed to alleviate the symptoms of a disease, or even cure the disease.

2.1 Introduction

One of the most significant ways in which cells communicate with each other is by the release and detection of extracellular signalling molecules, such as hormones, cytokines, and growth factors. Such molecules can often be released some considerable distance from their point of action, and have unspecific transport to the site of action. For example, a hormone might be released and carried by the bloodstream, in which it is transported to all parts of the body and washes around many types of cell, some of which will detect its presence, and many of which will not. Many cells will be unaffected by such a release of hormone, the specificity of effect being determined by the presence of specific receptor molecules present in the detecting cells (see Chapter 3; sections 3.1 and 3.4 in particular for detailed discussion).

Many of these extracellular signalling molecules are involved with and have been implicated in disease states, with either a defect being the cause of the disease, or the presence of the molecule propagating the manifestation of the disease. One of the most well studied and understood is the role of insulin in diabetes. This multifactorial disease may involve aberrations in the synthesis of the hormone, thus relaying no message to the target cells, or may involve a lack of detection by the target cells. Insulin and the signalling pathways it induces are further discussed in Chapter 9.

The role of hormones has been most well studied in animals, although such extracellular signals are widely used in the plant kingdom, and in fact as discussed in Chapter 1 (section 1.9), a plant hormone, or at least its effects, were studied by Darwin well over 100 years ago. Hormones have also been found to be important for unicellular eukaryotes, where the release of hormone-like molecules have been seen in the transfer of messages between individual organisms, in a similar manner used by mammals between tissues.

Although the distinction between hormones, cytokines, and growth factors appears to be a somewhat loose one, a rough division is given here and they are considered separately.

2.2 Hormones

Under the broad name of hormones, a term taken from the Greek meaning to arouse or excite, are several diverse types of molecule. Although the exact definition of a hormone appears to be somewhat vague, here the term is used to describe substances that are released into the extracellular medium by the cells of one tissue, to be carried to a new site of action, such as a different tissue, where they provoke a specific response. Hormones can be split into several broad classes:

- small water-soluble molecules;
- peptide hormones;

- lipophilic molecules which are detected by cell-surface receptors;
- lipophilic molecules which are detected by intracellular receptors.

However, what is common to all hormones is the fact that they have specificity, being released only when required, and detected only by cells which need to respond to them.

Small water-soluble molecules

These hormones share common characteristics in that they are all water soluble and cannot cross the plasma membrane of the cell, usually because they carry a charge at physiological pH and as such their detection requires the presence of specific receptors on the cell surface of the responding cell. They are usually released into the vascular system of the organism, to be carried freely to their site of perception. They are usually relatively small organic molecules and hormones which fall into this category include histamine and epinephrine (adrenalin) (see Fig. 2.1)

Histamine is produced in the mast cells and is responsible for the control of blood vessel dilation. Often topical drugs, for example for the treatment of bee stings, will contain anti-histamines which are used to counteract the release of histamine during an inflammatory response. Epinephrine, or as it used to be known, adrenalin, was classically known as the hormone released at times of panic, and has been referred to as the 'fright, fight, and flight' hormone. It is produced by the adrenal medulla and causes an increase in blood pressure and pulse rate, contraction of smooth muscles, an increase in glycogen breakdown in muscles and the liver, and an increase in lipid hydrolysis in adipose tissue, all factors which would ready the body in the situation where a greater energy requirement is envisaged, such as in fight and flight. In the majority of cases the effects of epinephrine are fast and, importantly, quickly

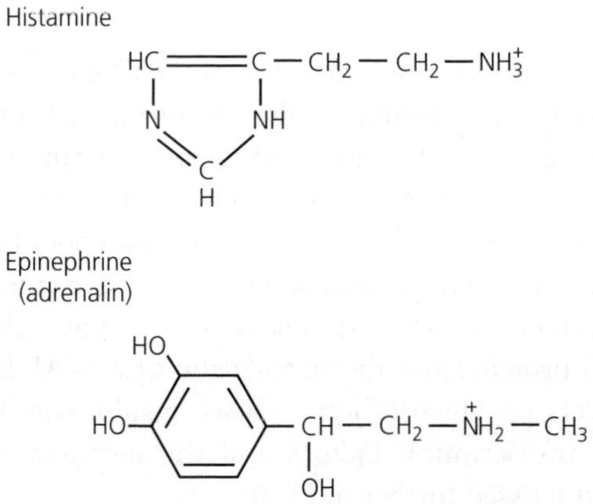

Histamine

$$HC = C - CH_2 - CH_2 - NH_3^+$$

Epinephrine
(adrenalin)

Fig. 2.1 The molecular structures of histamine and epinephrine (adrenalin): examples of small water-soluble hormones.

reversed. An organism would not wish to remain in a state of panic for too long.

Peptide hormones

Like histamine and epinephrine, the peptide hormones are water soluble and are carried to their site of action by the vascular system. Usually their release is rapid, as they are often stored within vacuoles in the cell, ready to be evicted from the cell by exocytosis. Such vacuoles will remain dormant in the cell until such time as the cell is stimulated, involving a signal transduction pathway (often involving a rise in calcium ion concentrations), when the vacuoles will be moved to the plasma membrane. Fusion of the vacuole membrane with the plasma membrane will cause the hormone to be released to the extracellular fluid (see Fig. 1.2b). The breakdown of peptide hormones is also rapid, usually by proteases in the blood and tissues. The two hormones which have been most extensively studied are glucagon and insulin, which in fact have antagonistic effects.

Insulin is produced by the β cells of the pancreas as a single preproinsulin polypeptide, which is converted to a single proinsulin polypeptide. This folds into its correct secondary and tertiary structure aided and stabilized by the presence of disulphide bridges. Once correctly folded active insulin is produced by proteolytic cleavage in two places causing removal of a substantial part of the middle of the polypeptide as illustrated in Fig. 2.2. The result of this is that insulin contains two polypeptide chains, an A chain of 21 amino acids and a B chain of 30 amino acids, held together by disulphide bonds and electrostatic forces.

■ The structure of insulin was identified by Dorothy Hogkins. Born 1910, she studied the structures of several biomolecules, including penicillin and vitamin B_{12}. She was awarded the Nobel Prize for Chemistry in 1964.

Once released, insulin stimulates the uptake of glucose by muscle and fat cells, increases lipid synthesis in the adipose tissues and causes a general increase in cell proliferation and protein synthesis. Detection of insulin is due to the presence on the surface of target cells of a multipolypeptide receptor (actually a tetramer) which contains intrinsic tyrosine kinase activity, and its activation leads to a cascade of events including the activation of phosphoinositide 3-kinase (PtdIns 3-kinase), activation of G proteins and the stimulation of a MAP kinase pathway. Among its effects on glucose homeostasis, insulin can also lead to the stimulation of transcription factors and the increase in selected gene expression, as discussed further in Chapter 9.

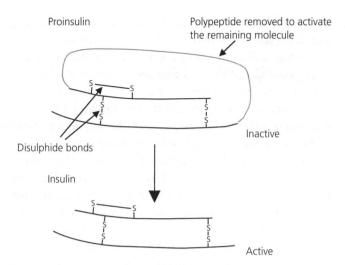

Proinsulin

Polypeptide removed to activate the remaining molecule

Disulphide bonds

Inactive

Insulin

Active

Fig. 2.2 The production of insulin
from proinsulin by the removal
of the linking polypeptide (shown
by the blue line).

Glucagon, like insulin and all peptide hormones, is also made from a precursor polypeptide. Active glucagon contains a single polypeptide chain of 29 amino acids. It is released from the α cells of the pancreas and leads to glycogen breakdown and lipid hydrolysis, allowing an increase in glycolysis and respiratory rates.

Other hormones in this class include follicle-stimulating hormone (FSH) and luteinizing hormone (LH). Both are like insulin in having two polypeptide chains. However, they are much bigger; FSH having an α chain of 92 amino acids along with a β chain of 118 amino acids. Both are produced by the anterior pituitary. FSH stimulates the growth of ovarian follicles and oocytes, and also increases the production of oestrogen. LH controls the maturation of oocytes and increases the release of oestrogen and progesterone.

Hormones often influence the release or action of other hormones, and here, the release of LH is under the control of another peptide hormone, this time a single-polypeptide hormone called LH-releasing hormone. This is produced by the hypothalamus and neurons. This type of interaction of different hormones and other extracellular signals, controlling each other's synthesis and secretion, is not unusual, and in fact interactions of this nature are exemplified by the action of cytokines and chemokines.

Lipophilic molecules which are detected by cell-surface receptors

As well as detection of the water-soluble hormones above, some receptors on the surfaces of cells also detect the presence of a group of hydrophobic, or lipophilic, molecules which act like hormones. The main group of signalling molecules here are the prostaglandins.

Prostaglandins are synthesized from arachidonic acid, a 20-carbon fatty acid. Arachidonic acid is often found covalently attached to the glycerol backbone of phospholipids which constitute the plasma membrane. The arachidonic acid is attached to the middle carbon, but can be hydrolytically released from lipids, for example, by the action of phospholipase A_2.

As the prostaglandins are derived from a molecule embedded in a lipid environment, it is not surprising that they are inherently hydrophobic in nature. Although prostaglandins are a relatively large group of chemicals they can be divided roughly into nine different classes. Although originally named after the organ where they were thought to be made, prostaglandins are in fact produced by most cells, with their action usually being local. Their synthesis can be inhibited by several anti-inflammatory drugs including aspirin and their effects range from control of platelet aggregation to the induction of uterine contraction.

Lipophilic molecules which are detected by intracellular receptors

Not all hormones are perceived by cell-surface receptors. Many are recognized by receptors inside the cell, and therefore such hormones need to be lipophilic, or hydrophobic, to enable them to penetrate the plasma membrane before perception. This class of hormones encompasses the steroid hormones, which include oestrogen, progesterone, androgens, and glucocorticoids, as well as the thyroid hormones and retinoids (see Fig. 2.3).

Steroid hormones are all derived from cholesterol and are synthesized and secreted by endocrine cells. Progesterone is synthesized by the ovaries and placenta and is involved in the development of the uterus in readiness for implantation of the new embryo as well as for the stabilization of early pregnancy and development of the mammary glands. Oestrogens, such as oestradiol, are also involved in the development of female sexual characteristics such as uterus differentiation and mammary gland function. Similarly, testosterone, which is produced by the testis, is responsible for the development and functioning of the male sex organs, as well as the development of less obviously useful male characteristics such as hair growth.

Mammals are not the only animals which use such hormones. For example, in insects and crustaceans a similar role in the development of sexual characteristics is fulfilled by a steroid-like compound called α-ecdysone.

Other steroid hormones include cortisol and vitamin D. Cortisol, a glucocorticoid, is produced by cells of the cortex of the adrenal gland and

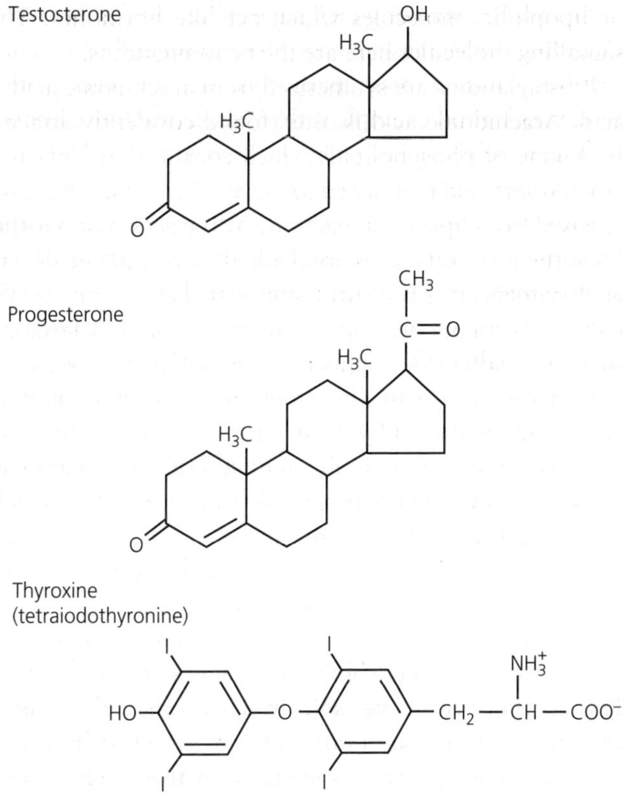

Testosterone

Progesterone

Thyroxine
(tetraiodothyronine)

Fig. 2.3 The molecular structures of testosterone, progesterone, and thyroxine: examples of lipophilic molecules detected by intracellular receptors.

controls the metabolic rates of many cells. It is formed from progesterone by three hydroxylation steps, at carbons 11, 17, and 21. Vitamin D is synthesized in the skin when it is exposed to sunlight. 7-Dehydrocholesterol (provitamin D_3) in the skin is lysed by ultraviolet light to previtamin D_3, an inactive form, which is activated by hydroxylation in the liver and kidneys with its conversion to calcitrol (1,25-dihydroxycholecalcitrol). Calcitrol is responsible for the control of Ca^{2+} uptake in the gut and lowering Ca^{2+} excretion by the kidneys. Lack of vitamin D in a child's development can lead to the development of the condition known as rickets. Here the cartilage and bone fail to calcify properly, leading to the malformation of the bones, and this often leaves the long bones of the patient bent. This is most noticeable in the legs of patients and was seen when children were forced to work long hours indoors or underground away from adequate sunlight, a situation which is fortunately now not common. Dietary vitamin D, such as derived from fish oils, can overcome the lack of its *de novo* synthesis in the skin and prevent the associated symptoms.

Thyroid hormones which have effects on the metabolism of many cells, including the increase in heat production and production of polypeptides involved in metabolic pathways, are derived from the amino acid tyrosine. A good example is thyroxine, otherwise known as tetraiodothyronine, which is produced by the thyroid gland. The structure of thyroxine is shown in Fig. 2.3.

Another vitamin which is involved in the synthesis of lipophilic hormones is vitamin A (also called retinol), from which the retinoids are synthesized. For example, retinoic acid is formed by the oxidation of the alcohol group of retinol to a carboxyl group. The effects of retinoic acid include an alteration of gene-expression profiles in the receptive cell.

Retinol is also the precursor of retinal, the light-sensitive group of rhodopsin, and therefore is instrumental in the photoperception of mammals (see section 10.2). It was thought that a deficiency in vitamin A could lead to an impairment in light detection of the eye, especially under low lighting conditions, a condition commonly called night blindness. This fact was used as propaganda during the second world war to persuade people to eat vegetables, particularly carrots.

Transport between cells of these lipophilic homones is not as simple as that seen for the water-soluble ones, as once released these lipophilic molecules are inherently insoluble in water. Therefore, in the bloodstream, they need to be stabilized, which is achieved by their association with specific carrier proteins. Dissociation from these carriers then needs to occur before they cross the plasma membrane and enter the cells.

In general, these hormones can persist in the circulation for hours or even days and are involved in long-term control. However, once inside the cells they are detected by receptors in the cytosol or in the nucleus (see section 3.2), often culminating in the alteration of specific gene expression, which again might lead to long-term effects in the cells and tissues.

2.3 Plant hormones

The existence of plant hormones was first postulated by Darwin, who showed by experimenting on the phototropism of coleoptiles that substances, yet to be identified, must be transported around plants. By 1928 Went had discovered auxin and now a wide range of substances have been placed under the rather broad umbrella of plant hormones. It is in fact a very loose term and other terms such as plant growth substances and phytohormones have been suggested, but never used extensively. In animals, the extracellular signalling molecules have been divided, albeit rather crudely, into classes such as hormones and cytokines: the term plant hormone embraces all the substances which act as extracellular signals in plants (**see Fig. 2.4** for the structures of some of the common ones). This means that they might exert their influence at a site that is some distance from their site of manufacture, or act within the tissue where they are made or even on the same cell. For example, ethylene has a very short distance of influence, whereas cytokinins can be transported from the roots to the leaves.

Auxin

Although the main auxin found in plants is indole-3-acetic acid (IAA; Fig. 2.4), several compounds derived from this molecule show auxin-like activity, including indoleacetaldehyde or indoleacetyl aspartate. IAA is manufactured in the leaves, particularly young leaves and in developing seeds from tryptophan or indole. Its transport appears to be primarily from cell to cell but it can also be found in the phloem. Its effects are diverse, ranging from the stimulation of cell enlargement and stem growth, stimulation of cell division and the differentiation of the phloem

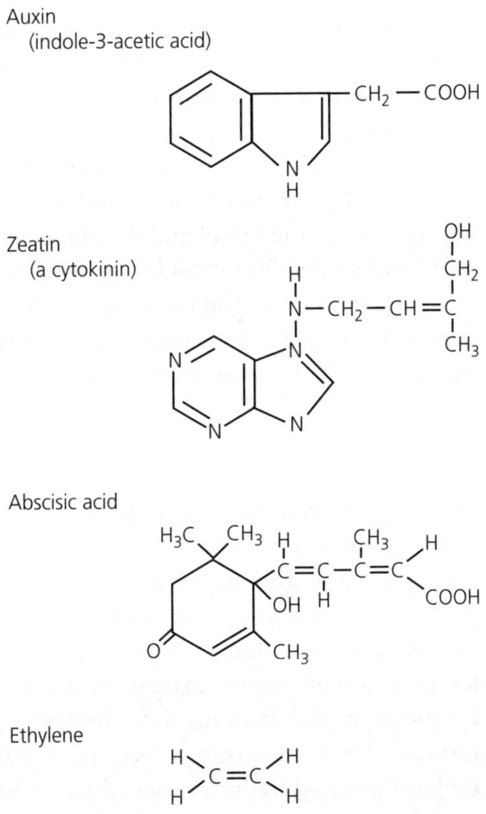

Fig. 2.4 The molecular structures of auxin (indole-3-acetic acid; IAA), zeatin, abscisic acid, and ethylene: examples of plant hormones.

and xylem vessels, to the mediation of tropistic responses to light and gravity. Even though auxin was identified a long time ago, and many of its effects have been well characterized, its perception—that is the receptor protein to which it binds—has yet to be identified unequivocally and understood. One of the main proteins being studied is known as auxin-binding protein 1 (ABP1), which has now had its structure solved, and with site-directed mutagenesis studies underway its mode of action should be revealed. However, its main cellular location appears to be associated with the endoplasmic reticulum, a place which is at odds with its perception of an extracellular hormone.

Cytokinins

Cytokinins (CKs) are derived from adenine in the root tips and in developing seeds, the most common CK being zeatin (see Fig. 2.4). Their transport is via the xylem system and their activity includes the induction of cell division, although this requires the presence of auxin, leaf expansion caused by cell enlargement, and the delay of leaf senescence.

Gibberellins

The most common biologically active gibberellin in plants is gibberellin A_1 (GA_1), although the term gibberellin encompasses a family of active compounds. Their structures are complex, based on what is known as the *ent*-gibberellane structure. This involves four ring structures with a fifth oxygen-containing ring bridging across. Gibberellins are synthesized from mevalonic acid in young shoots and in developing seeds while their transport is via both the xylem and the phloem vascular systems. The biological activity of gibberellins includes the stimulation of flowering, the induction of seed germination, and the stimulation of the synthesis of enzymes such as α-amylase. Their perception by cells also leads to cell division and cell elongation which manifests itself as the elongation of the stems.

Abscisic acid

Like the gibberellins, abscisic acid (ABA) is synthesized from mevalonic acid, but this time mainly in the roots and mature leaves, although most tissues are probably capable of producing it. Seeds can also synthesize ABA or indeed import it from the parent plant. ABA transport is carried out by both the phloem, down to the roots and by the xylem up to the leaves. It is synthesized particularly in times of water stress and one of the functions of ABA is to mediate stomatal closure, and so control transpirational water loss from leaves. Here ABA perception invokes a wide range of intracellular responses which include the production of hydrogen peroxide and nitric oxide, and the stimulation of calcium signalling. It also induces the synthesis of storage proteins in seeds and inhibits shoot growth. However, as with auxin, the receptor which perceives ABA has remained elusive.

Ethylene

What appears to be an odd compound to act as a signalling molecule is the gas ethylene (Fig. 2.4). However, other gaseous compounds that act as cell signals, such as nitric oxide (NO), have been discovered, firstly in animals and then in plants, as will be discussed in Chapter 8 (section 8.2). Ethylene is synthesized from methionine in most plant tissues, particularly in response to stress. Its transport is by diffusion although intermediates in its synthesis can be transported around the plant and lead to its appearance at a site distant from the original stimulus. One of the most commonly known roles for ethylene is in the induction of fruit ripening but it can also lead to shoot and root differentiation and growth, and to the opening of flowers, among many others.

Other plant hormones

Many other compounds come under this broad term of plant hormone, including jasmonates (discussed in section 6.8), polyamines, salicylic acids, and brassinosteroids. This last group of contains over 60 steroidal compounds, while salicylic acid has been implicated in thermogenesis and the production of pathogen-induced proteins.

Other extracellular plant signals, perhaps not classed as hormones, include the reactive species hydrogen peroxide and nitric oxide. Although first identified as instrumental signals in mammals, these compounds have now been found to be important signals in plants, controlling water loss from leaves by modulating stomatal apertures, and inducing the death of cells by a mechanism similar to that of apoptosis, referred to as programmed cell death (PCD), which is seen as part of the hypersensitive response.

▉ The hypersensitive response is a mechanism used by plants to fight pathogen attack. It often involves the death of cells in the immediate location of pathogen perception, and so limiting pathogen spread to the rest of the plant.

2.4 Cytokines

A group of peptide molecules which have profound effects on cells, but which are classed separately from the hormones, are the cytokines and chemokines. These extracellular signals have only come to light relatively recently, compared to the work carried out on hormones, and the rate at which new members of these families are being discovered is truly alarming, and no doubt new cytokines and chemokines are yet to be found. However, with the completion of the human-genome sequencing project, it should be possible to identify all of them, even if we don't know what they all do.

The cytokines are a group of peptide molecules which are produced by many cell types, but which have their effect on other cells within a short distance, or often even on the cells from which they originate. Hence, cytokine effects tend to be local, where they are referred to as being involved in paracrine or autocrine function. Under this definition come those molecules which are principally responsible for the coordination of the immune response of higher animals, and include the interleukin (IL) series, tumour necrosis factors (TNFs), and interferons.

At present there are more than 80 peptides which have been classed as cytokines. Many of the cytokines have names which are historic, in that they were classified as part of a growing family, for example the interleukin series, while others are named in logical way which denotes a function or characteristic, for example, FGF-a, which stands for fibroblast growth factor acidic. Often such naming leads to acronyms, which at least allows for the easy memorization of their names. Excellent examples here include Trail (TNF-related apoptosis-inducing ligand), Trance (TNF-related activation-induced cytokine) and April (a proliferation-inducing ligand).

The sizes of the cytokines is quite variable, with some being relatively small, weighing in at less than 10 kDa. Others are much larger, with individual subunits as large as 60–70 kDa. However, the majority seem to fall between 15 and 40 kDa. In most cases, the receptors which perceive the cytokines have now been identified, and the intracellular cascades which are invoked are being unravelled. The likely receptors and transduction components involved are discussed in the following chapters.

There is a massive interest worldwide in these peptides, not least because the elucidation of the mechanisms by which they function should lead to the identification of future drug targets, and as they are involved in a plethora of diseases they are extremely important.

Despite the wide range of cytokines, it is worth focusing on three families, namely the interleukins, interferons, and tumour necrosis factors.

Interleukins

The interleukin series has at the present time 18 members identified, designated IL-1 to IL-18. They span across several classes of cytokine, although IL-8 is classified outside of the cytokine series, being considered a chemokine (see below). Although IL-1 was named as one, it in fact exists in two distinct forms, IL-1α and IL-1β, which are coded for by separate genes. Both are produced as larger precursor molecules and a cleavage event produces the active extracellular form. IL-1α is produced as 271 amino acids which is cleaved to 159 amino acids while IL-1β is a 153 amino-acid peptide derived from a 269 amino-acid precursor. Therefore, similar to the peptide hormones, such as insulin, a major cleavage event has taken place, removing a substantial proportion of the polypeptide. Such cleavage events are common in the synthesis of the interleukin family. In the case of IL-1, the uncleaved cell-associated form still seems to retain biological activity. In fact, several of the cytokine family appear to have membrane-bound or membrane-associated forms, and are not released freely from the cell.

Several of the genes for the cytokines, for example IL-3, IL-4, IL-5, IL-13 along with granulocyte-macrophage colony-stimulating factor (GM-CSF), are grouped together on the same region of chromosome 5 in humans, or chromosome 11 in mice, suggesting that they originally arose through gene-duplication events. This is quite commonly seen where families of proteins exist. Evolution has allowed the copying of a successful protein, and then its subsequent subtle alteration to fulfil a new function or role, unable to be undertaken by the original protein. Such a process is then repeated, building up a family of subtly different but related proteins, each with slightly, but significantly, different roles.

Table 2.1 summarizes some of the characteristics of the interleukins, and indicates the receptors that have been identified.

Most of the interleukins are active in the monomeric state, but IL-5 exists as a homodimer of two 115 amino-acid polypeptides, while IL-12 exists as a heterodimer of variable-sized polypeptides. IL-8, otherwise known as neutrophil-activating protein (NAP-1) is also a homodimer but here the subunits are of variable length, 72 up to 77 amino acids, due to truncation of the N-terminal end of the polypeptides. It is interesting that the IL-8 molecules which contain the shorter versions of the subunits are more potent than the ones with the longer polypeptides.

Cytokines are produced by a variety of blood cells and cells involved in the immune response. IL-2 and IL-3 are for example produced mainly by helper T lymphocytes, while IL-6 is produced by a variety of cells including T cells, macrophages, and fibroblasts.

The biological responses of the interleukin family are also varied. IL-13, formerly known as P600, seems to induce the growth and differentiation of B cells and inhibits cytokine production of macrophages and their precursor cells, monocytes. IL-5 leads to the activation of eosinophil function, including chemotaxis and eosinophil differentiation. Some of the individual cytokines have a diverse range of biological activities, a good example being IL-6. It stimulates the differentiation of myeloid cell lines, acts as a growth factor on B cells, modulates the responses of stem cells to

Table 2.1 The interleukin family and some of their characteristics.

Cytokine	Amino acids		Molecular weight (kDa)	Receptor(s) used	Chromosomal location
	Precursor	Active form			
IL-1α	271	159	17.5	IL-1 RI, IL-1 RII	2q13
IL-1β	269	153	17.3	IL-1 RI, IL-1 RII	2q13–q21
IL-2	153	133	15–20	IL-2 Rα/β/γc, IL-2 R/β/γc	4q26–q27
IL-3	152	133	14–30	IL-3 Rα/βc	5q23–q31
IL-4	153	129	15–19	Combinations with IL-13 R	5q23–q32
IL-5	134	115	45	IL-5 Rα/βc	5q23.3–q32
IL-6	212	184	26	IL-6 R/gp 130	7p15–p21
IL-7	177	152	20–28	IL-7 R/γc	
IL-8	99	72–77	8–8.9	CXCR-1, CXCR-2	
IL-9	144	126	30–40	IL-9 R/γc	5q31–q35
IL-10	178	160	39 (dimer)	IL-10 R1/R2	1
IL-11	199	178	23	IL-11 Rα/gp130	19q13.3–q13.4
IL-12		197 and 306	70 (heterodimer)		
IL-13		132	9, 17	Four combinations: some with IL-4 R	5q31
IL-14			53		
IL-15	162	114	14–15	IL-15 Rα/IL2 Rβ/γc	
IL-16			14	CD4	
IL-17		136	15,22	IL-17 R	
IL-18		157	18	IL-18Rα, IL-18Rβ	

other cytokines and has effects on non-haematopoietic cells, including effecting the development of nerve cells.

These cytokines all work in a concerted way to coordinate the whole response. They can exert coordinating responses on the cells themselves or regulate the synthesis of each other. For example, IL-5 enhances the IL-4 effect on B cells, enhancing the IL-4-induced synthesis of IgE and the expression of CD23 while IL-11 has synergistic effects with both IL-3 and IL-4. On the other hand, IL-6 can induce the production of IL-2 in T cells but IL-10 inhibits the synthesis of several other cytokines, including IL-1, IL-6, IL-8, IL-10, and IL-12. Some cytokines produce similar responses. In the case of IL-13 and IL-4 no additive effect is seen if both of these cytokines are added together, and it is possible that they share a common receptor, or share the same signal transduction pathway. Likewise IL-15 and IL-2 may share common elements in their receptors.

An interesting interleukin is IL-1ra, which acts as an antagonist to IL-1 by competing for the receptor binding sites, so reducing the effect of IL-1. The production of IL-1ra may itself be under the control of IL-13.

Interferons

The interferons (IFNs) fall into two main groups. Type I includes IFN-α, IFN-β, and IFN-Ω while type II includes γ-interferon (IFN-γ). IFN-γ rose to prominence in the press as an anti-cancer agent and was branded as 'a wonder drug'. Although the interferons do have profound effects in some conditions, they are still not widely distributed by the medical profession, perhaps in some cases because of their terrific cost.

Like the interleukins, IFN-γ is a peptide. Here the polypeptide is 143 amino acids long and the interferon exists as a dimer in the active form, or even multimers. However, the peptide is made from the expression of a single gene on the long arm of chromosome 12 in humans, the sequence including a signal sequence of 23 amino acids. The α/β forms on the other hand are monomers, coded for by genes on chromosome 9.

Produced by T cells amongst other cells, IFN-γ's activity includes the induction of expression of the class II histocompatibility antigens on epithelial, endothelial, and connective tissue. It also acts as an macrophage-activating factor, causing specific gene expression in these cells. This leads to the enhanced ability of the cells to be cytotoxic to tumours and to kill parasites.

Tumour necrosis factors

Tumour necrosis factor-α (TNF-α), originally known as cachectin, is coded for by a single gene which in humans is found on chromosome 6, but in the soluble active form it is a homotrimer of 157 amino-acid subunits (17.5 kDa form). There is also a 26 kDa form that remains membrane-anchored.

TNF is produced by monocytes and macrophages. Its activity is closely coupled to that of IFN-γ and it has been shown to cause necrosis of tumours, hence its name, and to be cytotoxic to transformed cells *in vitro*.

TNF-β (also known as lymphotoxin-α, LT-α) is also a homotrimer and is encoded for by a gene that is close to the TNF-α gene. They both have similar biological activity and even bind to the same receptors, TNF-RI and TNF-RII. However, TNF-β can also form membrane-surface trimers with LT-β, and bind to the membrane protein herpes virus entry mediator (HVEM).

Other cytokines, chemokines, and receptors

Many of the small protein signalling molecules come under the grouping of chemokines. This is a large family, or superfamily, of peptides which are grouped together because of their structural relationships rather than any functional similarity. At present there are at least 36 peptides that can be grouped under the banner of chemokine.

There are two main sub-groups of the chemokine family, characterized by a particular structural motif containing four cysteine residues. The first sub-group is known as the CC chemokines and these contain the sequence:

$$\text{NH}_2\text{-X (10–11)-\textbf{Cys-Cys}-X (22–23)-Cys-X (15)-Cys-X (18–24)-\text{COOH}}$$

where the numbers in brackets denote the number of amino acid residues which are likely to be present at this point in the primary structure. Members of this group include Rantes (regulated on activation, normal T cell expressed and secreted), monocyte chemoattractant proteins (MCP1–4), macrophage inflammatory proteins (MIPs), and eotaxins. As the names suggest, like the cytokines, these peptides are commonly involved in immune responses.

The second class of chemokines is termed the CXC chemokines, and they contain the sequence:

$$\text{NH}_2\text{-X (6–12)-\textbf{Cys}-X \textbf{(1)}-\textbf{Cys}-X (23–24)-Cys-X (15–16)-Cys-X (15–53)-\text{COOH}}$$

where the first two cysteine residues are separated by one amino acid. This group contains peptides such as IL-8, monokine induced by IFN-γ (MIG), platelet factor 4 (PF4), and neutrophil-activating peptide 2 (NAP-2).

At least two chemokines fall outside of this CC or CXC classification. These are lymphotactin (LTN) and fractalkine, the latter classed as a CX_3C chemokine.

Chemokines have been alternatively referred to by the names intercrines, the small cytokine family, or scy, and the small, inducible secreted cytokines, or SIS. Many of these chemokines have been discovered not by their functional activity but by their sequence homology when their cDNAs have been cloned.

Many of the receptors for the cytokines have been found to contain some commonality in their structure and primary amino acid sequence. Four cysteine residues among a hydrophobic region along with a hydrophobic region encompassing runs of positively and then negatively charged residues have been found in several of the receptors cloned. Downstream of the receptor, the intracellular signalling on binding of the cytokine usually involves tyrosine phosphorylation, and this will be discussed in Chapter 4 (section 4.3).

2.5 Growth factors

The term growth factor here is used to define those compounds which have been shown to have specific functions in the regulation of the growth and differentiation of cells. Other works may well categorize members of this group under the other headings used above. This emphasizes the vagueness of the terms used to classify extracellular signals. At present over 50 known proteins which possess growth factor-like activity have been reported with at least 14 different receptor families involved in their detection.

Platelet-derived growth factors

Platelet-derived growth factors (PDGFs) are dimeric proteins which may contain two related polypeptides, A or B. The active growth factor has a molecular weight of between 25 and 29 kDa, and is made up of either two of the same subunits, that is AA or BB, or it may be a heterodimeric protein, AB. The individual subunits have been shown to have a molecular weight of 12–18 kDa and are coded for by two separate genes. They are held together by disulphide bonds, and in fact all the cysteine residues in the polypeptides are involved in either inter- or intra-molecular bonding. Interestingly, the receptors for PDGF are also dimeric, composed of either two identical subunits, αα or ββ, or two different ones, αβ, and the different forms of the receptor have different binding properties to the different forms of the PDGF (see Table 3.2). PDGF-BB, with two B subunits, can bind and activate all the receptor forms, whereas PDGF-AA, with two A subunits, can only activate the αα receptor. The heterodimeric PDGF-AB has an intermediate activity.

PDGF has been shown to induce both cell migration and cell proliferation and has been implicated in several disease states including fibrosis and arteriosclerosis.

Epidermal growth factor

A larger group of growth factors comes under the heading of epidermal growth factor (EGF). These include EGF itself, transforming growth factor-α (TGF-α), betacellulin, and heparin-binding EGF. They are very small peptides, rat EGF being only 5.2 kDa, although their precursors are very large. The EGF precursor is in fact a transmembrane protein of 1168 amino acids of which only 53 are cleaved off to create the active EGF molecule. Likewise TGF-α is only 6 kDa (50 amino acids), although there are larger members of the family—sensory and motor neuron-derived factor (SMDF) being 296 amino acids.

EGFs are characterized by the presence of at least one domain containing six cysteine residues which are involved in the formation of three disulphide bonds. These growth factors also appear to have several aromatic residues which are exposed to the aqueous medium. This is unusual in proteins which in general hide hydrophobic side chains in the interior of the folded polypeptide structure. With EGF it has been proposed that there are interactions between these exposed aromatic groups and the protein surface may contain an aromatic domain which is vital for the function of the protein. Proteins with domains related in sequence to EGF have been found in *Drosophilia* and in sea urchin embryos. The proposed signal transduction cascade leading from EGF receptors is discussed in Chapter 4 (section 4.3) and illustrated by Figure 4.8.

Fibroblast growth factor

Fibroblast growth factor (FGF) represents a family of molecules which are involved in the regulation of proliferation, differentiation, and cell mobility. In mammals the family comprises many members, including FGF-a, FGF-b, and FGFs 3–19.

Species differences

In general mammals are grouped together and assumed to be similar to each other, but here it should be noted that there appears to be no FGF-15 in humans. This highlights the inherent danger of assuming that all the animals in a group are the same at the molecular level.

The FGFs are structurally related proteins of 20–30 kDa, with the genes encoding them probably arising from gene duplication of an ancestral gene. However, as with the expression of many genes further isoforms can arise from alternative splicing or the use of alternative initiation codons for translation. Alternative post-translational modification leads to further diversification of the molecules within the family, where some are glycosylated, phosphorylated, or even methylated or cleaved.

Some of the FGF family have been identified as proto-oncogenes, where the oncogene variants lead to the transformation of cells, where proliferation is uncontrolled resulting in the formation of tumours.

2.6 Neurotransmitters

In the axon termini of the presynaptic cells of the nervous system there are storage vesicles called synaptic vesicles which contain neurotransmitters. Voltage-gated Ca^{2+} channels are opened on the arrival of an action potential, leading to a sharp increase in the concentration of intracellular Ca^{2+}. This in turn leads to exocytosis from the synaptic vesicles, releasing the neurotransmitters into the space between the nerve cells, with receptors on the postsynaptic cells detecting the presence of these compounds, leading to the propagation of the signal or the response.

These transmitters fall into two main groups. The first group is made up of a group of small molecules and contains neurotransmitters including acetylcholine, GABA or γ-aminobutyric acid, and dopamine. This group also includes some molecules already discussed above, such as epinephrine and histamine (see Fig. 2.5).

Also included here in this group of neurotransmitters are amino acids which can act as signals, such as glycine and glutamate, along with some derivatives of amino acids. For example, dopamine is derived from

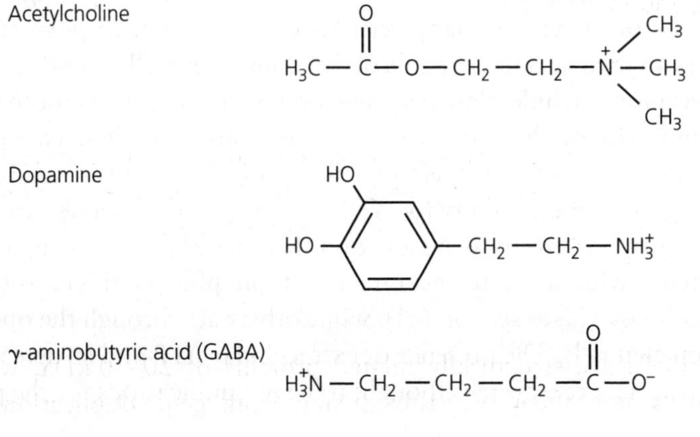

Fig. 2.5 The molecular structures of acetylcholine, dopamine, and γ-aminobutyric acid (GABA): examples of neurotransmitters.

tyrosine, serotonin is a derivative of tryptophan, and GABA is derived from glutamate. Also included here are nucleotide-derived compounds such as ATP and adenosine.

The second group are the neuropeptides. These include among their number substance P, β-endorphin, and vasopressin, the latter also being classed among the hormones.

■ Substance P was the first neuropeptide to be discovered.

The effects of the neurotransmitters are local, acting on the postsynaptic cell (across the synaptic cleft), or at least within a short distance. The receptors on the postsynaptic cell themselves fall mainly into two classes: G protein-linked or ligand-gated ion channels (see section 3.2).

2.7 ATP as an extracellular signal

Although ATP is well known as the molecule that acts as an ubiquitous supply of energy in cells, extracellular ATP can arise. ATP can be made in small amounts by the anaerobic metabolism of glycolysis, although this is not usually enough to sustain an active or busy cell, and therefore the majority of ATP is produced by aerobic respiration of mitochondria or from photosynthetic pathways in chloroplasts. Even though it is so crucial for the survival of the individual cells, extracellular release from cells can occur, for example, in cells involved in the secretion of neurotransmitters, or through storage granules from cells such as those of the adrenal medulla cells, or lymphocytes. Cell death and breakage can also lead to the release of intracellular ATP into the extracellular medium and therefore the local concentrations of ATP may be nanomolar or even micromolar. However, three different ectonucleotidases are responsible for the sequential hydrolysis of ATP to adenosine, so removing the ATP from solution. Even so, many cells have been shown to possess purinoceptors on their surface, capable of detecting ATP. Cells found to contain such receptors include platelets, neutrophils, fibroblasts, smooth muscle cells, and cells of the pancreas. These receptors fall into two groups, P_1 and P_2, depending on their specificity for the adenosine compound, P_2 having the highest affinity for ATP. P_2 receptors themselves can be subdivided into four groups depending on their action, some acting through G proteins with a commensurate effect on phosphatidylinositols via phospholipase C (see section 6.3), while others act through the operation of cation channels. The presence of extracellular ATP and its detection by the appropriate receptors is thought to have ramifications for the process of apoptosis, for example, and therefore should not be dismissed as a strange anomaly.

The slime mould *Dictyostelium discoideum* can exploit another nucleotide which is more normally associated with having a intracellular function as an extracellular signal; that is, the adenosine compound cAMP (see section 5.2 for discussion of cAMP as an intracellular signal). Here, cAMP is used for controlling differentiation and cellular aggregation. If a food source becomes scarce, cAMP is released into the extracellular medium and acts as a chemoattractant, signalling to the normally free-swimming cells to aggregate into a slug. This is also accompanied by changes in gene expression and culminates in the formation of a fruiting body.

2.8 Pheromones

Karlson and Lüscher first defined pheromones in 1959. They are substances which are excreted by an individual and which have their effect on another individual of the same species. The word pheromone comes from Greek, where *pherein* means to transfer and *hormon* means to excite.

Many bacterial species use pheromones in communication, the effects usually being seen if the cells grow to a particular density. Responses induced by pheromones include the production of light, or luminescence, the production of virulence factors, the development of fruiting bodies, and plasmid transfer.

The chemical structure of bacterial pheromones seems to be quite variant and includes amino acids, short peptides, proteins, and branched-chain fatty acids. One of the largest groups of pheromones are the so called AHLs, or N-acetyl-L-homoserine lactones. Many AHLs not only induce a response in another individual but also induce the genes for their own production. The transcriptional activators used in the AHL response are grouped together in what is called the LuxR superfamily of response regulators, of which at least 15 members of the family are known. Similarly, a superfamily of enzymes which produce AHLs has been defined, the 10 members of this family being known as the LuxI superfamily.

It is not just prokayotes that use pheromones in their organism-to-organsim communication, such systems are also used extensively by higher organisms. The so-called water mould, *Allomyces*, uses a compound called sirein as a sexual attractant. This compound is related to a cyclic organic compound called 2-carene which is found in pine resins. Another slime mould, *Achlya*, uses two steroid pheromones, one produced by the male and one produced by the female. Detection of male pheromone by the female is essential for the development of the female's sexual machinery, and likewise the male needs to detect the female's pheromone.

The sea anenome *Anthopleura elegantissima* uses a positively charged organic compound called anthopleurine. This is an interesting pheromone as it is distributed by a second species. Sea anenomes are eaten by sea slugs, and when this occurs the sea slug therefore also ingests the pheromone from the sea anemone. The pheromone is then released by the sea slug as it moves around and this acts as a warning to other sea anenomes that a predator is coming. The sea anemones near the advancing sea slug will then retract as a defence.

Much of the early work on pheromones was carried out by Adolph Butenandt, a German organic chemist. He worked with the silk-worm moth, *Bombyx mori*. This insect uses a compound which has been named bombykol, a long C_{16} unsaturated fatty acid. Much work has been done subsequently to study the pheromones of other insects.

Higher organisms also use pheromones in their silent communications, including fish, amphibians, and even mammals, including humans, and although it is impossible to discuss them all here in detail, the principles are the same: one individual communicating to another by means of chemical messengers, in a way analogous to the communication between cells in a single individual.

2.9 SUMMARY

- Cells commonly communicate by the release and detection of signalling molecules, often over relatively large distances.

- Released compounds have often been referred to as hormones, taken from the Greek meaning to arouse or excite.

- Other released molecules are now categorized separately and are known as cytokines, chemokines, or growth factors.

- Such classification does not appear to follow hard and fast rules, allowing the same compounds to be classified differently, a problem that might lead to confusion.

- It is important to remember that, like all signals, extracellular signals need to relay a specific message, to specific targets, at the correct time. That might require their transportation between cells within an

organism, or even between organisms. Either way, their recognition, and their ability to cause a defined response is the key to their success as messengers.

- Hormones are a diverse group of molecules, which may be small and water soluble, peptides, or lipophilic molecules.

- Lipophilic hormones are detected by either a cell-surface receptor or an intracellular receptor as seen with steroid hormones.

- Plant hormones are a very diverse group of chemicals, including among them compounds which are derived from amino acids, and lipids; they can even be gases, such as ethylene or nitric oxide.

- Cytokines are a group of peptides which include the interleukins, currently having 18 members in the group, the interferons and tumour necrosis factors.

- Cytokines can be classified into 16 functional or structural groups, with such molecules being important in the orchestration of the immune response and the development of the cells used in animal host defence.

- Chemokines are a large group of extracellular signals, often involved in control of the immune system.

- Chemokines can be mainly split into two famillies, and CC chemokines and the CXC chemokines.

- Growth factors are molecules which are involved in the regulation of the growth and differentiation of cells.

- The growth factor family is known to contain at least 50 proteins but, as with the cytokines and chemokines, undoubtedly more are yet to be discovered.

- Other extracellular signals include the energy-storage compound ATP and a compound usually associated with intracellular signalling, cAMP, as seen in control of the aggregation of the slime mould *Dictyostelium*.

2.10 FURTHER READING

Introduction

Baulieu, E.E. and Kelly, P.A. (1990) *Hormones: from Molecules to Disease*. Chapman and Hall London.

Le Roith, D., Shiloach, J., Roth, J., and Lesniak, M.A. (1980) Evolutionary origins of vertebrate hormones: substances similar to mammalian insulins are native to unicellular eukaryotes. *Proceedings of the National Academy of Sciences USA* 77, 6184–6188.

Hormones

Henry, H. and Norman, A.W. (eds) (2003) *Encyclopedia of Hormones*. Academic Press.

Cytokines

Arai, K.-I., Lee, F., Miyajima, A., Miyatake, S., Arai, N., and Yokota, T. (1990) Cytokines: coordination of immune and inflammatory responses. *Annual Review of Biochemistry* 59, 783–836.

Minami, Y., Kono, T., Miyazaki, T., and Taniguchi, T. (1993) The IL-2 receptor complex: its structure, function and target genes. *Annual Review of Immunology* 11, 245–267.

Moore, K., O'Garra, A., de Waal Malefyt, R., Vierra, P., and Mosmann, T. (1993) Interleukin-10. *Annual Review of Immunology* 11, 165–190.

Scott, P. (1993) IL-12: initiation cytokine for cell mediated immunity. *Science* 260, 496–497.

Thomson, A. (ed.) (2003) *The Cytokine Handbook*, 4th edn. Academic Press. An excellent review of the area, but 1800 pages.

Growth factors

Mason, I.J. (1994) The ins and outs of fibroblast growth factor. *Cell* 78, 574–552.

Oefner, C., D'Arcy, A. Winkler, F.K., Eggimann, B., and Hosang M. (1992) Crystal structure of human platelet-derived growth factor BB. *EMBO Journal* 11, 3921–3926.

Pheromones

Agosta, W.C. (1992) *Chemical Communications: the Language of Pheromones*. Scientific American Library, New York.

Kell, D.B., Kaprelyants, A.S., and Grafen, A. (1995) Pheromones, social behavior and the functions of secondary metabolism in bacteria. *Trends in Ecology and Evolution* 10, 126–129.

Wirth, R., Muscholl, A., and Wanner, G. (1996) The role of pheromones in bacterial interactions. *Trends in Microbiology* 96, 96–103.

ATP as an extracellular signal

El-Moatassim, C., Dornand, J., and Mani, J.-C. (1992) Extracellular ATP and cell signalling. *Biochimica Biophysica Acta* 1134, 31–45.

2.11 USEFUL WEB PAGES

Cytokines and chemokines www.rndsystems.com

Hormone levels www.il-st-acad-sci.org//data2.html

3 Detection of extracellular signals: the role of receptors

Chapter 2 discussed the array of signals that are used by cells to communicate to each other. However, regardless of how many signals are present on the outside of a cell, that cell will only respond if it has the capacity to recognize a signal, and then to respond to it. Therefore Chapter 3 discusses the mechanisms used by cells to perceive signals, and the immediate action taken when they do. Most signals are perceived at the cell surface by a variety of protein receptors, but some signals are able to penetrate the cell, and therefore need to be recognized on the inside.

As with dysfunction of the signals themselves, dysfunction of the receptors can, and does, lead to disease. For example, if the insulin receptor cannot recognize the presence of insulin, then a cell cannot take the appropriate action when insulin is released into the blood, and diabetes will result. Therefore, an understanding of the types of receptor, their mechanisms and actions, and how they lead to the signalling cascades inside the cell is vital to the understanding of cell signalling mechanisms.

The study of binding characteristics of receptors is discussed, as an understanding of the ability of a cell to bind to a signalling ligand will reveal the capability of a cell to respond, or not. Further, the capacity of a cell to recognize a signal may not be constant, with the number of available receptors changing over time. The mechanisms of how a cell may modulate its complement of receptors is also discussed. Cells can become less responsive to drugs, as well as endogenous signals, and therefore it is important to understand the way in which cells can change their array of receptors.

3.1 Introduction

Many signals to which a cell needs to respond reach it from the extracellular environment, whether that is via diffusion in the growth medium or actively through a vascular system, as discussed in Chapter 2 (section 2.1). Such extracellular signals may be at extremely low concentrations, but even when they rise to a level to which the cell needs to react, the concentration could still commonly be very low, perhaps in the order of 10^{-8} M. Therefore, if the cell is to respond to the presence of such a signal the cell must have the capacity to detect the signal molecule, even at these low concentrations, and have the capacity to act upon it. Secondly, as discussed in Chapters 1 and 12 (sections 1.1 and 12.2), cells are usually awash with a plethora of signalling molecules, and it is only to a selection of these that they need to respond.

The role of detection of signals arriving from outside of the cell is usually fulfilled by the presence of specific receptors, either on the cell surface or inside the cell (either in the cytoplasm or in the nucleus). However, three crucial critieria have to be met by the functioning of a receptor:

- a receptor has to have specificity, detecting only the signalling molecule (or range of molecules) which the cell wishes to perceive;
- the binding affinity of the receptor must be such that it can detect the signalling molecule at the concentrations at which it is likely to be found in the vicinity of the cell;
- the receptor must be able to transmit the message that the signalling molecule conveys to the cell, usually by the modulation of further components in a signalling cascade.

Therefore, receptors usually have a high binding affinity that is in the concentration range of that of their ligand, and binding of the ligand to the receptor will then stimulate the required intracellular response, usually via a complex signal transduction pathway. Detection of the signalling molecule has to be precise and a cell has to have the ability to have a repeated response to the same signal or even simultaneous responses to several signals.

The receptor must have the capacity to transmit the message on to the next component in the signalling cascade. Often this involves binding of a ligand on the outside of the membrane and an interaction with protein on the inside of the cell, the message being transmitted through the membrane. As discussed below, the membrane receptors typically have α-helices (perhaps one, often seven, sometimes more) passing through the membrane. Therefore, on ligand binding a series of molecular events will ensue. Ligand binding itself will induce a conformational change in the outer domain of the receptor; this change will be transmitted through the membrane-spanning helices, and this will induce a conformational change in the intracellular domain of the receptor. This will either activate or inhibit the receptor's intrinsic activity, if it has one, or perhaps disrupt the receptor's association with other proteins, such as a G protein. By this significant conformational change through the receptor, the ligand binding can effect events on the other side the membrane. Even if the receptor has no membrane location, as with steroid receptors, conformational change on ligand binding is still a key event.

3.2 Types of receptor

To further fulfil the requirements outlined above, a wide number of receptors have evolved to fill the vital role of the detection of extracellular signals. However, despite the vast array of extracellular molecules that need to be detected by a single cell, including hormones, cytokines, and chemokines (see sections 2.2–2.5), the majority of receptors fall into five classes (see Fig. 3.1):

- G protein-linked;
- ion channel-linked;
- containing intrinsic enzymatic activity;
- tyrosine kinase-linked;
- intracellular.

G protein-linked receptors

A receptor which is classed as a G protein-linked receptor, when activated by binding to its ligand, results in the activation of a G protein which conveys the message to the next component in the signal pathway. There is a vast number of receptors in this class, and they include those which have a specificity for hormones such as epinephrine, serotonin, and glucagon along with those listed in Table 3.1. These receptors are of great interest to the pharmaceutical industry as they can be targets of therapeutic interest. For example, drugs which have their action here include some anti-histamines, anti-cholinergics, inhibitors of the β-adrenergic receptor, and some opiates.

In general, the topology if these receptors is such that they contain seven regions of approximately 22–24 amino acids which form hydrophobic α-helices that span the plasma membrane (see Fig. 3.2). Therefore their structure is analogous to that described for bacteriorhodopsin and rhodopsin (see section 10.2 and Fig. 10.3) and they are commonly referred to as seven-spanning receptors. With G protein-linked receptors the N-terminal end of the polypeptide is on the exterior face of the plasma membrane but the C-terminal end is on the inside, which means that there are four cytoplasmic loop regions, the third of which is probably the site for G protein binding, along with the cytoplasmic C-terminal tail.

The class of G proteins involved in binding to these receptors and relaying the signal along the transduction pathway is known as the trimeric

Fig. 3.1 Three different types of cell receptor. (a) An ion channel-linked receptor. (b) A G protein-linked receptor. (c) An enzyme-linked receptor.

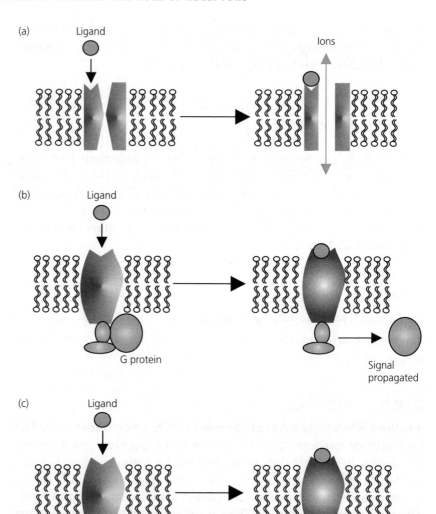

Table 3.1 A list of a few of the ligands which act through G protein-linked receptors.

Acetylcholine
Bradykinin
Calcitonin
Dopamine
Epinephrine (adrenalin)
Glucagon
Histamine
Leukotrienes
Lutropin
Neurotensin
Oxytocin
Parathyroid hormone
Prostaglandins
Retinal
Serotonin
Somatostatin
Thrombin
Thyrotropin
Vasopressin

G proteins, or heterotrimeric G proteins, and as their name suggests they are composed of three subunits of different sizes, termed α, β, and γ. This is an extremely important class of signalling protein and their functioning is further discussed in Chapter 5 (section 5.4).

As well as being activated by binding of the relevant ligand, the activity of these receptors is also altered by phosphorylation. Phosphorylation can be catalysed by cAMP-dependent protein kinase (PKA) or by a class of kinases known as G protein-coupled receptor kinases (GRKs). GRKs are known to phosphorylate these receptors on multiple sites, using threonine and serine residues as targets. Phosphorylation deactivates the receptor as

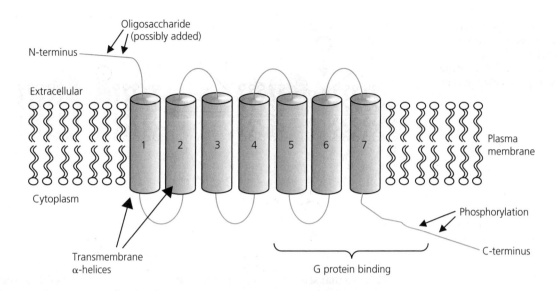

Fig. 3.2 The predicted structure of a G protein-linked receptor, showing the seven α-helices spanning the membrane and the likely G protein-binding region.

well as allows the interaction of the receptor with an inhibitory protein known as β-arrestin. Therefore, such mechanisms are involved in the termination of the signal and receptor desensitization. As will be discussed in Chapter 10 (section 10.2), one of the classical examples of such a system is seen with rhodopsin, used in light perception in the eye. One of the most studied examples of these kinases is β-adrenergic receptor kinase (βARK) and these kinases and their actions will be further discussed in Chapter 4 (section 4.2).

Ion channel-linked receptors

These receptors are often involved in the detection of neurotransmitter molecules and often these receptors are referred to as transmitter-gated ion channels. Binding of the ligand to the receptor changes the ion permeability of the plasma membrane as the receptor undergoes a conformational change which opens or closes an ion channel, allowing the efflux or influx of specific ions. However, this is only a transient event, the receptor returning to its original state very rapidly.

These receptors make up a family of related proteins, but with one distinguishing feature being that they contain several polypeptide chains which pass through the membrane. For example, the acetylcholine receptor is composed of two identical polypeptides which contain acetylcholine-binding sites along with three different polypeptides, giving an α, α, β, γ, δ subunit structure (see Fig. 3.3). These five polypeptides are therefore encoded by four separate genes but interestingly they show a large degree of homology suggesting that they probably arose from a gene-duplication event. In the membrane the proteins are arranged in a ring, in a similar

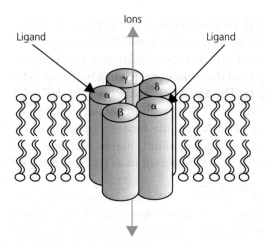

Fig. 3.3 The predicted structure of an ion channel-linked receptor.

fashion to that seen with the gap junctions (see Fig. 1.5), and therefore there is a water-filled channel which runs through the middle from one side of the membrane to the other, allowing the passage of ions. The ions which pass through the hole made by the acetylcholine receptor are usually positively charged, such as Na^+, K^+, or Ca^{2+} as the presence of negatively charged amino acids at the ends of the hole bestows some selectivity to the channel.

Receptors in this class with different ligand-binding specificities, and subsets within these groups of receptors, all contain polypeptides with high levels of sequence similarity, and are encoded by genes with high levels of homology or by genes where the expression involves alternative splicing events. Therefore a vast array of receptors can be made using combinations of these different gene products. Often the presence of a particular receptor subset is tissue-specific. Ion channel-linked receptors, are the targets of many drugs, such as barbiturates used in the treatment of insomnia, depression, and anxiety.

Alternative Splicing

Alternative splicing is where different exons within a gene are used, in a variety of combinations, when the gene is expressed, so that one gene may give rise to many different transcipts and therefore many different proteins. A good example of this mechanism of gene expression is seen with antibody production.

Receptors containing intrinsic enzymatic activity

The receptors containing intrinsic enzymatic activity make up a quite heterogeneous class of receptors which is characterized by the presence of a catalytic activity integral to the receptor polypeptide, and it is this catalytic activity that is controlled by the ligand-binding event. The ligand-binding domain is found on the extracellular side of the membrane, with usually a single span of the membrane, leading to the catalytic domain on the cytoplasmic side. The catalytic activity may be a guanylyl cyclase (as discussed further in section 5.5), a phosphatase, or a kinase. The receptors may contain a serine/threonine kinase activity and are often referred to as receptor serine/threonine kinases, or they may contain a tyrosine kinase activity and are referred to as receptor tyrosine kinases or RTKs.

RTKs represent a class of the receptors containing intrinsic enzymatic activity which has been studied extensively. Here with RTKs, ligand binding leads to activation of the kinase activity of the receptor which causes phosphorylation of the receptor itself on tyrosine residues. This leads to the creation of new binding sites, particularly for proteins which contain protein-binding domains, for example, SH2 domains (see section 1.7). Such binding proteins may be adaptor proteins, such as GRB2 of mammals or Drk in *Drosphilia*, which contain both SH2 and SH3 domains. On phosphorylation of the receptor (by itself), the new binding sites are recognized by the binding protein, and binding of such adaptor proteins to the receptor stimulates the formation of protein complexes. Such complexes may also include guanine nucleotide-releasing factors; for example, Son of Sevenless (Sos). Here, the result would be the activation of a G protein. This in turn may lead to a transduction cascade which could include further kinases, such as MAP kinases. Such signalling is further discussed in Chapter 4 (section 4.4 and Fig. 4.8) and Chapter 9 (see Fig. 9.7).

However, phosphorylation by the receptor is not confined only to autophosphorylation and other proteins may also be phosphorylated, leading to further propagation of the signal. For example, the IRS1 protein is phosphorylated on multiple sites by the insulin RTK as discussed in Chapter 9.

So far, over 50 RTKs have been identified and these can be grouped into at least 14 different families. Most classes of RTK are monomeric in nature, but some share the insulin receptor's tetrameric topology; that is, they have an $\alpha_2\beta_2$ structure held together by disulphide bonds. Some RTKs contain cysteine-rich extracellular domains while others contain extracellular antibody-like domains. However, RTKs all seem to share the characteristics of having their N-terminal ends on the extracellular side of the membrane and the polypeptides only cross the membrane once, through the use of a single hydrophobic α-helix.

Although in the inactive state these receptors are often monomers, on activation they are often found as dimers, the ligand binding leading

Table 3.2 The binding specificities of the PDGF receptor, showing how dimerization allows for subtleties of ligand binding.

Type of receptor	Isoforms of ligand that are recognized
αα	AA AB BB
αβ	AB BB
ββ	BB

to the dimerization event. However, some receptors are constitutively dimers/tetramers, as seen with the insulin receptor. Interestingly, the insulin receptor is coded for by one gene which leads to the formation of one mRNA, but the protein product undergoes post-translational modification, including a cleavage event. This leads to the formation of two polypepedides, α and β, where the two polypeptide chains are held together by the formation of cystine or disulphide bridges. This structure then dimerizes to form a tetrameric receptor structure (see Chapter 9 for more discussion of this receptor). Having said that, dimerization is a common theme in the activation of RTKs. The dimer may be a homodimer, containing two identical receptor subunits, or a heterodimer where the complex is composed of two different subunits from the same receptor family. Subunit structure may also involve other, or accessory, proteins. An example here is the involvement of the protein gp130. The formation of heterodimers allows the creation of a wider diversity of receptor specificities as seen with PDGF receptors where the receptor dimers can be αα, ββ, or αβ, each having a different specificity to isoforms of growth factor (see Table 3.2).

Receptors linked to separate tyrosine kinases

Several receptors do not themselves contain a tyrosine kinase domain, but on activation by ligand binding they cause the stimulation of a tyrosine kinase. Such kinases are normally resident in the cytoplasm of the cell, but will recognize and bind to an activated receptor, which on ligand binding has adopted a new conformation. This class of receptors is commonly referred to as the cytokine receptor superfamily, as they are involved commonly in the recognition of cytokines and growth factors.

As discussed above, binding of the ligand to the receptor often induces dimerization to occur. Here, for example, IFN-γ binding leads to homodimerization of its receptors. Other ligands, on the other hand, cause heterodimerization on binding to the receptor protein, or dimerization may involve accessory proteins, for example gp130. However, the receptor for TNF-β forms trimers (receptor trimers are further discussed in Chapter 11; section 11.5) while other ligands can lead to the formation of heteroligomers of three different polypeptides.

Once activated, the receptor in this group needs to recruit and activate the relevant protein kinase, and it is the soluble protein tyrosine kinases called Janus kinases (JAKs) that are often involved. These kinases, which contain two catalytic domains, enable the propagation of the signal along the signal transduction cascade, and are discussed further in Chapter 4 (section 4.3).

Intracellular receptors of extracellular signals

Not all extracellular signalling molecules are detected on the surface of the cell by plasma membrane-borne receptors. Many very important signals are released by cells but the receptors for their perception are inside the target cells, not on their surface. These include receptors for steroid hormones, thyroid hormones, retinoids, fatty acids, prostaglandins, and leukotrienes, the receptors for which are all intracellular. Steroid hormones are derived from cholesterol and include cortisol, vitamin D, and steroid sex hormones. The amino acid tyrosine is the base for the thyroid hormones while the retinoids are derived from vitamin A: such signals are discussed further in Chapter 2 (section 2.2). These signalling pathways are also an important target for many therapeutic regimes, with steroid-based drugs being used commonly.

Intracellular ligand binding for these extracellular signalling molecules means that they have to move through the plasma membrane and commonly these signals are small and hydrophobic, hence allowing them to get access to the intracellular receptors. However, as these molecules are readily soluble in the hydrophobic environment of the membrane, allowing their free passage into the cell, they are therefore inherently insoluble in the aqueous fluids outside the cells, such as the bloodstream. Therefore, these extracellular signalling molecules need their solubility in water to be increased, which is facilitated by their association with specific carrier proteins. Dissociation from the carrier occurs before the signalling molecules can enter the cell.

The receptors found inside cells are commonly referred to as the intra-cellular receptor superfamily or steroid hormone receptor superfamily. The activation of such receptors often leads to effects in the nucleus of the cell, commonly the alteration of transcription rates of specific genes, or sets of genes. Therefore, the signal can either enter the cell and bind to a cytoplasmic receptor that then moves to the nucleus to have an effect, or the signalling molecule itself can move directly to the nucleus and bind to the receptors there. Both scenarios exist in cells, and receptors for these extracellular signals can be found either in the cytoplasm or in the nucleus of the cell.

There are numerous nuclear receptors in cells. There are known to be 48 different nuclear receptors in humans, 21 in the fly *Drosophila melanogaster*, and hundreds in the nematode *Caenorhabditis elegans*, highlighting the importance of this type of signalling to cells. Its is certainly not an oddity, but an important part of the signalling that takes place in an organism. The steroid hormone receptor superfamily represents, in fact, the largest known family of transcription factors described for eukaryotes. However, these include putative receptors identified through the use of sequence similarity analysis for which the ligand specificity has not yet been determined, so-called orphan receptors, as well as numerous isoforms of known receptors.

Orphan receptors

With the advent of modern molecular biological techniques and whole-genome sequencing, once a consensus sequence has been determined for a class of proteins, such as receptors, other putative members of a protein family may easily be identified in the genome of interest. With receptors, if no ligand is known to bind to such proteins that have been identified, they are termed orphan receptors. However, biochemical and functional-genomic analysis is required to determine if the novel protein actually does anything, what it might bind, or in some cases if it even exists.

The structures of the intracellular receptors are very conserved and it is thought that nearly all these receptors have evolved from a common ancestral receptor, the oestrogen receptor. In general, the amino acid sequences show that these receptors can be divided into several domains. At the N-terminal end of the polypeptide is a variable region known as the A/B domain, which contains an activation function 1 (AF-1) region which is a transcription activator and is involved in gene activation. The length of the A/B domain is very variable between receptors, ranging from less than 50 to over 500 amino acids in humans. The next domain along the polypeptide is the C domain or DNA-binding domain (DBD), often containing two zinc fingers. This domain is relatively short, being around 60 amino acids, and is responsible for DNA recognition and also partly for dimerization of the receptor. The next domain, the D domain, consists of a variable hinge region and may contain sequences responsible for the localization of the receptor to the nucleus. This domain is also relatively short in most cases, and can be as little as 18 amino acids.

Ligand binding is the responsibility of a large E domain (often referred to as the ligand-binding domain (LBD). This region is approximately 200–250 amino acids in length and is also responsible for association with heat-shock proteins (Hsps), as well as being involved in receptor dimerization. The E domain also contains an activation function 2 (AF-2) region which is needed for the transcriptional activation brought about by these receptors. At the C-terminal end is the last region, called the F domain, but not all of the intracellular receptors seem to contain this region and no defined function has been assigned to this area of the protein.

The structure of intracellular receptors also often involves other proteins. The receptors that are found in the cytoplasm exist there as an inactive complex, which involves proteins known as the heat-shock

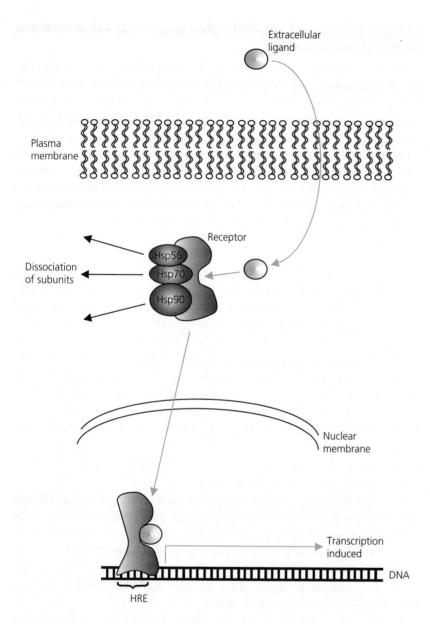

Fig. 3.4 A schematic representation of the detection of lipophilic hormone molecules by an intracellular receptor. HRE, hormone response element.

proteins (Hsps), usually Hsp90, Hsp70, and Hsp56. On binding of the ligand to the receptors, the latter undergoes a conformational change and dissociation from the inhibitory heat-shock protein complex (see Fig. 3.4). Receptor proteins may then enter the nucleus through nuclear pore complexes which span the nuclear membrane. Once in the nucleus binding of the receptor can occur to specific sequences of bases on DNA, so altering the rates of transcription of specific genes, and therefore these receptors can be classified as transcription factors, even though they were originally found in the cytoplasm. The receptors bind to the DNA as either homodimers or heterodimers.

The receptors that are already located in the nucleus in an inactive state can also be associated with heat-shock proteins, in particular Hsp90, Hsp70, and Hsp40, along with other proteins. Such protein–protein binding interactions probably control the access of the steroid to the ligand-binding site.

Intracellular receptors control the expression of genes once they are activated, but for there to be a controlled cellular response to the presence of a ligand there must be a specific gene or set of genes which are targeted by these receptors. So, as well as characterizing the receptors themselves, the region of DNA recognized by the receptors is also important to determine. The region of the DNA that is the target for intracellular receptors is called the hormone-response element (HRE). For example, for glucocorticoid receptors the HRE has been found to be two short imperfect inverted repeats with three nucleotides separating them. Other HREs have been found to be similar to this pattern, although of course the exact nucleotide sequence involved bestows the specificity on the receptor–DNA interaction and therefore the specificity of the cellular response to the original stimulus (i.e. presence of the specific ligand).

Ligand binding is not the only mechanism by which the activity of the intracellular receptors is modulated. Phosphorylation also alters the activity of these receptors, phosphorylation taking place usually on serine or threonine amino acids in the N-terminal domain or in the DNA-binding domain. Phosphorylation may alter the ability of the nuclear receptor to bind its interacting proteins, or modulate DNA binding itself, or may be involved in the turnover of the receptor.

It should be noted that activation of an intracellular receptor does not always increase the rate of expression of a gene, as receptor activation can lead to decreased transcription rates too. Further, the genes which are expressed might encode proteins that themselves alter the rates of transcription, a point further discussed in Chapter 12. So the overall response to the activation of an intracellular receptor could be complex and very profound to the workings of the cell.

A study of receptors in the cytoplasm and nucleus will therefore highlight that it is not only the receptors on the plasma membrane that are important for ligand binding, and control of cellular activity, but also those receptors present inside the cell. Many of these intracellular receptors, such as those in the nucleus, are already present in the cell organelle where the response is to take place, so bypassing the need for the complex pathways used by plasma membrane receptors. These receptors often control transcription of many genes, and therefore are in direct control of the future cellular complement of proteins, and hence future cellular activity.

3.3 Ligand binding to their receptors

Ligands that bind to receptors can be classified under general headings. A ligand which is termed an agonist is one which binds to the receptor and results in the activation of that receptor. Alternatively, a ligand referred to as an antagonist will bind the receptor but not result in activation of the receptor, and further the presence of an antagonist may interfere and stop the action of an agonist. However, a receptor may be active in the absence of a ligand, in which case it would be said to be constitutively active, a situation seen with some oncogenes.

Receptors are, and need to be, highly specific for their ligand and usually have a high affinity for that ligand. However, in molecular terms, the binding site of the receptor can in many ways be viewed as being like an active site of an enzyme. It is a specific local environment, determined by the presence of specific amino acids which are held in a three-dimensional orientation by the other amino acids within the protein. It is the make-up the binding site that determines both the specificity and affinity for the ligand. The individual forces involved in holding the ligand on to the receptor are generally weak, being ionic attractions, van der Waals forces, hydrogen bonding, or hydrophobic interactions.

Ligand binding is usually a reversible reaction, often allowing the receptor to be used and reused over a long period of time. Therefore the binding reaction can be written as follows, where L means ligand and R means receptor.

$$L + R \leftrightarrow LR$$

Therefore, as with enzyme kinetics, where it is useful to determine the concentration of substrate at which the reaction proceeds at half the maximal rate, i.e. K_m, one of the calculations that is useful when characterizing ligand binding is the determination the concentration of ligand at which half of the receptors are bound, with half the receptors in the unbound state. This value is called K_d and can be defined by the following equation:

$$K_d = \frac{[R][L]}{[RL]}$$

where [R] is the concentration of receptor, [L] is the concentration of ligand, and [RL] is the concentration of receptor bound to ligand as a complex. The lower the K_d value the higher the affinity of the receptor for its ligand. Usually the K_d values approximate to the physiological concentrations of the ligand, allowing the receptor to have the highest sensitivity to changes in the ligand concentration in the concentration range usually found. This of course would be ideal for the cell to respond quickly and efficiently as ligand concentrations fluctuate.

To perform the calculation above, the amount of ligand actually bound to the receptor needs to be determined experimentally. Ligand binding can usually be studied using a ligand that has been labelled or tagged, and therefore the binding to its receptor can be followed. To visualize the ligand binding on, for example, a cell-surface receptor, a fluorescent label may be used in conjunction with a fluorescence microscope, a laser-scanning confocal microscope, or a fluorescence-activated cell sorter (FACS). However, as fluoresecence is hard to quantify accurately, to quantify ligand binding a radiolabel is usually employed. The most common radiolabels are ^{131}I and [125]I. However, these isotopes have extremely short half-lives, approximately 8 days and 60 days respectively. Further, the presence of a large iodine molecule may well interfere with the ligand–receptor interaction. Alternatively, ^{3}H can be used as a label in many instances; one advantage being its long half-life of approximately 12 years.

If we consider a cell-surface receptor, a usual experiment would quantify the amount of ligand bound as the concentration of the ligand was increased. This would result in the total binding curve as depicted in Fig. 3.5.

Binding measurements under the conditions adopted usually mean that binding has come to equilibrium. However, it should be noted that the speed of binding may be affected by the pH of the solution and/or the temperature of the reaction. Furthermore, some of the larger ligands might well have binding which is very slow and therefore care needs to be exercised if a true measure of binding is to be obtained.

Once the ligand binding has been quantified, the total amount of ligand bound does not necessarily mean that all the ligand has bound to the receptor, and the amount bound will almost certainly include an element of non-specifically bound ligand. The contribution of the non-specific binding to the

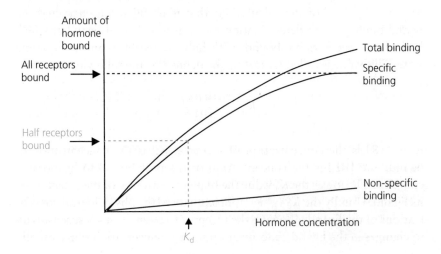

Fig. 3.5 A plot of the amount of hormone bound versus the hormone concentration, showing the total binding and the proportions made up by the specific binding and the non-specific binding.

total is usually estimated by the inclusion of control experiments in which a huge excess of unlabelled ligand, approximately 100-fold excess, is added. This means that the high-affinity receptor sites will be saturated with unlabelled ligand and therefore any labelled ligand remaining bound will be due to it binding to non-specific sites. The non-specific binding is in general linear in a concentration-dependent manner. Once these values are subtracted from the total binding curve the binding specific to the receptor can be seen (see Fig. 3.5). This binding starts off being very much greater than the non-specific binding, due to the high affinity of the receptors, but as in typical enzyme kinetics, the receptors become saturated and the binding curve tails off to a maximum, beyond which no more ligand binding can occur, despite the addition of more ligand. Therefore, from this graph, the total specific binding sites can be estimated as can the K_d value.

Estimating the binding characteristics from a curve can be difficult and not very meaningful, and therefore, in an analogous way to the analysis of enzyme activity, the ligand-binding data need to be manipulated mathematically to give a linear relationship. Such analysis should give a better insight into the characteristics of the binding: for example, whether the binding shows any cooperativity; that is, whether the binding of the second ligand is more or less favourable because the of bound first ligand. The two common methods employed are derived from those developed by Hill and Scatchard. The original development of the Hill plot was to analyse the binding of oxygen to haemoglobin, but the same rationale can be used here.

The Scatchard plot is derived by dividing the concentration of bound ligand by the concentration of free ligand and then subsequently plotting this against the concentration of bound ligand, that is:

$$\frac{[\text{Bound ligand}]}{[\text{Free ligand}]} \text{ versus } [\text{Bound ligand}]$$

As illustrated in Fig. 3.6, from this plot the K_d value can be obtained, as the slope of the line will be $-1/K_d$ and the maximum of amount of ligand bound (B_{max}) can be determined by extrapolating the line to cross the x axis.

Although theoretically Scatchard analysis should give a straight line, this is quite often not the case. Disruption of the ligand's interaction with the receptor, for example by the interference of the label used on the ligand, may result in a curvature of the line. The Scatchard analysis may not be linear for reasons other than artifactual ones. The binding of the ligand to the receptor may show what is termed negative cooperativity, where binding of the first ligand to the receptor causes the receptor to have a reduced affinity for the second ligand. Alternatively, there may be more than one type of receptor present, each with different binding characteristics, or the interaction of an intracellular subunit with the receptor, such as a G protein, may reduce its affinity for the ligand. If the Scatchard analysis is done

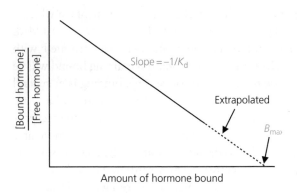

Fig. 3.6 A typical Scatchard plot showing the extrapolation of the slope to the x axis.

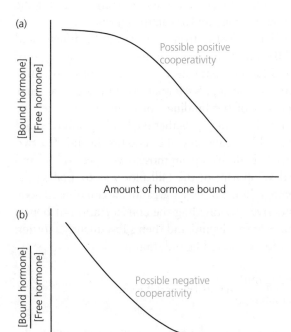

Fig. 3.7 Typical Scatchard plots where the ligand binding is not straightforward. A curve is obtained for both positive cooperative binding (a) and negative cooperative binding (b).

when any of these factors is involved the resultant line will be curved and lie under the expected straight line (Fig. 3.7b).

Alternatively the line obtained may be curved but lie above the expected linear plot (Fig. 3.7a). This may be an indicator of positive cooperativity, where binding of the first ligand increases the receptor's affinity for a second ligand, very much the same as in the binding of oxygen to haemoglobin. Such effects may be due to conformational changes within the protein on ligand binding and may involve more than one polypeptide: for example, interactions of receptor molecules which would be possible with receptor dimerization or complex formation.

3.4 Receptor sensitivity and receptor density

The concentration, or density, of receptors on the surface of a cell is not necessarily constant, and in fact rarely is, and it is very apparent that a cell may become more sensitive or less sensitive to a given concentration of extracellular ligand.

An increase in ligand sensitivity, or sensitization, may occur by an increase in the amount of a receptor on the cell surface. Here a cell is maximizing its chance of detecting the ligand and so responding to it. A real increase in the receptor available can be achieved by the synthesis of new receptor molecules, their recruitment from intracellular stores, such as from vesicles, and a decrease in the rate of removal of the receptor from the cell surface, with one or more of these methods being responsible in any particular cell.

In many cases, if a cell has been exposed to a specific ligand, and has shown a given response but shortly afterwards has a second exposure to that same ligand, the second response seen is very much reduced. The cells are said to have become refractory to the second dose of ligand, as shown in Fig. 3.8. If the cell retains a normal response against other ligands, that is if only one receptor seems to be involved in the refractory state, the phenomenon is called homologous desensitization. An example of this is seen with the β_2-adrenergic receptor and its response to adrenalin. The receptor becomes desensitized extremely rapidly, and it has been shown that both Mg^{2+} and ATP are required, suggesting the involvement of a phosphorylation step. It is now clear that there are in fact two separate phosphorylation events. Firstly, the β_2-adrenergic receptor causes a rise in intracellular cAMP and subsequent activation of cAMP-dependent protein kinase (PKA), leading to phosphorylation of the receptor on a serine residue, and hence disruption of activation of a G protein by the receptor. Secondly, once activated, the receptor becomes a target for a specific kinase, β-adrenergic receptor kinase, which phosphorylates the receptor on several threonine and serine residues towards the C-terminal

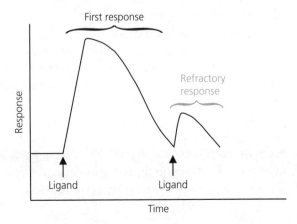

Fig. 3.8 A plot of the hormonal response with time showing that the application of a second dose of the same amount of ligand, e.g. a hormone, does not necessarily invoke the same magnitude of response as the first dose.

end of the polypeptide. This phosphorylated polypeptide then binds to a protein called β-arrestin, which stops the receptor activating its associated G protein, and so prevents propagation of a response. It is interesting to note that rhodopsin, which shows structural similarity to the β-adrenergic receptors, also undergoes a similar desensitization involving an arrestin type of mechanism (discussed further in section 10.2).

As β$_2$-adrenergic receptors can be phosphorylated in response to rises in intracellular cAMP levels, which are controlled by many other receptors as well, β$_2$-adrenergic receptors may well be desensitized in response to one of these other receptors binding to its respective ligand. Such down-regulation is referred to as heterologous desensitization, as it is not caused by the presence of the ligand normally associated with that receptor.

As well as direct control of receptor function by phosphorylation, desensitization of a cell to a ligand may also occur by removal of the receptor from the cell surface. A common sequence of events on ligand binding may be as follows. Once a cell-surface receptor has been activated by a particular ligand and has transmitted its message to the next element of the signal transduction cascade, for example a G protein, the receptor is internalized into the cell through endocytosis. The membrane-bound receptor and ligand complex become an integral part of the vesicle which is formed during endocytosis and will become part of an endosome. On the cell surface the receptor faces outwards, that is its ligand-binding site is on the outside of the cell, but in the formation of the vesicle the receptor ends up facing the inside of the endosome and there is exposed to the environment inside the endosome. This is usually an acidic environment, unlike that to which the receptor would have been exposed on the outside of the cell, which generally is neutral. The change in pH on arriving at the endosome often causes a change in the conformation of the receptor, and so alters its affinity for the ligand, usually reducing it. Therefore, often the receptor–ligand complex dissociates. The receptor-binding site is now empty, enabling the receptor to potentially be reused. It only needs to be transported by the vesicular system of the cell, this time back to the plasma membrane where it can again be used to detect the extracellular presence of the ligand. The ligand which has been left behind in the endosome is usually delivered to lysosomes where it is degraded. Hence, cells can effectively remove ligands from extracellular fluids and cause ligand-induced signals to be turned off, unless of course the ligand continues to be released from its source and continues to activate the receptors as they are recycled.

However, not all receptor–ligand complexes dissociate to allow the recycling of the receptor back to the plasma membrane and in many cases both the receptor and ligand are transported to the lysosomes and destroyed. To maintain the receptor concentration on the cell surface will therefore require *de novo* protein synthesis.

Transcytosis

Sometimes a receptor may bind a ligand at one point on the plasma membrane, be internalized and but then subsequently be relocated in a different region of the plasma membrane, so transporting its ligand across the cell. This is a process referred to as transcytosis, and is seen in cells lining the gut in mammals.

The process of internalization of receptors is commonly preceded by their relocalization in the plane of the membrane into clusters (Fig. 3.9),

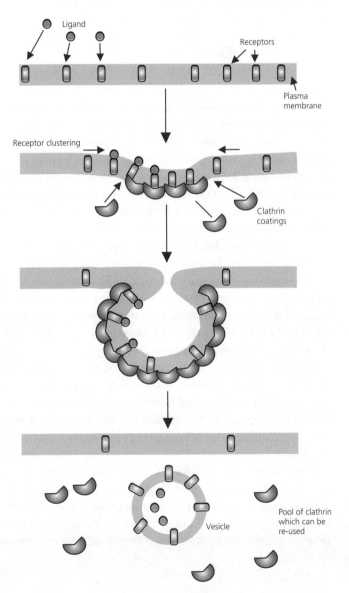

Fig. 3.9 The process of endocytosis. Receptors firstly move to the site of invagination and then a vacuole is formed, aided by the presence of clathrin molecules. The capture of ligand by the receptor is shown on the left-hand side of the figure, while on the right empty receptors are shown.

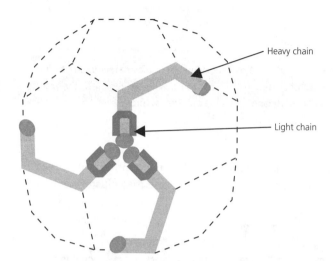

Heavy chain

Light chain

Fig. 3.10 The likely quaternary structure of the clathrin protein, showing how the heavy chains and light chains align, giving the appearance of a Manx flag. The overall shape encloses a sphere.

a process called capping. This is followed by invagination of the membrane and the formation of what is known as a coated pit. This is so called because the cytoplasmic side of the vesicle that is forming is covered, or coated, with a protein, in this case clathrin. Clathrin is composed of two polypeptides, a light chain and a heavy chain, and three of each come together to form a structure known as a triskelion. This is a three-legged structure that can polymerize further to form a basket-like matrix which will form a scaffold around the vesicle (Fig. 3.10).

The clathrin probably performs two main roles. First, it propagates the formation of the invagination of the membrane and stabilizes the vesicle formation, as the clathrin complex naturally takes up a concave shape. Secondly, it may be involved in the capture of the receptor molecules. Many receptors contain on their cytoplasmic side a short stretch of four amino acid residues, which acts as a signal for endocytosis. This short polypeptide region is recognized by a group of proteins known as adaptins. These adaptin proteins recognize both the receptor and the clathrin protein and will facilitate the association of the receptor with clathrin, and hence the uptake of the receptor into the cell. Different adaptins will recognize different receptors and so coordinate the internalization process, giving specificity to the receptor–clathrin association.

Once formed the vesicles will shed their clathrin coat, as illustrated in Fig. 3.9. This process probably involves ATP and heat-shock proteins such as Hsp70. The control of uncoating is also thought to involve the concentration of Ca^{2+} in the cell, where local rises in concentration may be encountered as the vesicle is transported deeper into the cell.

3.5 SUMMARY

- The detection of extracellular signals and the transmission of those signals into the cell is the responsibility of proteins known as receptors.

- Receptors are commonly found on the plasma membrane of the cell (see Fig. 3.11), where they are ideally placed to be in contact with their extracellular ligand.

- Steroid receptors, however, are found inside the cell, with the ligand itself transversing the membrane.

- Membrane receptors fall into several groups:

  - G protein-linked receptors, which lead to activation of the trimeric class of G proteins;

  - ion channel-linked receptors leading to changes in ion movements across the membrane;

  - receptors which contain intrinsic enzyme activity such as the receptor tyrosine kinases;

  - receptors which recruit separate enzymes, for example, tyrosine kinases such as the JAK proteins.

- Ligand binding to the receptor can be viewed as a reversible event, quantified by the K_d value; that is, the concentration of ligand at which half the receptors are bound. However, the binding of the ligand to a cell usually includes two elements: specific binding to the receptor and non-specific binding.

- Binding curves can be linearized by the use of Scatchard analysis, where the slope of the line is defined as $-1/K_d$. Such analysis can indicate if any cooperativity is involved in receptor–ligand binding.

- A cell's sensitivity to the concentration of extracellular ligand can be varied by altering the density of receptors available at its surface.

- New receptor molecules can be synthesized and routed to the membrane or the response can be down-regulated by capping and endocytosis, resulting in the internalization of the receptor molecules. Such internalized receptors can be returned to the membrane for the further perception of ligand, or alternatively destroyed.

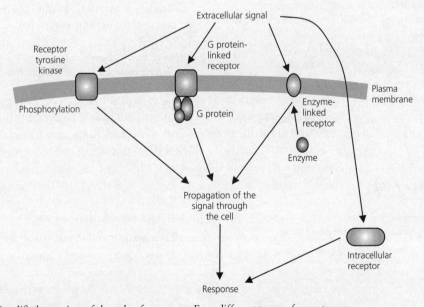

Fig. 3.11 A simplified overview of the role of receptors. Four different types of receptor are shown in blue, but all lead to signalling pathways and responses.

3.6 FURTHER READING

Types of receptor

Evans, R.M. (1988) The steroid and thyroid hormone receptor superfamily. *Science* **240**, 889–895.

Heldin, C.-H. (1995) Dimerization of cell surface receptors in signal transduction. *Cell* **80**, 213–223.

Lemmon, M.A. and Schlessinger, J. (1994) Regulation of signal transduction and signal diversity by receptor oligomerization. *Trends in Biochemical Sciences* **19**, 459–463.

Laudet, V. and Gronemeyer, H. (2002) *The Nuclear Receptors Factsbook*. Academic Press, London.

McEwan, I. (ed.) (2004) The nuclear receptor superfamily. *Essays in Biochemistry*, vol. 40. Portland Press, London. A collection of essays on the topic, written by experts in the field.

Russell, D.W. and Mangelsdorf, D.J. (eds) (2003) Nuclear receptors. *Methods in Enzymology* **364**. A series of articles on the topic.

Strader, C.R., Fong, T.M., Tota, M.R., and Underwood, D. (1994) Structure and function of G protein-coupled receptors. *Annual Review of Biochemistry* **63**, 101–132.

Tsai, M.-J. and O'Malley, B.W. (1994) Molecular mechanisms of action of steroid/thyroid receptor superfamily members. *Annual Review of Biochemistry* **63**, 451–486.

Turner, A.J. (ed.) (1996) *Amino Acid Neurotransmission*. Portland Press, London.

Ligand binding

Hill, A.V. (1913) The combination of haemoglobin with oxygen and carbon monoxide. *Biochemical Journal* **7**, 471–480.

Scatchard, G. (1949) The attraction of protein for small molecules and ions. *Annals of the New York Academy of Science* **51**, 660–672.

Receptor sensitivity and receptor density

Anderson, R. (1992) Dissecting clathrin-coated pits. *Trends in Cell Biology* **2**, 177–179.

Benovic, J.L., Mayor, F., Staniszeski, C. Lefkowitz, R.J., and Caron, M.G. (1987) Purification and charaterisation of the β-adrenergic receptor kinase. *Journal of Biological Chemistry* **262**, 9026–9032.

Brodski, F.M., Hill, B.L., Acton, S.L., Nathke, I., Wong, D.H., Ponnambalan, S., and Parham, P. (1991) Clathrin light chains: array of protein motifs that regulate coated-vesicle dynamics. *Trends in Biochemical Sciences* **16**, 208–213.

Keen, J.H. (1990) Clathrin and associated assembly and disassembly proteins. *Annual Review of Biochemistry* **59**, 415–438.

Lefkowitz, R.J. (1993) G-protein-coupled receptor kinases. *Cell* **74**, 409–412.

Nathke, I.S., Heuser, J., Lupas, A., Stock, J., Turck, C.W., and Brodsky, F.M. (1992) Folding and trimerisation of clathrin subunits at the triskelion hub. *Cell* **68**, 899–910.

Palczewski, K. and Benovic, J.L. (1991) G-protein-coupled receptor kinases. *Trends in Biochemical Sciences* **16**, 387–391.

Pearce, B.M. (1989) Characterisation of coated-vesicle adaptins: their assembly with clathrin and with recycling receptors. *Methods in Cell Biology* **31**, 229–246.

3.7 USEFUL WEB PAGES

Nuclear receptor resource
http://nrr.georgetown.edu/NRR/nrrhome.htm

Nuclear RDB www.receptors.org/NR/

Ligand binding http://alto.compbio.ucsf.edu/ligbase

4

Protein phosphorylation, kinases, and phosphatases

Phosphorylation is a central theme of cell signalling. Signalling often involves the alteration of the activity of a protein, seen as either an increase or a decrease. A protein's ability to interact with another protein can also be altered, either allowing or stopping a signalling cascade. Such alterations of a protein's function is often, although not always, brought about by the addition or removal of a phosphate group: phosphorylation and dephosphorylation. Chapter 4 discusses the mechanisms used to carry out this protein alteration.

Phosphorylation occurs at many points in cell signalling cascades. It may be right at the beginning, at the level of the receptor, or it might be at the end, with the phosphorylation of a transcription factor for example. Alternatively, phosphorylation may occur at a point in between, and there are also cascades of phosphorylation events. The end of a signalling response also may involve phosphorylation or dephosphorylation.

As with most signalling mechanisms, dysfunction of phosphorylation can lead to disease, but further, a vast number of drugs are directed against the modulation of phosphorylation of proteins in cells. Therefore it is important to understand the mechanisms involved in phosphorylation, and many of the main proteins involved are discussed this chapter.

4.1 Introduction

The arrival of an extracellular signal at a cell's surface and its detection by its receptors must lead to further events inside the cell if the cell is going to respond to the presence of that ligand. The activation of a receptor allows it to interact with and/or activate several different types of protein which lead to a wide range of intracellular signalling cascades, as will be discussed in subsequent chapters. However, a crucial event in all cell signalling pathways is the modification of the activity of enzymes or the alteration of the function of control factors.

An enzyme's specific activity may be altered by an alteration in its conformation; that is, the folding of its polypeptide chain. Sometimes this involves the alteration of the primary structure of the protein, where parts of the polypeptide might be cleaved off for example, but much more commonly this is seen as an alteration in its tertiary structure, or for multi-polypeptide enzymes it may also involve the quaternary structure of the protein. The altered spatial arrangement of the active-site amino acids will then reduce or increase the substrate binding and/or the catalytic action of the protein.

There are several ways of facilitating the change in the conformation of a protein. A change in pH may alter the interaction capabilities of amino acids and so disrupt the way that the polypeptide is held in three dimensions. Such changes were discussed in the previous chapter (section 3.3), especially seen when receptors are internalized and a conformational change leads to dissociation of the ligand from the receptors. A change in the reduction/oxidation environment of a protein has been suggested recently to be able to have profound changes on the way a protein might function, as further discussed in Chapter 8 (section 8.4). However, the most common way of modifying protein structure is by the addition, or removal, of one or more phosphate groups to the primary amino acid sequence of the polypeptide, processes known as phosphorylation and dephosphorylation respectively, as shown schematically in Fig. 4.1.

For phosphorylation to be an effective control mechanism, allowing the activity of an enzyme to be both increased and decreased, the overall reaction has to be reversible (Fig. 4.1). If, for example, the need for glycogen breakdown was suddenly increased due to a necessity to supply the muscles with glucose, and hence energy, the enzyme responsible, phosphorylase, will become phosphorylated with a concomitant increase in its activity. However, on cessation of the muscle's activity, the requirement for glucose would drop and therefore the rate of glycogen breakdown would need to be reduced. A quick and energetically favourable way to do this would be dephosphorylation of the phosphorylase, thus lowering its activity, rather than for example destruction of the phosphorylase polypeptide

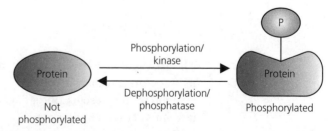

Fig. 4.1 The phosphorylation and dephosphorylation of the protein, the former being catalysed by a kinase, and the latter by a phosphatase. Note: phosphorylation usually causes a conformational change in the protein.

to remove its activity (the mechanism of control of glycogen metabolism is further explored in Chapter 5; see section 5.2 and Fig. 5.1).

The coordinated control of metabolic pathways also requires that futile cycles are avoided. Again using glycogen storage as an example, to facilitate glycogen usage phosphorylase is phosphorylated, causing an increase in its activity, as already mentioned. However, there would appear to be little point in increasing glycogen breakdown if synthesis of glycogen continued unabated, or in fact increased due to the rise in the concentration of intracellular glucose resulting from the enhanced glycogen breakdown. Indeed, this is coordinated. The enzyme responsible for synthesis of glycogen, glycogen synthase, is phosphorylated at the same time as phosphorylase, but in the case of the synthase phosphorylation causes a lowering of activity of the enzyme, and hence slowing the production of glycogen. Therefore, by phosphorylation of the two key enzymes in the pathway at the same time, both the breakdown is increased and the synthesis decreased, resulting in the necessary effect (Fig. 4.2); no futile cycle ensues.

Futile cycle

A futile cycle is when the production of a compound and the breakdown of the same compound are carried out at the same time, resulting in little change in the steady-state concentration of the compound. Such cycles would be wasteful to cells, and mechanisms are usually in place to avoid this happening.

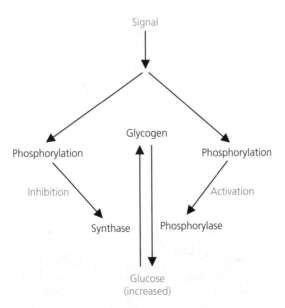

Fig. 4.2 Phosphorylation controls both glycogen breakdown and synthesis (see also Fig. 5.1).

Some enzymes that are involved in metabolic pathways and which are controlled by phosphorylation and dephosphorylation are listed in Table 4.1.

The addition of phosphate groups to proteins, and their removal, are enzyme-catalysed events. The enzymes which catalyse protein phosphorylation are known as protein kinases and there is a reciprocal group of enzymes which carry out dephosphorylation, called phosphatases. However, phosphorylation usually only takes place on one of three amino acids in the primary sequence of the polypeptide, on either a serine, threonine, or tyrosine (Fig. 4.3), although as we shall see there are exceptions. Even so, the amino acid targets of the kinases are used in their

Table 4.1 Some of the enzymes which are controlled by phosphorylation/dephosphorylation and the metabolic pathways in which they are involved. CoA, coenzyme A.

Enzyme	Metabolic pathway
Phosphorylase kinase	Glycogen usage
Glycogen phosphorylase	Glycogen usage
Glycogen synthase	Glycogen storage
Acetyl-CoA carboxylase	Lipid metabolism
Lipase	Lipid breakdown
Isocitrate dehydrogenase	Citric acid cycle in prokaryotes

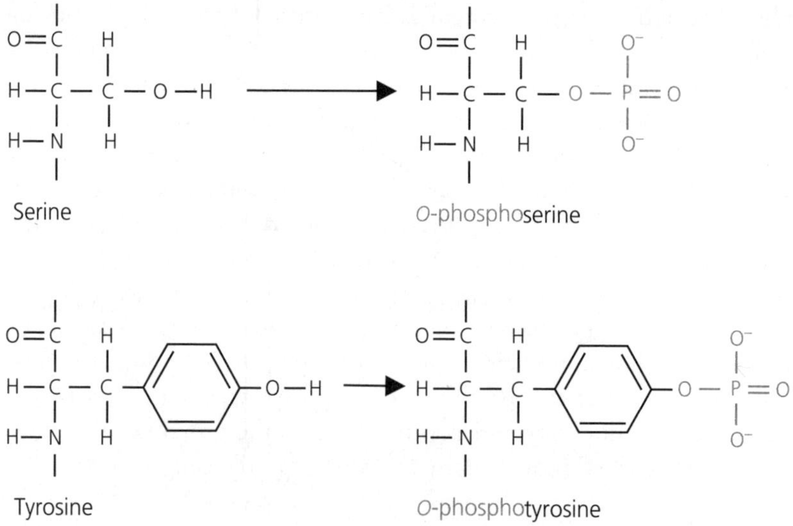

Fig. 4.3 The phosphorylation of serine and tyrosine yields the altered residues phosphoserine and phosphotyrosine.

classification, being grouped according to which amino acid they are specific for. The two main groups are:

- serine/threonine kinases which add the phosphate to serine and/or threonine
- tyrosine kinases which only use tyrosine as acceptors of the phosphate.

Kinase families

Other groupings of the kinases include the splitting of the kinases into five families: AGC which includes PKA, PKG, and PKC; CMGC which includes the MAP kinases and cyclin-dependent protein kinases; PTKs which are the proein tyrosine kinases; CaMKs which are the calcium/calmodulin-dependent protein kinases; and a group for others not able to be included in the groups above.

The phosphate group itself is supplied by ATP, the third phosphoryl group (γ) of the chain within ATP being transferred to the hydroxyl group of the acceptor amino acid, with the subsequent release of adenosine diphosphate (ADP); (the structure of ATP is shown in Fig. 5.3). Phosphorylation usually takes place inside the cell and here ATP is generally abundant, being supplied mainly by the mitochondria in eukaryotes. Dephosphorylation is the simple removal of the phosphoryl group from the amino acid with the regeneration of the hydroxyl side chain and the release of orthophosphate. Each enzyme catalyses their relevant reaction in an irreversible manner, although the overall reaction of phosphorylation and the subsequent return of the protein to its original state is effectively accompanied by the hydrolysis of ATP and hence results in a favourable free-energy change.

Phosphorylation of the protein is not restricted to a single site on the polypeptide chain and indeed a protein may be phosphorylated by more than one kinase, allowing in many cases the convergence of several signalling pathways. Glycogen synthase can be phosphorylated by protein kinase A (cAMP-dependent protein kinase or PKA), phosphorylase kinase, and Ca^{2+}-dependent kinase. Each phosphorylation event might have a different effect on the protein, or may have no effect at all. For example, phosphorylation of serine 10 on cAMP-dependent protein kinase appears to serve no purpose as phosphorylation seems to bestow no alteration of activity on this polypeptide. Using bioinformatic analysis, once the amino acid sequence of a protein is known it is relatively easy to predict if a protein can be phosphorylated by particular kinases, but that does not indicate the effect on the protein of the phosphorylation, or whether it actually takes place in the cell (see discussion below).

Bioinformatic analysis

Bioinformatic analysis refers to the use of computers in biological studies. Computers have allowed rapid and elaborate analysis of biological systems, and in particular the analysis of DNA, RNA, and protein sequences. Many annotated databases of sequences are now available on the internet. For example GenBank (www.ncbi.nlm.nih.gov/).

The phosphorylation of a polypeptide can alter the structure, and perhaps the enzymatic activity, of that molecule and this might be achieved for several reasons. The added phosphoryl group adds negative charge to an enzyme which can disrupt electrostatic interactions or may be instrumental in the formation of new interactions. Similarly the phosphoryl group can form hydrogen bonds which may favour a new conformation. The free-energy change involved in phosphorylation may also help to push the equilibrium from one conformational state to another.

Phosphorylation is ideal as a means of regulation in response to a cellular signal as it can occur in under a second, as can dephosphorylation, therefore making the system ideal for the fast interaction often needed in the alteration of metabolic rate. Conversely, the process may have kinetics over a matter of hours, which might be needed in other physiological conditions. Further, one of the basic needs of a signalling system in the cell is the ability to have amplification of a signal. As discussed in Chapter 1 (section 1.5), a few molecules arriving on the outer surface of the cell might need to alter the activity of many enzyme molecules on the inside. The activation of a single kinase molecule will result in the phosphorylation of many proteins and so the process of phosphorylation plays a major role in the amplification of intracellular signals.

If kinases are to control the activity of enzymes within the cells, then they themselves have to be under some sort of control. The modulation of a cell's activity may be through various different routes, including alteration of calcium concentrations, cAMP, guanosine 3′,5′-cyclic monophosphate (cGMP), and inositol phosphate metabolism. These different signalling pathways appear to end commonly with the activity of kinases with their own degree of specificity towards the enzymes which they control. However, not all kinases are controlled by second messengers, a classic case of messenger-independent protein kinases being the casein kinases. These enzymes are widely distributed throughout the plant and animal kingdoms where they are used for the phosphorylation of acidic proteins. In fact, some kinases might themselves be controlled by a phosphorylation event. For example, phosphorylase kinase is phosphorylated by cAMP-dependent protein kinase, while mitogen-activated protein kinase (MAPK) cascades show a series of phosphorylation events controlling a series of kinases.

Some kinases are specific for one protein, for example, phosphorylase kinase which phosphorylates phosphorylase, while others have a more generic action, where they are capable of phosphorylating many proteins. For example, protein kinase C has many protein substrates. The specificity of the kinase is determined by the specific amino acid sequence either side of the target amino acid residue which is destined to receive the phosphoryl group. In many cases consensus sequences for the sites of phosphorylation for different kinases have been identified using bioinformatics, but the appearance of a consensus sequence in the protein's primary sequence does not automatically mean that the site will be phosphorylated. Many potential phosphorylation sites will be buried deep inside the structure of the protein and not be accessible to kinase action, and will never be used. Using such information, once the gene encoding a protein has been cloned and sequenced, a researcher can more easily predict whether that protein can be phosphorylated by a variety of kinases. With such predictions, biochemical experiments can be carried out to determine which kinases actually modify the protein, and to determine the effects of such phosphorylations.

Consensus sequence

If the sequences of several related proteins are known, bioinformatic analysis allows for the sequences to be aligned and an 'average' or consensus sequence to be determined, incorporating the common features found in all the sequences known. Positions in the sequence where the identity of a single amino acid cannot be determined in a consensus, usually because it can be a variety of amino acids across a group of sequences, are simply marked as an X.

Although the kinases have been separated into these two broad classes, the actual catalytic sites within most of them seem to be quite well conserved, and contain common characteristics, among these enzymes. Most protein kinases contain a catalytic core domain of approximately 250 amino acids in size, which is relatively conserved between them. Within this domain there appears to be two particular regions that have had consensus sequences or signature patterns assigned to them. The first region is located at the N-terminal extremity of the catalytic region. It is characterized by a lysine residue believed to be involved in ATP binding, which is close to a stretch of glycine residues. In the central part of the catalytic region the second conserved region identified contains an aspartic acid residue believed to be important in the catalytic activity of the kinase. However, the amino acids around this residue appear to differ for the two classes of kinase and therefore separate two signature patterns can be deduced.

Not all kinases contain these regions. Interestingly, the signature pattern specific for the tyrosine kinase catalytic site has homology to some bacterial phosphotransferases. These are thought to be evolutionarily related and in fact do contain some structural homology to protein kinases.

The identification of protein signatures and consensus sequences will of course aid in the identification of kinase-like active sites in newly discovered proteins, but, like all consensus searches, just because it is found does not necessarily mean that it is active and bestows any functionality on the protein. It has in fact been estimated that the human genome may contain as many as 2000 genes coding for different kinase polypeptides, leaving us many that are yet to be discovered.

4.2 Serine/threonine kinases

The kinases which preferentially phosphorylate the amino acids serine or threonine within polypeptides, and therefore come under the classification of serine/threonine kinases, encompass a large group of phosphorylating enzymes, including cAMP-dependent protein kinase (PKA), cGMP-dependent protein kinase (cGMP), protein kinase C (PKC), Ca^{2+}/calmodulin-dependent protein kinases, phosphorylase kinase, pyruvate dehydrogenase kinase, and many others. Not all these can be treated in detail here, but important representative examples will be examined which should give an overall picture of the activity of these ubiquitous enzymes.

Oncogenes commonly encode for proteins which contain a cell signalling function, as discussed in Chapter 1 (section 1.8), and it is no surprise that several oncogenes appear to function because they encode proteins which contain serine/threonine kinase activity. These include the products of the genes *mil*, *raf*, and *mos*. Such activity highlights the importance of phosphorylation by these kinases in the control of cellular functions, and also highlight the importance of such functionality being tightly controlled.

Raf

Raf, the protein encoded by the *raf* gene, is an extremely important kinase found at the top of mitogen-activated protein kinase cascades. See later chapters for further discussion, for example Chapter 9.

---

cAMP-dependent protein kinase

cAMP-dependent protein kinase, otherwise referred to as protein kinase A, PKA, or cAPK, is widespread in eukaryotes, being found in animals, fungi, and as a slightly different form in plants.

Many processes within the cell are controlled through the activation of PKA. These include the regulation of the rates of metabolism and the control of gene expression. Examples of the former include the control of phosphorylase kinase, and hence activation of phosphorylase. This results in the increased breakdown of glycogen. As discussed above and in Chapter 5 (section 5.1), the activation of PKA not only causes the activation of phosphorylase kinase and so phosphorylase, it also catalyses the concomitant phosphorylation and deactivation of glycogen synthase, so preventing a futile cycle (see also Fig. 5.1).

Gene expression can also be controlled through phosphorylation by PKA, causing the activation of transcription factors such as CRE-binding protein (CREB). CREB in its active state binds to CRE (cAMP-response element) regions of the DNA. Genes containing such control elements include those which encode enzymes of gluconeogenesis in the liver.

As its name suggests, cAMP-dependent protein kinase is controlled by the levels of cAMP present in the cell; see further discussion on the production of cAMP in Chapter 5 (sections 5.2 and 5.3). In the inactive state—that is, when the levels of cAMP are low—cAMP-dependent protein kinase is found as a tetramer containing two catalytic subunits (C) and two regulatory subunits (R). If the concentration of cAMP rises it causes activation of the PKA. cAMP binds to the regulatory subunits, causing a conformational change in the structure of the protein and an alteration in the affinity of the regulatory subunits for the catalytic subunits, which causes the complex to dissociate. The regulatory subunits remain as a dimer with the release of the two active monomeric catalytic subunits, as shown in Fig. 4.4.

It is thought that virtually all of a cell's responses to cAMP are mediated by the activity of the catalytic subunits of cAMP-dependent protein kinase. As a monomer this catalytic polypeptide has a molecular weight of around 41 kDa, but at least three isoenzyme forms have been identified.

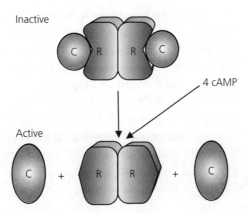

Fig. 4.4 The activation of cAMP-dependent protein kinase, showing the dissociation caused by the binding of cAMP. C, catalytic subunit; R, regulatory subunit.

The isoforms C_α and C_β in mammals differ by less than 10% when their amino acid sequences are compared and they seem to be highly conserved between species. However, unlike C_α, which appears to be expressed constitutively in most cells, the expression of C_β is tissue-specific. The catalytic core of these PKA subunits also shares homology with other known kinases and includes defined regions for peptide binding, ATP binding, and a catalytic site (also see discussion above). Such molecular studies of enzymes have revealed details of the mechanisms involved. Analysis of the catalytic regions of these PKA polypeptides, both by the use of fluorescence analogues of ATP and by studying the modification of lysine residues by acetic anhydride in the presence and absence of MgATP, have shown that lysine residues at amino acid positions 47, 72, and 76 are important for the functionality of these proteins. Sequence comparisons with other kinases showed that lysine 72 is totally invariant and site-directed mutagenesis has shown the vital importance of this lysine in the catalytic cycle of kinases. To the N-terminal side of the lysine lie three highly conserved glycine residues, found at positions 50, 52, and 55, which are probably involved in the binding of the phosphate in the nucleotide. Interestingly, point mutations in an analogous glycine-rich region may be responsible for the constitutive activity of the *v-erbB* onco-gene product, which resembles the EGF receptor but is in a permanently turned-on state. However, it should be noted that the *v-erbB* oncogene protein also lacks the extracellular EGF-binding domain as well.

Once activated, PKA needs to be able to phosphorylate the next component in the cell signalling cascade. Therefore, if it has many potential targets, it should be possible to identify some commonality among such targets. cAMP-dependent protein kinase typically phosphorylates peptides that contain two consecutive basic residues, usually arginine, which lie at positions 2 and 3 towards the N-terminal end of the protein from the site of phosphorylation. This would give the consensus sequences as:

$$\text{-Arg-Arg-X-Ser-X-} \quad \text{or} \quad \text{Arg-Arg-X-Thr-X-}$$

The residue designated as X between the Arg doublet and the phosphorylation site is usually a small amino acid while the other amino acid designated as X is usually hydrophobic in character. The arginines may also be replaced by lysines, and therefore the consensus sequence above does not make a hard and fast rule.

As well as the lysines identified above, e.g. lysine 72, carboxyl groups within the sequence of the catalytic subunit may well be responsible for this specificity, so allowing the recognition of peptide substrates. Carboxyl groups, particularly at position 184, an aspartate, and 91, a glutamate, may also be responsible for ligation to the Mg^{2+} in the MgATP substrate. Like lysine 72, these two groups appear to be invariant throughout the kinase domains studied so far. Further carboxyl groups at position 170, and six at the C-terminal end of the protein, have also been implicated in the recognition of the peptide substrate.

As with all enzymes, kinetic analysis of the catalysis can be determined and here for cAMP-dependent protein kinase such studies have found that the enzyme usually has a K_m in the region of 10–20 μM substrate, with a V_{max} of 8–20 μmol/min per mg. The first step in the binding of the target peptide sequence to the enzyme probably involves ionic interactions between the peptide's arginine residues and the enzyme which is followed by recognition of the amino acid, e.g. serine, which will ultimately accept the phosphate group. Binding of MgATP, which supplies the phosphate needed for phosphorylation, to the catalytic subunit enhances the binding of the peptide, with large conformational changes across the enzyme being induced by substrate binding.

The catalytic subunit of cAMP-dependent protein kinase also contains two phosphorylation sites itself, one at threonine 197 and another at serine 338, and once phosphorylated the phosphate groups are not readily removed by phosphatases. A serine residue at position 10 can also be autophosphorylated; that is, the phosphate group is added by PKA itself, although it is not known if this has any physiological role.

Other modifications to PKA can also occur, besides phosphorylation. For example, the catalytic subunit can be covalently modified by the myristoylation of the N-terminal end. It was thought that this might serve as a signal for translocation of the subunit to the plasma membrane, an event with which myristoylation is often associated, but after such modification the catalytic subunit remains soluble, and myristoylation appears not to be essential for catalytic activity either.

Therefore, much is known about the functioning and mechanisms involved with the activity of the catalytic subunits. But what of its controlling peptide, the regulatory subunit? In the presence of low concentrations of cAMP, the regulatory subunits of PKA have the function of binding to the catalytic subunits, so rendering them inactive. As with many signalling proteins, there is more than one isoform of this peptide. Two main groups of the regulatory subunits have been found, called type I and type II, which differ in their amino acid sequences but also in their function. Type II subunits can be autophosphorylated while type I subunits are not autophosphorylated, but do contain a binding site for MgATP which binds with relatively high affinity. Isoforms of the different types of subunit have been cloned, and expression patterns have been found to differ too. Some isoforms seem to be expressed fairly widely while others are expressed in a tissue-specific way. Furthermore, the expression of some isoforms is inducible while others appear to be expressed constitutively.

In general, the regulatory subunit exists as a dimer, with all the subunits sharing the same structural features (Fig. 4.5). These include two

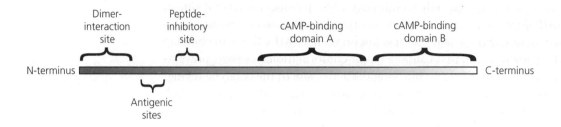

Fig. 4.5 The domain structure of the regulatory subunit of cAMP-dependent protein kinase.

consecutive gene-duplicated sequences at the C-terminal end of the molecule which are the cAMP-binding sites. Type I subunits are held together covalently by two disulphide bonds with the two polypeptide chains running antiparallel. In type II, the subunits are held together by interactions between the N-terminal amino acids in the chains. Another general feature of the subunits is what is referred to as the hinge region towards the N-terminal end, which is sensitive to proteolytic cleavage. This region encompasses the amino acids which are most likely to be involved in the interaction with other proteins as well as being a highly antigenic part of the molecule.

The most important part of the regulatory subunit is a region which controls the activity of the catalytic subunit. The hinge region contains an amino acid sequence, known as the peptide-inhibitory site, which resembles that of the substrate for the catalytic subunit. In type II regulatory subunits there is an autophosphorylation site here, while type I subunits have a sequence which contains the two arginines needed for recognition by the catalytic-subunit active site but which lacks the serine or threonine that would normally accept the phosphate group. This so-called pseudophosphorylation site contains an inert alanine or glycine residue instead. Several lines of evidence, including limited proteolytic cleavage, affinity studies of the regulatory subunit to the catalytic subunit, and site-directed mutagenesis, support the idea that the regulatory subunit maintains the catalytic subunit in an inactive state, by pseudophosphorylation or by the inhibitory site of the regulatory subunit occupying the peptide-binding site of the catalytic subunit, so preventing the binding of the correct protein substrate.

It can be seen therefore that here in PKA a sequence which mimics the true substrate of the enzyme can inhibit by competing for the active site of the enzyme. In the inactive state, the conformation of the subunits is such that binding to the pseudo-substrate—that is, the regulatory subunit—is preferred, but on activation this is dissociated, allowing the binding of the true substrate. In PKA the inhibitory site and the catalytic site are on separate polypeptide chains, but this general principle of a regulator region controlling the activity of the catalytic region is used by several other protein kinases. However, in other kinases both sites are usually part of a single polypeptide chain. This includes the related kinase, cGMP-dependent kinase, as well as other kinases such as myosin light-chain kinase (MLCK) and protein kinase C. With MLCK it is the binding of Ca^{2+}/calmodulin that causes a major conformational change, preventing the inhibitor site from blocking the active site of the catalytic region. In many kinases a mechanism more akin to that of the type II PKA regulatory subunits is used, in that the regulatory site is actually autophosphorylated. Enzymes using this type of mechanism include cGMP-dependent kinase and Ca^{2+}/calmodulin-dependent kinase II.

In cAMP-dependent protein kinase the conformational change needed to release the inhibitor site from the catalytic active site is induced by the binding of cAMP. Each regulatory site has two cAMP-binding sites, both of high affinity. These two sites, A and B, show high sequence homology to each other but their binding to various cAMP analogues varies. They also show high sequence homology to catabolite gene-activator protein (CAP) of *Escherichia coli*. This protein is involved in the cAMP-dependent regulation of the lactose operon, turning on gene expression when the cells are depleted of glucose and allowing them to survive on a new carbon source. In CAP it was found that two amino acid residues were of crucial importance to the functioning of the protein: an arginine which interacts with the negative charge of the phosphate of cAMP and a glutamine residue which hydrogen bonds to the ribose ring. These two residues are found in PKA regulatory subunits as well as in the cGMP-binding sites of cGMP-dependent protein kinase.

Studies where parts of the protein have been deleted and the functionality analysed have shown that the N-terminal region of the regulatory subunit, along with the cAMP-binding site B, can be removed and still the polypeptide is able to bind to the catalytic subunit in a cAMP-dependent manner. Removal of the cAMP-binding site A has also been shown to produce a molecule which is cAMP-dependent and can still bind to the catalytic subunit. Furthermore, removal of either cAMP-binding site seems to have little effect on the affinity for cAMP of the other binding site in isolated regulatory subunits. However, in reality, both cAMP-binding sites probably participate in activation of type I PKA. Both sites need to be occupied for dissociation of the native holoenzyme and binding of the two cAMP-binding sites shows cooperativity. cAMP probably binds to the site B first, causing conformational changes which increase the accessibility of the A site to cAMP. Binding at the A site causes a further conformational change, altering in particular the hinge region, so resulting in the activation of the enzyme.

To be inactivated, once the need for the phosphorylation catalysis of PKA has passed, the enzyme needs to be reassociated into its original tetrameric form. For the type II regulatory subunit-containing enzymes reassociation of the cAPK complex probably involves the use of phosphatases, allowing the inhibitor site to once again enter the catalytic site and cause deactivation of the enzymes activity. However the reassociation for regulatory subunit type I-containing enzymes involves the binding of MgATP. This binding shows positive cooperativity which probably ensures that the holoenzyme is a tetramer and that trimers with only one catalytic subunit are not formed.

The N-terminal end of the regulatory subunits is also the target for phosphorylation, in fact by several kinases, including protein kinase CK2 and glycogen synthase kinase. It may be that phosphorylation here is involved in the association of the regulatory subunits with other proteins.

PKA is usually found as a soluble enzyme in most cells, but it has been found that some forms of the kinase are associated with the membrane fractions of cells. Such associations are probably mediated by the regulatory subunits binding to an integral membrane protein. The catalytic subunits do not appear to be involved here, as they can be released and found in the soluble fractions of cells following activation.

cGMP-dependent protein kinase

cAMP is not the only nucleotide that is responsible for the activation of a kinase. cGMP is also an important cell signalling molecule in cells (see Chapter 5 for further discussions; section 5.5). The levels of cGMP in cells, like that of cAMP, can be used to control phosphorylation, often via cGMP-dependent protein kinase—otherwise referred to as cGPK or PKG—an enzyme with many similarities to PKA. cGMP-dependent protein kinase has been found to be abundant in many mammalian tissues, including smooth muscle, heart, lung, and brain tissues. Although mainly cytosolic in location, following cell subfractionation analysis the enzyme has also been found in particulate fractions.

cGPK, such as that purified from lung and heart, has been found to be a dimer of identical 76 kDa subunits, with the holoenzyme having a molecular weight of approximately 155 kDa. The two subunits are held together in an antiparallel arrangement by disulphide bonds. Just as was found in PKA, this enzyme too has an inhibitor sequence. In this case the orientation of the dimer allows the inhibitor site of the regulatory domain of one subunit to act as the inhibitor of the catalytic domain of the other subunit.

Sequencing of cGPK has revealed that it can be thought of as being six segments, making up four functional domains. The segment at the N-terminal end of the polypeptide contains the sites used to maintain the dimer structure, a hinge region as in cAMP-dependent protein kinase, and the inhibitor site which will undergo autophosphorylation. Also like cAMP-dependent protein kinase, the next two segments have a high sequence homology to the CAP protein of *E. coli* and make up the two cGMP-binding domains of the polypeptide. The rest of the molecule is the catalytic domain, with the fourth and fifth segments showing high homology to other protein kinases.

cAMP-dependent protein kinase and cGMP-dependent protein kinase share very similar preferences for the peptide sequences which are phosphorylated. The cGPK enzyme typically phosphorylates a serine, or threonine, in the peptide which lies at positions 2 and 3 towards the C-terminal end of the protein from two consecutive basic residues, usually arginine. Therefore, like cAMP-dependent protein kinase, the consensus sequences would be:

$$-\text{Arg-Arg-X-Ser-X-} \quad \text{or} \quad -\text{Arg-Arg-X-Thr-X-}$$

However, in cGPK, phosphorylation is enhanced by the presence of a proline residue N-terminal to the Ser/Thr phosphorylation site and a basic residue on the C-terminal side. Hence, such differences give cGPK a different substrate specificity to cAMP-dependent protein kinase.

Among other similarities with PKA is the fact that both nucleotide-binding sites appear to be needed to be occupied for activation. However, unlike cAMP-dependent protein kinase, cGPK does not dissociate on activation.

Although activated primarily by cGMP, cGPK can potentially also be controlled by other mechanisms. It can bind for example to cAMP and cyclic inosine monophosphate (cIMP), albeit at much higher concentrations than that seen with cGMP, while cGPK has also been shown to be phosphorylated. However the physiological relevance of such mechanisms is not clear, but would allow for the convergence of signalling in cells.

Protein kinase C

One of the most important protein kinases is one called protein kinase C, or PKC. PKC was originally thought to be just a single protein, but more recent research has revealed that it is in fact a family of closely related protein kinases. These different polypeptides are encoded for by different genes, or in some cases are derived from the alternate splicing of a messenger RNA (mRNA) transcript from single genes. However, not all cells express all the variants, but cells often express more than one form.

The protein kinase C proteins can be broadly split into two families. The first to be cloned was the group containing the α, βI, βII, and γ subspecies, but later cloning revealed a group containing δ, ε, and ζ subspecies. The genes for the α, β, and γ forms have been located to different chromosomes, with the variation in the βI and βII forms coming from alternative splicing from one gene. Other isoforms described include η, θ, and λ, making a total of at least 10 isoforms that have been described.

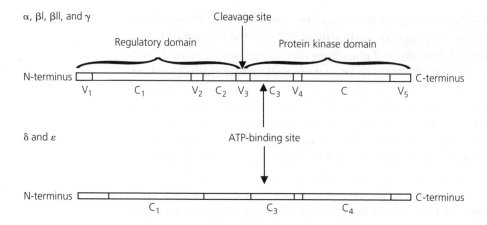

Fig. 4.6 The domain structure of the two families of protein kinase C.

In general the protein kinase C enzymes are monomeric in nature, having between 592 and 737 amino acids, which give molecular weights of between 67 and 83 kDa. The two groups of PKCs each have a common structure although distinct differences are also seen between the two families. The first group, containing the α, βI, βII, and γ forms, have four conserved regions (C_1–C_4) along with five variable regions (V_1–V_5). The second group, containing, δ, ε, and ζ, lacks the C_2 conserved region, although overall their molecular weights appear to be similar across the families.

The polypeptides can be divided roughly into two domains, a regulatory domain and a protein kinase domain (Fig. 4.6). The regulatory domain contains the C_1–C_2, and V_1–V_2 regions of the first group, but only the C_1 region of the δ/ε group. The C_1 region of both groups contains a highly cysteine-rich region which resembles the consensus sequences of a cysteine–zinc DNA-binding finger, although there appears to be no DNA binding. The second domain contains the rest of the molecule and is referred to as the protein kinase domain. Sequence homology between this region and other protein kinases has been reported, and it is the C_3 region that contains an ATP-binding sequence. This motif is repeated in the C_4 region although it is probably not used.

X-ray absorption studies suggest that protein kinase C may contain four zinc ions within its structure, which are coordinated mainly by sulphur atoms supplied by cysteine residues. However, it is likely that these zinc ions serve a role in the structural stability of the polypeptide rather than having a direct role in the catalytic cycle.

As with all cell signalling components, protein kinase C has to have its activity controlled. PKC can be activated in a wide range of manners, including by proteolytic cleavage. In many cases its activity can also be controlled by intracellular Ca^{2+} concentrations which is how it derived its name; that is, as a calcium-dependent protein kinase. However, some isoforms are Ca^{2+}-independent, such as δ, η, θ, and ε.

Of particular importance is the fact that protein kinase C can also be controlled by phospholipids and in particular by diacylglycerol (DAG). DAG is derived from inositol phosphate metabolism simultaneously with inositol trisphosphate (InsP$_3$) by the action of PLC. InsP$_3$ formation leads to an increase in intracellular calcium, and therefore the action of PLC has two potential ways of controlling PKC activity, through the release of Ca^{2+} and the formation of DAG. Inositol metabolism is discussed in further detail in Chapter 6 (section 6.3). The response by PKC to DAG is not homogeneous across the isoforms of PKC; the exact effect differs between isoenzyme forms, due to the fact that some seem to lack a DAG-binding site, for example ζ and λ. Furthermore, it has been suggested that some PKC subspecies become activated at different times during a cellular response, orchestrated by a series of phospholipid metabolites such as DAG, arachidonic acid, or other unsaturated fatty acids.

Activation of some PKC forms can also be through the phosphoinositide 3-kinase (PtdIns 3-kinase) pathway. This kinase, as discussed in Chapter 6 (section 6.3), forms phosphorylated inositides, which are now thought to be extremely important in many pathways, such as insulin signalling to GLUT4 glucose transporters (see Chapter 9). Phosphoinosides formed are able to activate a kinase known as 3-phosphoinositide-dependent kinase (PDK1), and this kinase has been found to be able to phosphorylate PKC, particularly isoforms ζ and λ.

Chemical stimulants, such as the tumour-promoting phorbol esters, for example phorbol 12-myristate 13-acetate (PMA), otherwise known as 12-O-tetradecanoyl-phorbol-12-acetate (TPA), are widely used in the laboratory to cause activation of protein kinase C, as they work by their action as DAG analogues. However, their exact effect differs between isoenzyme forms because, as mentioned above, some lack the DAG-binding site. Caution is needed when using phorbol esters in experimental work, not only due to their toxicity to the users but also because they are only very slowly metabolized from the cells, unlike DAG. New stimulators which are more specific than phorbol esters, such as Sapintoxin A, can also be used in the laboratory to activate protein kinase C.

On looking at the molecular mechanism of the enzyme, it can be seen that here, as in cAMP-dependent protein kinase, it appears that protein kinase C has an inhibitor site, or pseudophosphorylation site, near the N-terminus of the polypeptide. Again, this site contains an inert alanine instead of the threonine or serine which would normally be found in the substrate. Binding of PKC activators such as Ca^{2+}, DAG, and phosphatidylserine will cause a conformational change within the protein, releasing this pseudophosphorylation site and resulting in an activation of the kinase.

Activation of PKC by proteolytic cleavage can occur through the action of enzymes such as calpain. Cleavage of PKC occurs within the variable

V_3 region, releasing a catalytically active fragment. It may be that the active form of PKC is the target for calpain, which itself is active in micromolar Ca^{2+} concentrations, although unlike the γ form not all subspecies of PKC are susceptible to rapid cleavage; for example, subspecies α is relatively resistant. It is possible that cleavage of protein kinase C is in fact not part of its physiological activation, but perhaps the first step to its degradation and removal from the cell.

Having discussed the activation of PKC, and the presence within it of a pseudophosphorylation site as part of its enzymatic mechanism, it is necessary to investigate the likely substrate targets of this enzyme. Under physiological conditions, once activated, protein kinase C preferentially phosphorylates a polypeptide on a serine or threonine residue which is found in close proximity to a C-terminal basic residue. Therefore, the consensus sequence would thus be:

-(Ser/Thr)-X-(Arg/Lys)-

Additional basic residues, on either the C-terminal or N-terminal side of the target amino acid, may enhance the V_{max} and reduce the K_m of the phosphorylation reaction.

The exact role, or roles, of PKC within the cell however remains surprisingly obscure. Phosphorylation experiments *in vitro* have shown that a large variety of polypeptides become phosphorylated by PKC and it has been implicated in the activation of Ca^{2+} ATPases and Na^+/Ca^{2+} exchangers, controlling Ca^{2+} levels within the cell as well as phosphorylation of receptors such as the EGF receptor and the IL-2 receptor. Clearly, future work will establish the exact role of each isoform and how they fit into the complex web of cellular signals.

Ca^{2+}/calmodulin-dependent protein kinases

One of the major ways of controlling the functioning of enzymes and metabolic pathways in a cell is by the alteration in the intracellular concentration of Ca^{2+}, $[Ca^{2+}]_i$, as discussed further in Chapter 7 (see section 7.1). Many kinases have been found to be dependent on or regulated by Ca^{2+}, including protein kinase C (discussed above), myosin light-chain kinase and phosphorylase kinase. Besides these more specific enzymes there is a group of kinases which are controlled by Ca^{2+} called the multifunctional calcium/calmodulin-dependent protein kinases. Several classes of these have been identified but the most characterized is known as calcium/calmodulin-dependent protein kinase II. This enzyme is also referred to as CaM kinase II, type II CaM kinase or sometimes simply as just kinase II.

Although kinase II is widespread throughout several tissues, it is most abundant, up to 20–50 times more concentrated, in brain and neuronal tissue compared to non-neuronal tissues. The enzyme prepared from brain tissue consists of an α subunit of approximately 50 kDa and a protein doublet (β/β′) of approximately 60 kDa. Both types of subunit are able to bind to calmodulin and are autophosphorylated in a Ca^{2+}/calmodulin-dependent manner. Both subunits therefore contain both regulatory and catalytic domains. The α and β/β′ subunits are very closely related, the N-terminal halves of the two subunits being 91% identical, with over three-quarters of the other half also being found to be identical when the genes were cloned from rat brain. The holoenzyme has been reported to have a molecular weight of 500–700 kDa in most tissues that have been investigated but a smaller complex of 300 kDa has been found in liver. The holoenzyme complex probably assumes a dodecamer structure in which two hexameric rings are stacked on top of each other. The ratio of α subunits to β/β′ subunits seems to vary enormously between tissues while other subunits of approximately 20 kDa have been reported in spleen tissues. Isoenzymes cloned include δ and γ subunits which are of similar molecular weight to the β/β′ subunits and which appear to have a wide tissue distribution. The sequences of these isoforms show that they are again closely related to the α subunits.

The subunits of this kinase can be split into three distinct functional domains. The N-terminal section is the catalytic domain, with the centre section encompassing the regulatory domain. A domain essential for the formation of the holoenzyme complex is at the C-terminal end of the polypeptides, known as the association domain (see Fig. 4.7). Proteolytic cleavage can release an active catalytic domain, and here homology to other calmodulin-regulated kinases is seen, such as phosphorylase kinase. The fact that proteases can cleave here suggests the presence of a hinge region between the catalytic and regulatory domains.

As with other kinases, the regulatory domain contains an autoinhibitory site, as discussed above, but here the regulatory domain also has a calmodulin-binding site. Binding of Ca^{2+}/calmodulin to the enzyme, a process which shows positive cooperativity, will cause an overall conformational change in the kinase's three-dimensional structure, which releases the autoinhibitory site from the active site and allows kinase to have full activity. In the inactive state the autoinhibitory site not only

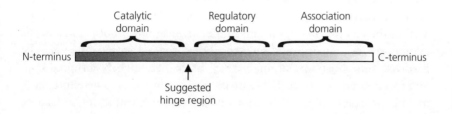

Fig. 4.7 The domain structure of calcium/calmodulin-dependent protein kinase II.

prevents peptide-substrate binding but also prevents the binding of the donating ATP. The autoinhibitory site involves autophosphorylation, as seen with other kinases, and it is thought that this autophosphorylation potentiates the effects of the second messenger. A rapid spike of intracellular Ca^{2+} can activate the enzyme not only allowing it to phosphorylate its normal substrate but also causing autophosphorylation, which may slow down the calmodulin dissociation from the enzyme. This will have a consequence of lengthening the time that the kinase has activity, even after the Ca^{2+} concentrations have returned to their basal levels.

The subcellular location of the enzyme also seems to vary, sometimes being membrane-associated, sometimes being soluble, and sometimes being bound to the cytoskeletal structures of the cell.

As with all kinases, there are likely to be specific targets for their action. The substrates for kinase II are quite wide ranging, and include tryptophan hydroxylase, glycogen synthase from skeletal muscle, synapsin I, proteins associated with the microtubules, ion channels, and transcription factors. The specific part of proteins targeted—that is, the target peptide specificity—in some cases seems to be similar to that identified for cAMP-dependent protein kinase or myosin light-chain kinase, but in other cases the peptides that are phosphorylated are distinct from those recognized by other kinases. In general, it has been found that an arginine three residues towards the N-terminal end of the polypeptide from the serine or threonine which is to accept the phosphate group seems to be essential. It has also been suggested that acidic residues within two residues of the phosphorylated amino acid or a hydrophobic residue on the C-terminal side of the phosphorylation site might also be important.

As discussed above, kinase II is not the only multifunctional calcium/calmodulin-dependent protein kinase, and others differ from kinase II substantially. Kinase I, for example, is a monomer with a molecular weight of only approximately 40 kDa. Kinase I and kinase III (an elongation factor-2 kinase), also seem to have a much narrower substrate specificity than kinase II. In fact, kinase I has a substrate specificity more closely related to cAMP-dependent protein kinase than to kinase II.

A fourth Ca^{2+}/calmodulin-dependent kinase, kinase IV, is expressed primarily in brain tissue, and in the thymus, but has also been seen in the spleen and testis. However, kinase IV appears to be absent from several tissues studied. Kinase IV is expressed as two splice variants, and has been purified from rat brain as (two bands on gels from sodium dodecyl sulphate polyacrylamide gel electrophoresis (SDS-PAGE), of 65 and 67 kDa. Within the cell it is mainly found in the nucleus, but it does reside in the cytoplasm too. Again, as discussed above, the activation of kinase IV involves autophosphorylation, here of its Ser/Thr-rich N-terminal end, and it has been shown to be involved in the control of gene expression, mainly through its phosphorylation and activation of transcription factors, for example CREB.

Although calcium kinases are clearly regulated by the calcium signalling, some of them are now known to be part of a cascade. A CaM
kinase kinase, itself activated by elevated Ca^{2+} ion concentration, can
activate both kinase I and kinase IV through phosphorylation. CaM
kinase kinase exists in two forms, α and β, which are found in the cytoplasm and nucleus respectively. Interestingly, CaM kinase kinase can also
phosphorylate and therefore control protein kinase B, allowing divergence of the signalling.

G protein-coupled receptor kinases

An important class of kinases has been discovered which are involved in the
down-regulation of the receptors which act through the trimeric class of
G proteins. These cytosolic kinases are known as G protein-coupled receptor kinases (GRKs), of which the best studied are probably β-adrenergic
receptor kinase (βARK) and rhodopsin kinase (the latter is discussed
further in Chapter 10, section 10.2).

At least six GRKs are known, of which two are classed as βARKs.
Towards the C-terminal ends of these proteins are Pleckstrin homology
(PH) domains which are capable of binding the $\beta\gamma$ subunit complex from
the trimeric G protein. It was originally thought that the $\beta\gamma$ subunits of
these G proteins contributed little to cell signalling pathways, but clearly
here they are involved in kinase activation. An interaction between
the G protein subunits and the kinase results in the translocation of the
kinase to the membrane where it can interact with and phosphorylate
the receptor. With kinases GRK2 and GRK3 lipids are also seen to
stimulate this activity, including lipids such as phosphatidylserine,
although the activity was inhibited by the lipid phosphatidylinositol
bisphosphate ($PtdInsP_2$) at high concentrations. It has been suggested
that lipids and $\beta\gamma$ G protein subunits may compete for the same binding
site on GRK-type kinases.

Further to regulation by lipids and G proteins, β-adrenergic receptor
kinase has been shown to be phosphorylated by protein kinase C. Again,
translocation of GRK to the membrane was enhanced along with an
increase in kinase activity. Phosphorylation of the kinase was towards the
C-terminal end of the GRK polypeptide.

Of the GRKs which are not classed as βARKs, GRK4 is expressed most
highly in the testis. GRK4 is actually a family of four GRKs which arise
from alternative splicing of the mRNA, this splicing involving exons II
and XV. Meanwhile, GRK5 appears to be a unique member of the family
of GRKs. It is capable of phosphorylation of rhodopsin, m2 muscarinic
cholinergic receptor, and β(2)-adrenergic receptor, and it is inhibited by
heparin.

What is being recognized and phosphorylated by these kinases? In the
β(2)-adrenergic receptor the phosphorylated residues are all located within

a 40 amino-acid stretch at the extreme C-terminal end of the receptor polypeptide. GRK5 will phosphorylate threonines at positions 384 and 393, and serines at 396, 401, 407, and 411 while GRK2 will phosphorylate serines 396, 401, and 411 and threonine 384. A similar phosphorylation pattern is seen with rhodopsin. It is important to note that it is only the activated receptors which are phosphorylated, so the conformation of the protein must be very important in the kinase–receptor recognition.

The role of the kinase in the control of the receptor is thought to work in the following way. Once the receptor has bound ligand, the receptor undergoes a conformational change and the G protein is activated. The Gα transmits its signal to, for example, adenylyl cyclase while the βγ complex activates the GRK. GRK will phosphorylate only those receptors which are bound to ligand and deactivate the receptor, stopping its further activation of G proteins. Further, phosphorylation of the receptor allows its interaction with a protein known as β-arrestin, which further inactivates the receptor. Therefore the receptor is desensitized and the cell shows adaptation to a prolonged exposure to ligand (see also Chapter 3, section 3.4). This process is reversible to allow the receptor to be subsequently used when needed.

Protein kinase B

A serine/threonine kinase that, although it was discovered 10 years ago, has not had much prominence in cell signalling literature, is protein kinase B (PKB). PKB is otherwise known as c-Akt, and is the cellular homologue of a virally encoded oncoprotein v-Akt. However, like many proteins, its importance is being recognized and it is now known to be used in a variety of signalling pathways, including those of glycogen metabolism and in response to insulin and growth factors.

PKB is a 57 kDa protein which exists in three isoforms in mammals; α, β, and γ (also known as Akt1, Akt2, and Akt3). The PKB family of proteins shows some homology to PKA and PKC, hence the adoption of its name. When looked at in more detail it can be seen that PKB contains a PH domain (see Chapter 1, section 1.7) at the N-terminal end, a kinase domain in the middle, and a regulatory domain at the C-terminal end.

Activation is thought to be through the PH domain, where it interacts with 3-phosphoinositide lipids created in the membrane by the action of PtdIns 3-kinase (see Chapter 6, section 6.2). Activation also involves translocation of PKB from the cytoplasm to the plasma membrane, and once there it is phosphorylated. One site of phosphorylation of PKB is in the kinase domain, while another is in the regulatory domain. The phosphorylation of the kinase domain is thought to involve an enzyme called 3-phosphoinositide-dependent kinase (PDK1), and PKB autophosphorylation is also thought to occur.

AMP-activated protein kinase

A protein kinase that has recently been shown to be important in a variety of pathways is 5′-AMP-activated protein kinase (AMPK). In the active state AMPK phosphorylates, and inactivates, several metabolic enzymes which are involved in the synthesis of fatty acids and cholesterol, and AMPK has also been shown to be important for insulin signalling (discussed in Chapter 9) where inhibition of AMPK in the presence of glucose activates insulin secretion. The mammalian kinase is a heterodimer which consists of a catalytic α subunit and non-catalytic β and γ subunits, which exist in different isoforms and aid in the control of the enzyme. The presence of either control subunit increases kinase activity, but both of these non-catalytic subunits have been found to be needed for full activity of the kinase, the subunits having a positive and synergistic effect on the control of the enzyme. Experimentally, AMPK can be activated by the addition of 5-amino-4-imidazolecarboxamide riboside (AICAR). Physiologically, AMPK is controlled as its name suggests by 5′-AMP, but also by phosphorylation. The target for phosphorylation within AMPK is threonine 72, phosphate being added by AMPK kinase (AMPKK).

Interestingly, the mammalian AMPKs are related to the SNF1 kinase (sucrose non-fermentor kinase), although the control of the two enzymes by their respective interacting proteins (mammalian β and γ, and yeast Sip1p/Sip2p/Gal83p proteins and Snf4p) seem to be subtly different.

Haem-regulated protein kinase

Haemoglobin

Although haemoglobin is known as an extremely important protein, it should be noted that it contains a haem prosthetic group. It is this haem group which is crucial to its functionality and therefore haem and protein need to come together to produce a working haemoglobin molecule.

It is not only the enzymes in a metabolic pathway that are under the control of specific kinases; protein synthesis can also be regulated in this way. The classical example of this is the work done in reticulocytes where it has been shown that globin synthesis is regulated by haem. It would be a waste for a cell to produce a large quantity of globin polypeptides if there is no protohaem available for completion of the holoenzyme. This mechanism is regulated by phosphorylation of one of the protein synthesis initiation factors, eIF-2, used to bring Met-tRNA (methionyl-transfer RNA) to the ribosome to start the new polypeptide chain. If phosphorylated, this initiation factor forms a stable complex with a guanine nucleotide-exchange factor (GEF), preventing the exchange of GDP for GTP on the initiation factor and so preventing another round of protein synthesis initiation. However, if haem is present, it inhibits the phosphorylation event and allows protein synthesis and formation of holoproteins.

Phosphorylation of eIF-2 is catalysed by a kinase known as haem-regulated protein kinase or haem-controlled repressor (HCR). The purified kinase has a molecular weight of approximately 95 kDa under denaturing conditions, but shows an apparent molecular weight of about 150 kDa under non-denaturing conditions. The target sequence recognized by the kinase appears to be -Leu-Leu-Ser-Glu-Leu-Ser-, where the first serine is the site of phosphorylation. The only known substrate for the enzyme is the initiation factor eIF-2, although the enzyme also undergoes autophosphorylation. Both phosphorylation events are inhibited by the presence of haem. The activity of the kinase is also affected by the state of its thiol groups. If its thiol groups are oxidized or modified the enzyme is found in the active state even in the presence of haem. Other factors might also be involved in the activation of this kinase, and will no doubt be revealed by future research.

A similar control of translation is seen by an interferon-induced kinase, suggesting that phosphorylation might be a commonly used pathway for the control of protein synthesis.

Plant-specific serine/threonine kinases

Although the text above has been devoted to the discussion of kinases as characterized in animal tissues, it should be noted that plants and fungi also use protein phosphorylation in their control of cellular functions, and in fact plants have been found to contain several kinases which are strongly analogous to those described above and are under similar control mechanisms. However, several plant-specific kinases have also been identified and partially or completely purified. One of the best characterized is a kinase which phosphorylates the light-harvesting chlorophyll a/b complex (LHC). LHC is phosphorylated on at least one threonine residue, activating it when exposed to low levels of red light. The kinase, found in thylakoid membranes, has an approximate molecular weight of 64 kDa, and its activation probably involves the redox state of plastoquinone.

With growing evidence that plants also contain kinases analogous to those found in animal tissues, one of the exciting areas of research now is the search for animal-like proteins and control mechanisms in plants. For example, several effects of phorbol esters have been seen in plant tissues, suggesting the presence of a protein kinase C-like enzyme. It is not only with kinases of course, but many areas of similarity in cell signalling can be seen between animals, plant, and organisms of other kingdoms, and research in one area will no doubt inform the future of research elsewhere.

4.3 Tyrosine kinases

The second major class of protein kinases includes those which add a phosphoryl group to tyrosine, as opposed to serine or threonine. Phosphoryl addition to tyrosine is less common than the modification on serine or threonine and an analysis of the phosphoamino acid content of a cell in 1980 revealed that only approximately 0.05% was phosphotyrosine. Even though much more is known about phosphorylation events now, some 25 years later, this still suggests that tyrosine phosphorylation is not that prevalent. Such numbers should not belittle their importance, however, and tyrosine kinases are vital components of many cell signalling pathways. The tyrosine kinases which carry out this phosphorylation are themselves found in two broad groups; those which are part of a receptor and those which are soluble.

The general phosphorylation site which has been characterized as the target for these kinases has a lysine or an arginine residue seven amino acids to the N-terminal side of the tyrosine. An acidic amino acid such as aspartate or glutamate is quite often found three or four residues also to the N-terminal side of the tyrosine, giving the consensus sequences:

-(Lys/Arg)-X-X-(Asp/Glu)-X-X-X-Tyr   or
-(Lys/Arg)-X-X-X-(Asp/Glu)-X-X-Tyr

However, as with most of this type of signature, there are exceptions which do not seem to fit into this neat pattern.

Receptor tyrosine kinases

Receptor tyrosine kinases (RTKs) can be split into 14 groups characterized by their general structural patterns, although they all share the same basic topology. They all possess an extracellular ligand-binding domain, a single transmembrane domain, and a cytoplasmic domain which contains the kinase activity. These receptors are also discussed in Chapter 3 (section 3.2), and later, for example, in Chapter 9.

In all cases, binding of the ligand to the extracellular binding site of the receptor causes a conformational change to the protein which activates the kinase activity on the cytoplasmic side of the receptor. Phosphorylation by this kinase activity then leads to intracellular signalling by the activation of signalling pathways. Activation of RTKs leads to the involvement of many other proteins in such pathways, including, among many others, PtdIns 3-kinase as part of the inositol pathway, GTPase-activating protein involved in G protein signalling, and mitogen-activated protein kinase (MAP kinase) cascades.

Ligand binding to RTKs may in many cases also involve receptor dimerization. For example, the EGF receptor family of kinases form dimers on activation. Where the receptor exists in different isoforms, the dimers can be made up of two of the same polypeptide or may be a mixture of the isoforms. A good example of this is seen with the platelet-derived growth factor (PDGF) receptor, which can have dimers that are two of the same subunit, αα or ββ, or one of each subunit as a heterodimer, αβ, with each combination having a different specificity towards the precise ligand that can be bound (see section 3.2). However, the insulin receptor is a tetramer having a $\alpha_2\beta_2$ comformation. Although this is a tetramer, it is effectively a dimeric-type structure, with each part of the 'dimer' comprising of two others (see Chapter 9).

Activation of RTKs commonly also involves autophosphorylation; that is, the addition of phosphate groups to tyrosine residues on the cytoplasmic side of the receptor. This autophosphorylation may be intramolecular, where the polypeptide chain phosphorylates itself, or in many cases autophosphorylation is intermolecular, where one polypeptide in a dimer phosphorylates the other and vice versa. This latter case is seen for example with the fibroblast growth factor (FGF) receptor family.

It is often autophosphorylation of the receptor itself that is the signal which enables the message to be relayed further down the pathway. The addition of the phosphate group creates a binding site for other cytoplasmic proteins. As discussed in Chapter 1 (section 1.7) proteins that contain SH2 domains are able to recognize and bind to phosphotyrosine residues. Therefore creation of such covalent modifications within proteins creates binding site for these SH2 domains. Observations that have arisen from the use of synthetic peptides, which can be used to block the interactions of polypeptides by binding to recognition sites and stopping the binding of the true protein partner, have helped to unravel such mechanisms. These studies have also shown that other amino acids are also involved in the polypeptide interactions, other than the phosphotyrosine. For example, for the binding of PtdIns 3-kinase a methionine residue is required three residues to the C-terminal side of the phosphotyrosine. Residues to the N-terminal side of the tyrosine seem to be less important than those on the C-terminal side. The importance of the creation of these binding sites by autophosphorylation may well be to create high-affinity binding sites for the proteins to be phosphorylated, allowing the efficient phosphorylation of proteins which might be in a very low abundance within the cell, and once bound to the receptor protein kinase the bound protein will itself be phosphorylated on tyrosine so altering its activity. Alternatively, it may be the simple act of binding of the protein to the receptor that alters the conformation of that protein, so allowing it to relay the message onwards without the need for it to be phosphorylated. A good example here is the use of the linker or adaptor protein. Adaptor proteins commonly

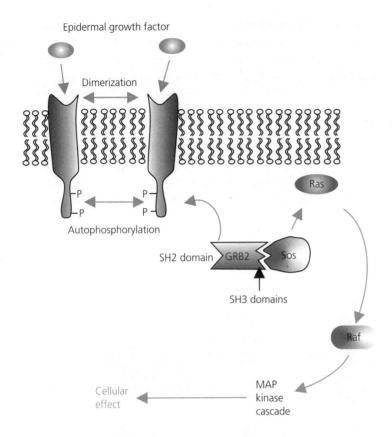

Fig. 4.8 The signalling cascade associated with epidermal growth factor detection as an example of signalling from RTKs. Binding of the growth factor to the receptor causes dimerization and autophosphorylation of the receptor. Phosphotyrosines formed create binding sites for SH2 domains of the GRB2 adaptor protein. This protein is found associated with Sos (Son of Sevenless), a guanine nucleotide releasing protein through SH3 domains. The new receptor–GRB2–Sos complex causes activation of the G protein Ras which leads to the activation of a kinase cascade and ultimately the cellular effect.

contain SH2, and perhaps SH3, domains. Examples of such adaptor proteins are GRB2 in mammals and Drk in the fly *Drosophila*. Shown as a scheme in Fig. 4.8, adaptor proteins are often associated with guanine nucleotide-releasing proteins (GNRPs) and the interaction of the adaptor with an activated RTK through SH2 domains and newly formed phosphotyrosine residues on the RTK leads to the association of the GNRP with membrane proteins and the activation of its respective G protein. This is discussed further in Chapter 9.

In several of the classes of tyrosine receptor kinases common areas of homology have been characterized, which will be useful in the identification of new kinases and the assigning of kinase activity to proteins which may at present have no function assigned to them.

In many of these receptor kinases, it is not only tyrosine which can be phosphorylated but also serine or threonine. In the insulin receptor family serine/threonine phosphorylation occurs on the β subunit of the receptor. However, the exact significance of this is yet to be determined.

Cytosolic tyrosine kinases

Kinases which contain tyrosine phosphorylation capacity are not only membrane bound, as the receptor kinases, but may also be soluble and reside in the cytoplasm of the cell. One such group of kinases are the Janus kinases, otherwise known as the JAKs. Janus in Roman mythology

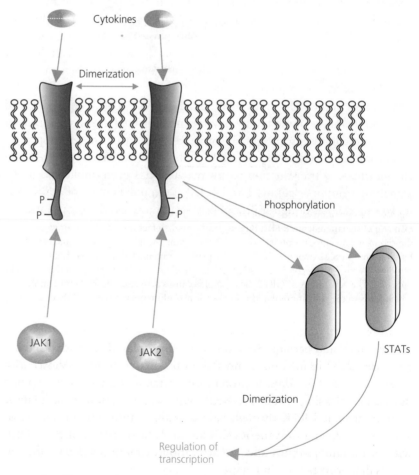

Fig. 4.9 Signalling cascade proposed to lead from cytokine receptors. Binding of the cytokine, such as IFN-γ, leads to dimerization of the receptor, and the recruitment of soluble tyrosine kinases, JAKs. Phosphorylation of the receptors, JAKs, and other proteins, such as STATs, takes place. STAT proteins themselves dimerize and carry the signal to the nucleus where transcription is regulated.

was the god of doors and gateways, and signified beginnings which ensured good endings. The statue of Janus has two heads which are seen to be gazing in opposite directions. Janus kinases contain tandem but non-identical catalytic domains, but they lack both SH2 and SH3 domains. They are involved in the signal transduction pathways that lead from the plasma membrane to deep within the cell, such as those which are cytokine-induced, for example in leucocytes and lymphocytes. The Janus kinases associate with an activated receptor, that has undergone a conformational change on binding to its ligand, and hence once activated themselves JAKs cause the phosphorylation of their respective receptor protein.

Once activated JAKs phosphorylate cytoplasmic proteins, such as STAT (*signal transducers and activators of transcription*) proteins, which leads to the activation of transcription (Fig. 4.9). As these cytosolic STAT proteins contain SH2 domains, once they are phosphorylated by JAKs they can dimerize by association of the new phophotyrosine of one STAT protein with the SH2 domain of another. Such activated dimers then cause the alteration of transcription observed.

Members of the JAK family include JAK1, JAK2, JAK3, and Tyk2, and different members of the family have been implicated in message transfer from different receptors. For example, activation of the erythopoietin receptor leads to activation of JAK1, while JAK1 associates with JAK2 in the interferon-γ transduction pathway. JAK3 has been implicated in the signalling from interleukins 2 and 4. JAK3 is a polypeptide of approximately 120 kDa, although it probably exists as at least three variants, but other Janus kinases are slightly bigger.

4.4 Mitogen-activated protein kinases

Many external factors, including cytokines and growth factors, can lead to the phosphorylation and activation of a family of kinases, other than the Janus kinases. This other class of kinases are often referred to as extracellular signal-regulated kinases (ERKs) or are commonly known as mitogen-activated protein kinases (MAP kinases). Activation of these serine/threonine kinases can lead to their translocation to the nucleus and subsequent phosphorylation and activation of transcription factors, leading, for example, to promoted growth, differentiation, and altered gene-expression profiles.

MAP kinases are themselves activated by phosphorylation but in an unusual manner; that is, they are phosphorylated on tyrosine and threonine through the action of specific threonine/tyrosine kinases known as MAP kinase kinases (MAPKKs). This MAPKK enzyme is alternatively

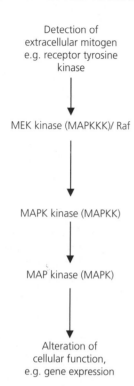

Fig. 4.10 A simplified schematic representation of the MAP kinase cascade.

known as MAP/ERK kinase (MEK). The sequence commonly recognized and phosphorylated within the MAP Kinase by MAPKK is -Thr-Glu-Tyr-, where both the threonine and tyrosine residues receive phosphoryl groups leading to the activation of the MAP kinase. However, the glutamate residue may be variant in some cases, giving the consensus target for the MAPK kinase as -Thr-X-Tyr-.

The MAPKK is itself also regulated by phosphorylation, catalysed this time by MAPKK kinase, otherwise known as MAPKKK or MEK kinase (MEKK). This is in fact a group of related serine/threonine kinases. One of the proteins which can catalyse this phosphorylation is Raf, the product encoded by the *raf* gene. The MAPKKKs (MEK kinases) from mammalian systems show conservation of the catalytic domains but interestingly they show divergence of a regulatory domain towards the N-terminus. A typical MAP kinase cascade is illustrated in a simplified form in Fig. 4.10.

The initial signal for this cascade of phosphorylations may originate in the activation of a receptor tyrosine kinase (RTK) activity, or, in lower organisms, in the activation of a two-component system as described below (section 4.5). Activation of the MAP kinase cascade can also involve the activation of G proteins, in some cases through the action of a trimeric G protein, either the Gα subunit or in other cases the G$\beta\gamma$ dimer. Alternatively activation of the MAP kinase cascade may be through a monomeric G protein such as Ras, as shown in Fig. 4.8 and Fig. 9.7. MAP kinases have also been shown to be activated by reactive oxygen species (ROS) as discussed in Chapter 8 (section 8.3), but whether this is a direct effect or not needs to be established.

To show how MAP kinases fit into a pathway, the activation by epidermal growth factor serves as a good example, as illustrated above in Fig. 4.8. The tyrosine kinase receptor leads to formation of binding sites for, and the activation of, an adaptor protein, GRB2, which contains both SH2 and SH3 domains. The SH2 domains are responsible for the adaptor protein binding to the receptor once the receptor has been autophosphorylated on tyrosine residues while the SH3 domains are involved in the interaction of the adaptor protein with a guanine nucleotide-exchange factor and the subsequent activation of Ras. Once Ras has been converted to the GTP-bound and active form it is able to cause the activation of the kinase Raf and phosphorylation catalysed by Raf leads to activation of the rest of the cascade.

As well as the specific example above, it have been found that MAP kinase pathways show remarkable conservation across species and even kingdoms. Although the details are different, the principles of a MAP-KKK activating a MAPKK and so on are used in mammals, yeast, and plants. Selected examples are shown in Fig. 4.11a.

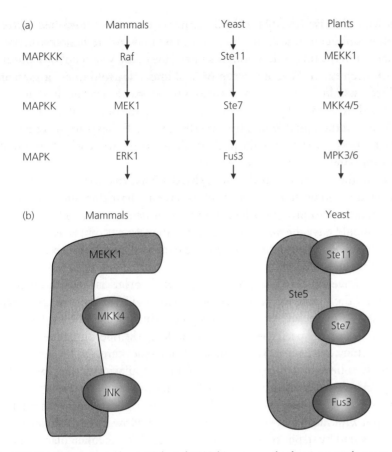

Fig. 4.11 MAP kinase cascades. (a) Selected MAP kinase cascades from mammals, yeast, and plants, showing the similarity and conserved nature of these pathways. (b) MAP kinase cascades are held by scaffold proteins, some of which act only in this capacity, as with the yeast Ste5, while others have scaffold function and have a kinase domain. JNK, c-Jun N-terminal Kinase.

It has been postulated that other proteins are also involved in the MAP kinase pathway, some which are themselves kinases but others of which are proposed to have a function as scaffold proteins rather than possessing kinase activity (Fig. 4.11b). An example of the latter is the protein Ste5 from the yeast *Saccharomyces cerevisiae* which contains a zinc-finger-like conformational domain and has been shown to interact with other components of the MAP kinase pathway. The existence of such scaffold proteins is really important, even though they often appear to have no obvious inherent function. The same kinase components have been seen to function in more than one pathway and more than one pathway seems to have the same end effect in some cases. The existence of such a scaffold structure would indeed be an attractive hypothesis to explain the relative lack of interaction between different MAP kinase pathways in the same cell.

At least six distinct MAP kinase pathways have been identified in yeast and it is difficult to see how the signals, the messages, remain distinct with such similar and related components. Surely, they should just interact with each other? The existence of a physical scaffold and kinase complexes would help explain these observations. However, if there is a defined scaffold and physical architecture then the potential for the amplification of the signal would be severely limited. This is a pay-off that cells might have evolved to allow defined messages to reach their targets appropriately.

It should be noted here that some MAP kinase cascade components can act as scaffold proteins but themselves have a kinase domain, as seen with the mammalian protein MEKK1 and the yeast protein PbS2. These proteins would have the dual role of keeping the correct components together as well as relaying the signal down the pathway (Figure 4.11b).

Once the MAP kinase has been phosphorylated and therefore activated it must have a role in phosphorylating further proteins, causing an alteration in their activity. Transcription factors are known to be targets for MAP kinase with phosphorylation causing an increase in the rate of transcription of specific genes; for example, in the pheromone response of yeast. However, there is a translocation issue here. MAP kinases are usually found in their inactive state in the cytoplasm of cells, while their transcription factor targets are in the nucleus. On activation, the MAP kinases are phosphorylated and then often translocated into the nucleus, where, while active, they can have their effect. However, MAP kinases are deactivated by dephosphorylation, and once this has taken place they are recognized by the trafficking mechanisms of the cell and re-translocated back to the cytoplasm, where they are ready in the inactive state for a further round of signalling. However, it should be noted that there are also cytosolic targets of some MAP kinases, so not all signals here lead to the nucleus.

As discussed, like all signalling pathways, once turned on the response has to be turned off again, and here the reversal is by dephosphorylation. The removal of the phosphate group is catalysed by phosphatases. A group of phosphatases whose synthesis is inducible by growth factors and stimuli of the MAP kinase pathway has been shown to be involved in switching-off of the pathway itself. These phosphatases seem to be specific for MAP kinase while amino acid sequences appear to have some homology to a dual-specificity tyrosine/serine phosphatase from vaccinia virus, VH1. One such phosphatase, CL100, also shows some homology to another dual-specificity phosphatase cdc25. Another phosphatase, PAC-1, has been found to be localized to the nucleus, where MAP kinase has its effect.

Other likely candidates to be involved in the cessation of the MAP kinase cascade are less-specific phosphatases such as PP2A or CD45.

4.5 Histidine phosphorylation

Although the most common phosphorylations of proteins are based on the amino acids serine, threonine, or tyrosine they are not the only amino acids to participate in this type of reaction and so cause an alteration in the activity of the protein and hence the propagation of a cellular signal. Histidine and aspartate are found to be phosphorylated in bacteria, and in some higher organisms. These are relatively high-energy phosphoamino acids and are involved in what has been termed the two-component signalling system. Detection of a stimulus leads to phosphorylation of the detector protein on a histidine residue, using ATP as the donor for the phosphate group, and the phosphoryl group is subsequently transferred to a second protein, the response regulator, which becomes phosphorylated on an aspartate residue. These events lead to an alteration in the functioning of this latter protein. For example, it may have enhanced DNA-binding capacity and so alter DNA transcription rates.

In general, the first phosphorylation step is an autophosphorylation event, so the proteins must contain kinase and target domains. Histidine kinases belong to a group of kinases which in fact share in common the presence of several conserved domains. Five main domains have been identified but any particular protein may not contain all of them, and may also contain other functional domains. However, most contain a region in which there is a conserved histidine which receives the phosphoryl group. Two glycine-rich regions, termed G1 and G2, contain the kinase activity as well as phosphatase activity and nucleotide-binding ability, while two other regions, termed N and F, which are conserved have also been identified. The initiating stimulus can be detected by either a receptor region which is part of the kinase polypeptide itself or a separate polypeptide which activates the kinase by a protein–protein interaction. An integral receptor domain would probably be extracellular with the polypeptide having a transmembrane region and the other domains would reside on the other, intracellular, side of the membrane. However, other histidine kinases are soluble and are found in the cytoplasm of the cells. These histidine kinases are terrific examples of modular proteins, as different domains seem to be glued together to give the catalytic activity and specificity required.

The response regulators, which receive the phosphoryl group on to their aspartate residues, share a common domain containing two aspartate residues and a conserved lysine residue. Phosphorylation will lead to an alteration of the protein's function. The protein can be subsequently reinstated back to the inactive form by an intrinsic phosphatase activity or by the activity of a separate phosphatase enzyme.

Although in many cases the kinase domain and the response domains reside on separate polypeptide chains, there are many examples where both parts are on the same protein. These are referred to as hybrid kinases. Therefore, histidine kinases can be grouped into two main classes: hybrid and non-hybrid.

Systems in bacteria which use this two-component type of signalling pathway include those which monitor osmolarity, temperature, pH, and cell density. In fact, over 50 such systems have been identified.

Using the conserved sequences of the histidine kinases as a basis, researchers have looked for such kinases in higher organisms. Such a strategy has revealed sequences homologous to histidine kinases in the fungus *Neurospora crassa*, in the slime mould *Dictyostelium discoideum*, and in the yeast *S. cerevisiae*, while several proteins sharing homology with non-hydrid and hybrid histidine kinases have been found in the higher plant *Arabidopsis thaliana*. Here, for example, the histidine kinases have been found to be instrumental in the recognition of, and signalling induced by, ethylene. Ethylene has been found to bind to the membrane-spanning domain of at least one of them, while the kinase and response domains reside in soluble domains on the intracellular region of the protein. Interestingly, these kinases have been shown to dimerize, but some have a cellular location in the endoplasmic reticulum.

It is very likely that histidine and aspartate residue phosphorylations will be discovered to be more widespread and abundant than first thought in eukaryotic systems, and to be involved in responding to or even sensing many signals. For example, a two-component system has been found to be involved in oxidative stress responses in yeast. However, there is little evidence of sequences encoding histidine kinases in mammalian genomes, including the human genome.

4.6 Phosphatases

No treatment of phosphorylation would be complete without a discussion on how the phosphoryl group is removed to reverse the cycle—that is, dephosphorylation—and return the protein to its previous state, be that active or inactive (refer to Fig. 4.1). Dephosphorylation is carried out by a group of enzymes called phosphatases. Like the kinases, these again are split broadly into two groups; those which remove the phosphoryl group from serine or threonine residues, the serine/threonine phosphatases, and the ones which remove the phosphate groups from tyrosine residues, the tyrosine phosphatases. In a similar fashion to the kinases, the number of phosphatases coded for in the human genome has been estimated, research suggesting that it could be as many as 1000.

As with the kinases, although most of the research in this area has been carried out on animals, plants and fungi also contain phosphatases analogous to those characterized, although some plant-specific phosphatases have also been found, such as the LHC phosphatase associated with the thylakoid membranes.

Serine/threonine phosphatases

Several different types of these enzymes have been identified, referred to as PP1, PP2A, PP2B, PP2C, PP4, and PP5. Interestingly, it has been estimated that a cell's total serine/threonine phosphatase activity is approximately equal to a cell's total serine/threonine kinase activity, suggesting that a balance of these activities exists within cells. Such a balance would make much sense, as otherwise the activation of a pathway may occur in preference to the inactivation, as the reversal, or dephosphorylation, might not be able to keep pace with overwhelming kinase activity. Indeed, further to this, the cellular concentrations of the major forms of these two enzymes are also comparable.

PP1 has a broad substrate specificity and in mammals two closely related isoforms have been identified, PP1α and PP1β. These two enzymes arise from alternative splicing of the same gene. The catalytic subunits, of 37 kDa, are under the control of two thermostable proteins that are able to inhibit the activity of PP1: inhibitor 1 and inhibitor 2. PP1 appears to be involved in the regulation of glycogen metabolism, where it is associated with a glycogen-binding protein (R_{G1}). This glycogen-binding protein is itself under the control of phosphorylation. Phosphorylation of this latter protein by cAMP-dependent protein kinase makes it unable to bind to the catalytic subunit of the phosphatase, showing that the phosphatase PP1 is under the control of the cAMP signalling pathway. Inhibitor 1 is also only active in inhibiting the PP1 when it is in the phosphorylated state, again phosphorylated by cAMP-dependent protein kinase. Therefore the action of the cAMP pathway is twofold; that is, in the deactivation of PP1 and thus the prevention of dephosphorylation of proteins which are phosphorylated by kinases which are turned on directly or indirectly by the rise in cAMP.

PP2A is a trimeric protein and shows some similarity to PP1. PP2A has two identified isoforms of the catalytic subunit, PP2Aα and PP2Aβ, but here they are separate gene products. The other subunits involved are a structural subunit (A) of 65 kDa and a regulatory third subunit (B) which shows a degree of variability. At least 13 genes have been found encoding variants of the B subunit, and splice variants of these have been reported too. Roles for PP2A include the control of DNA replication and apoptosis.

PP2B, also known as calcineurin, is a calcium-dependent enzyme. In the presence of Ca^{2+}/calmodulin the activity of PP2B is increased. PP2B exists as a heterodimer of a catalytic subunit (A subunit) and a calcium-binding subunit (B subunit), which contains four Ca^{2+}-binding regions.

Therefore, the effect of Ca^{2+} may be either direct or through the action of calmodulin. One form of the enzyme that was purified had a molecular weight of approximately 80 kDa, composed of a 60 kDa A subunit and a 20 kDa B subunit. However, the A subunit seems to be quite variable between tissues which may account for reported substrate-specificity differences. The subcellular location of PP2B may be dictated by the B subunit which may be myristoylated, allowing a tight association with membranes.

PP2C, a 42–45 kDa monomeric protein, on the other hand, appears to be a Mg^{2+}-dependent enzyme, while PP5 is primarily found in the nucleus and contains sequences which are reminiscent of those found with proteins associated with RNA and DNA binding.

Study of the roles of phosphatases, like that of many other signalling components, has been greatly enhanced by the use of inhibitors. PP1 and PP2A have been shown to be inhibited strongly by okadaic acid, a fatty acid produced by dinoflagellates. Okadaic acid has very little effect on PP2B, and has no effect on PP2C or tyrosine phosphatases. A similar inhibitory pattern is seen with tautomycin from *Streptomyces*.

Tyrosine phosphatases

When, in 1988, Tonks and colleagues published the partial sequence of the first protein tyrosine phosphatase (PTP 1B) it was found that not only did the enzyme lack homology with serine/threonine phosphatases but that another such enzyme had already been cloned a few years earlier. However, that protein, CD45, had been assigned no function. Obviously such findings have sparked a wave of activity in this area of biochemistry and it is now known that there are two main classes of tyrosine phosphatase: those which are intracellular and those which are receptor-linked, very much as is seen with the tyrosine kinase families.

The intracellular phosphatases reside, as the name suggests, on the inside of the cell, with the activity usually associated with the cytoplasm, and they are soluble proteins. PTP 1B can be thought of as the model for this family of phosphatases. The N-terminal end of the single polypeptide contains the catalytic domain which removes the phosphate group from the target protein. It has been found that all of the enzymes in the protein tyrosine phosphatase family contain the following consensus sequence in their catalytic centre.

-(Ile/Val)-His-Cys-X-X-Gly-X-X-Arg-(Ser/Thr)-

Of particular importance here is the cysteine, as this residue is essential for catalysis. This cysteine is at position 215 in PTP 1B. It is also this cysteine which is able to be modified, so removing catalytic activity, by the

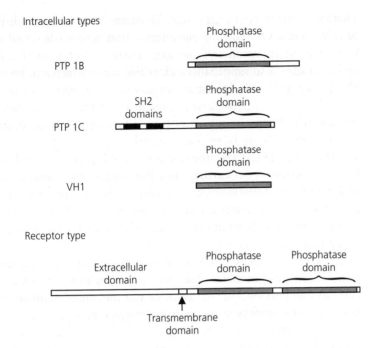

Fig. 4.12. The domain structure of several protein tyrosine phosphatases.

inhibitors pervanadate and hydrogen peroxide, as discussed in depth in Chapter 8 (section 8.4).

The C-terminal end of the polypeptide contains a signal to direct the protein to its intracellular location. PTP 1B appears to be localized to the cytoplasmic side of the endoplasmic reticulum, but removal of the C-terminal end relocates the enzyme into the cytosol. Thirty five amino acids at the C-terminal end of the protein are particularly hydrophobic, and allow membrane association. Relocation of the enzyme also alters its functional activity and it is thought that regulation of the phosphatase may be associated with a regulation of its location in the cell. Alternative splicing of the mRNA for some PTPs might well be responsible for controlling their cellular location.

Upon this basic structure for intracellular PTPs, others also contain additional features (Fig. 4.12). For example, PTP 1C also contain two protein-binding SH2 domains and, as discussed above for the receptor tyrosine kinases, these domains will be used to locate and bind phosphotyrosine residues. Such localization would aid the catalytic action of the enzyme. The SH2 domains might also be crucial to maintain the enzyme's intracellular location.

Another addition to the basic PTP 1B structure that has been identified is the addition of a domain which resembles cytoskeletal-associated proteins. For example, PTPMEG1 has an area which shares some homology with erythrocyte protein band 4.1. Again this is probably important in the maintenance of intracellular location of the phosphatase. Tyrosine phosphorylation is thought to be important in the polymerization of tubulin, a main element of the cytoskeleton, and tyrosine phosphatases would be vital in this regulatory mechanism.

The smallest and simplest PTP identified is one which contains the phosphatase domain and nothing else (Fig. 4.12). This protein, VH1, is coded for by the vaccinia virus. This phosphatase interestingly also shows activity to serine-phosphorylated sites as well as tyrosine-phosphorylated sites.

Just as PTP 1B can be seen as a model for the intracellular PTPs, CD45 can be used as a model for the receptor-linked tyrosine phosphatases. The CD45 protein has an extracellular domain, a transmembrane domain, and a domain containing phosphatase activity on the other side of the membrane; that is, in the cytoplasm (Fig. 4.12). CD45 is actually a glycoprotein, with the N-terminal extracellular domain containing O-linked glycosylation.

Unlike the intracellular PTPs, the receptor-linked family nearly all have two phosphatase domains (Fig. 4.12), although exceptions to this have been found. However, using mutations of the amino acids thought to be crucial for the catalytic action of these receptor phosphatases, mainly conserved cysteine residues, it is thought that the second catalytic site is probably inactive in removing the phosphoryl group from tyrosines. However, it could have a function in bringing about the association of the enzyme with its substrate, in a similar manner to the SH2 domains seen elsewhere.

Tyrosine phosphatases, like the tyrosine kinases, are probably regulated by phosphorylation although the precise mechanisms in many cases have yet to be identified. It may be that some phosphatases are constitutively active, ever ready to switch off pathways activated by kinases. Interestingly, reactive oxygen species, particularly hydrogen peroxide, have been found to inactivate these phosphatases, in particular PTP 1B. Oxidation of the active-site cysteine to a sulphenyl-amide group is thought to be responsible for the loss of activity of this enzyme. This is further discussed in Chapter 8 (section 8.4).

Several inhibitors have been identified which are useful in identifying the activity of tyrosine phosphatases. Molybdate and particularly orthovanadate are inhibitors of all tyrosine phosphatases while other compounds have been used as diagnostic tools for certain PTPs. Such molecules include zinc ions, ethylenediaminetetra-acetic acid (EDTA), and spermine.

4.7 Other covalent modifications

Although the discussion above has concentrated on the covalent addition of a phosphoryl group and its subsequent removal from a polypeptide as a means of regulating activity, it is not the only covalent modification

which is seen in the control of enzyme function. Many proteins are proteolytically cleaved, turning them from an inactive precursor into an active
enzyme. The classical case here is in the blood-clotting cascade where one
factor cleaves and activates another, which then goes on to cleave the next
and so on, leading to great amplification of the initial signal. Such action,
however, unlike phosphorylation, is not readily reversible as once the
polypeptide has been cleaved it is not able to be rejoined. An example of
relevance to cell signalling here is the cleavage activation of protein kinase C,
and the cleavage and hence formation of many cytokines (see section 2.4).

A more reversible covalent modification is adenylation. Here the
adenylyl group (AMP) from ATP is added to a target tyrosine residue of
the polypeptide, releasing inorganic pyrophosphate, the breakdown of
which probably helps to drive the reaction. The reversal releases AMP,
restoring the enzyme to its former condition. An example of this is seen in
E. coli where glutamine synthase in the cell's nitrogen metabolism is controlled in this way. Adenylation inhibits the enzyme's activity.

Many proteins are covalently modified by the addition of a large
hydrophobic group which allows the protein to become associated with a
membrane, for example, a myristoyl group. The myristoyl group is a C_{14}
fatty acid which is added to the N-terminal end of the polypeptide. The
donor for myristoylation is myristoyl-CoA, while the enzyme which catalyses the reaction is N-myristoyl transferase. For this reaction to take place,
the target polypeptide must have a glycine residue at its N-terminal with
a typical consensus sequence being:

-Gly-X-X-X-(Ser/Thr)-Y-Y-

where X is any variety of amino acids but Y is a basic amino acid.

Palmitoylation involves the addition of a palmityol group, again often
allowing a protein to become membrane-associated. The palmityol group
is a C_{16} fatty acid, donated from palmitoyl-CoA. It is usually added to a
cysteine residue. An example of a protein which has undergone this type of
modification is rhodopsin. Other covalent modifications of large hydrophobic groups include farnesylation which is the addition of a farnesyl group, a
C_{15} fatty acid. The donor in this case is farnesyl pyrophosphate. This process
takes place at the C-terminal end of the polypeptide where a consensus
sequence of -Cys-Y-Y-X is found. Cysteine again is the target for the modification, Y would be aliphatic in nature and X could be any amino acid.
Typically, the -Y-Y-X sequence would subsequently be removed and the end
of the polypeptide would also be methylated. An example of a signalling
protein which is altered in this way is the monomeric G protein Ras.

The last modification of this type to be considered here is the addition
of the C_{20} fatty acid geranylgeranyl, added in a process called geranyl-
geranylation. Again, like farnesylation, the pyrophosphate is the source

of the fatty acid and again it is attached to the C-terminal end of the polypeptide. Here, however, the consensus sequences needed at the C-terminus are -Cys-Cys, -Cys-Cys-X-X, or -Cys-X-Cys. One or both cysteine groups could be modified. An example of a protein here is the monomeric G protein Rab.

Finally, reactive species, such as those referred to as the reactive oxygen species (ROS) or reactive nitrogen species (RNS), can cause the modification of amino acids in proteins. Oxidation of cysteine groups can lead to the formation of cystine disulphides, which can then be re-reduced back to reverse the effect. In plants such reversible reduction is used in the regulation of some enzyme activities, with thioredoxin being used as the reductant in the reaction. The subsequent re-reduction of the thioredoxin is light-driven. In animal systems such mechanisms have also been proposed. Alternatively, ROS might lead to the formation of oxidized derivatives of the cysteine thiol group, such as seen in the formation of the sulphenyl-amide group in the phosphatase PTP 1B. Alteration to the amino acids in these ways leads to changes in the conformational shape of the protein, and usually in its activity. Such oxidized forms of the thiol may react further with glutathione, to form glutathionated forms of the protein, where perhaps the activity has been modified. RNS can likewise modify proteins, usually by a process referred to as nitrosylation. Here the nitric oxide attacks the thiol group to form an -SNO group, again with ramifications for the three-dimensional structure of the protein and its activity.

4.8 SUMMARY

- Most signalling pathways involve phosphorylation, an extremely important biochemical mechanism used in the control of proteins (Fig. 4.13).

- Phosphorylation is the addition of one or more phosphoryl groups to certain amino-acid side groups on a polypeptide chain.

- Dephosphorylation is the removal of one or more phosphoryl groups from certain amino-acid side groups on a polypeptide chain.

- Phosphorylation events are classified as either serine/threonine phosphorylations or tyrosine phosphorylations, depending on the target amino acids.

- The enzymes which add the phosphoryl groups are known as kinases, with the reverse reaction being catalysed by phosphatases.

- The human genome probably contains over 1000 genes for kinases and phosphatases, and other genomes also encode numerous examples.

- Serine/threonine kinases include protein kinase C, cAMP-dependent protein kinase, Ca^{2+}/calmodulin-dependent protein kinase, and cGMP-dependent protein kinase, which in general have wide substrate specificities.

- cAMP-dependent protein kinase exists in an inactive tetrameric state which dissociates on activation—that

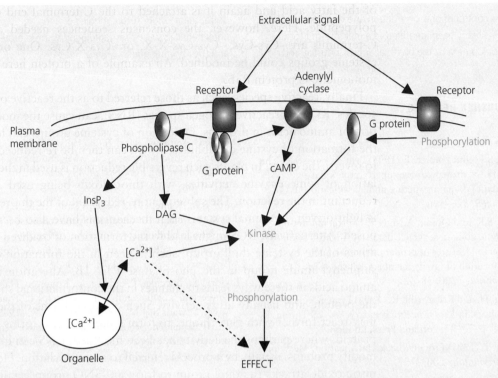

Fig. 4.13 A schematic representation of the central role of phosphorylation in cell signalling.

is, on binding to cAMP—to release two catalytic subunits.

- Protein kinase C, which is activated by the presence of Ca^{2+} ions, is in fact two families of protein kinases, each family containing several members, which probably have subtle differences in both their activation and specificity.

- Another important group of Ca^{2+}-controlled kinases are the Ca^{2+}/calmodulin-dependent protein kinases, again with several isoforms existing.

- The other major class of kinases are the tyrosine kinases, including receptor kinases which are membrane-bound receptors. Binding to their ligand results commonly in autophosphorylation of the receptor along with phosphorylation of intracellular proteins.

- Other tyrosine kinases include the soluble Janus kinase (JAK) family.

- A group of sequential kinases are those of the MAP kinase cascades.

- MAP Kinase pathways lead commonly from receptors which bind growth factors, or indeed insulin, and may involve the G protein Ras and the kinase Raf at the early part of the transduction pathway.

- The two-component systems of prokaryotes and some higher organisms are unusual in that they involve the attachment of phosphoryl groups to histidine and aspartate, and the enzymes are more commonly referred to as the histidine kinases. Histidine kinases fall into two main groups, the hybrid and non-hybrid histidine kinases. Such systems may be more widespread than first thought, although probably not found in mammals.

- The reversal of kinase action is performed by phosphatases and again these are usually either serine/threonine-specific or tyrosine-specific, the latter being either cytosolic or receptor-linked. Again several isoforms of each are known to exist.

- The control of phosphorylation and dephosphorylation has already been identified as a major target for drug development by the pharmaceutical industry, and no doubt many of our future medicines and agrochemicals will be targeted against these proteins and their homologues.

4.9 FURTHER READING

General

Hanks, S.K. and Hunter, T. (1995) Protein kinases 6. The eukaryotic protein kinase superfamily: kinase (catalytic) domain structure and classification. *FASEB Journal* 9, 576–596.

Hanks, S.K. and Quinn, A.M. (1991) Protein kinase catalytic domain sequence database: identification of conserved features of primary structure and classification of family members. *Methods in Enzymology* 200, 38–62.

Hidaka, H. and Kobayashi, R. (1994) Protein kinase inhibitors. In *Essays in Biochemistry*, vol. 28 Tipton, K. (ed.), pp. 73–97. Portland Press, London.

Hunter, T. (1991) Protein kinase classification. *Methods in Enzymology* 200, 3–37.

Hunter, T. (1995) Protein kinases and phosphatases: the yin and yang of protein phosphorylation and signalling. *Cell* 80, 225–236.

Kholodenko, B.N. (2003) Four dimensional organisation of protein kinase signaling cascades: The roles of diffusion, endocytosis and molecular motors. *Journal of Experimental Biology* 206, 2073–2082.

Kikkawa, U., Kishimoto, A., and Nishizuka, Y. (1989) The protein kinase C family: heterogeneity and its implications. *Annual Review of Biochemistry* 58, 31–44.

Knighton, D.R., Zheng, J., Ten Eyck, L.F., Ashford, V.A., Xuong, N.-H., Taylor, S.S., and Sowadski, J.M. (1991) Crystal structure of the catalytic subunit of cyclic adenosine monophosphate-dependent protein kinase. *Science* 253, 407–414.

Sun, H. and Tonks, N.K. (1994) The co-ordinated action of protein tyrosine phosphatases and kinases in cell signalling. *Trends in Biochemical Sciences* 19, 480–485.

cAMP-dependent protein kinase

Hanks, S.K., Quinn, A.M., and Hunter, T. (1988) The protein kinase family: conserved features and deduced phylogeny of the catalytic domains. *Science* 241, 42–52.

Øgreid, D. and Døskeland, S.O. (1981) The kinetics of the interaction between cyclic AMP and the regulatory moiety of protein kinase II. *FEBS Letters* 129, 282–286.

Smith, S.B., White, H.D., Siegel, J., and Krebs, E.G. (1981) Cyclic AMP-dependent protein kinase I; cyclic nucleotide binding, structural changes and release of the catalytic subunits. *Proceedings of the National Academy of Science USA* 78, 1591–1595.

Taylor, S.S., Buechler, J.A., Slice, L., Knighton, D., Durgerian, S., Dingheim, G.E., Neitzel, J.J., Yonemoto, W.M., Sowadski, J.M., and Dospmann, W. (1989) cAMP dependent protein kinase: a framework for a diverse family of enzymes. *Cold Spring Harbor Symposia* 53, 121–130.

cGMP-dependent protein kinase

Lincoln, T.M. and Corbin, J.D. (1983) Characterisation and biological role of the cGMP dependent protein kinase. *Advances in Cyclic Nucleotide Research* 15, 139–192.

Takio, K., Wade, R.D., Smith, S.B., Krebs, E.G., Walsh, K.A., and Titani, K. (1984) Guanosine cyclic 3'-5'-phosphate dependent protein kinase: a dimeric protein homologous with 2 separate protein families. *Biochemistry* 23, 4207–4218.

Protein kinase C

Coussens, L., Parker, P.J., Rhee, L., Yang-Feng, T.L., Chen, E., Waterfield, M.D., Francke, U., and Ullrich, A. (1986) Multiple, distinct forms of bovine and human protein kinase C suggest diversity in cellular signalling pathways. *Science* 233, 859–866.

House, C. and Kemp, B.E. (1987) Protein kinase C contains a pseudosubstrate prototype in its regulatory domain. *Science* 238, 1726–1728.

Naor, Z., Shearman, M.S., Kishimoto, A., and Nishizuka, Y. (1988) Calcium independent activation of hypothalamic type I protein kinase C by unsaturated fatty acids. *Molecular Endocrinology* 2, 1043–1048.

Nishizuka, Y. (1988) The molecular heterogeneity of protein kinase C and its implication for cellular regulation. *Nature* **334**, 661–665.

Ono, Y., Fujii, T., Ogita, K., Kikkawa, U., Igarashi, K., and Niskizuka, Y. (1988) The structure, expression and properties of additional members of the protein kinase C family. *Journal of Biological Chemistry* **263**, 6927–6932.

Ca²⁺/calmodulin-dependent protein kinases

Anderson, K.A. and Kane, C.D. (1998) Ca²⁺/calmodulin-dependent protein kinase IV and calcium signaling. *Biometals* **11**, 331–343.

Hanson, P.I. and Schulman, H. (1992) Neuronal Ca²⁺/calmodulin-dependent protein kinases. *Annual Review of Biochemistry* **61**, 559–601.

Payne, M.E., Schworer, C.M., and Soderling, T.R. (1983) Purification and characterisation of rabbit liver calmodulin dependent glycogen synthase kinase. *Journal of Biological Chemistry* **258**, 2376–2382.

Soderling, T.R. (1999) The Ca²⁺-calmodulin protein kinase cascade. *Trends in Biochemical Sciences* **24**, 232–236.

Woodgett, J.R., Davison, M.T., and Cohen, P. (1983) The calmodulin dependent glycogen synthase kinase from rabbit skeletal muscle: purification, subunit structure and substrate specificity. *European Journal of Biochemistry* **136**, 481–487.

G protein-coupled receptor kinases

Benovic, J.L., Mayor, F., Staniszeski, C. Lefkowitz, R.J., and Caron, M.G. (1987) Purification and charaterisation of the β-adrenergic receptor kinase. *Journal of Biological Chemistry* **262**, 9026–9032.

Debburman, S.K. Ptasienski, J., Benovic, J.L., and Hosey, M.M. (1996) G protein coupled receptor kinase GRK2 is a phospholipid dependent enzyme that can be conditionally activated by G protein beta/gamma subunits. *Journal of Biological Chemistry* **271**, 22552–22562.

Lefkowitz, R.J. (1993) G-protein-coupled receptor kinases. *Cell* **74**, 409–412.

Palczewski, K. and Benovic, J.L. (1991) G-protein-coupled receptor kinases. *Trends in Biochemical Sciences* **16**, 387–391.

Winstel, R., Freund, S., Krasel, C., Hoppe, E., and Lohse, M.J. (1996) Protein kinase crosstalk: membrane targetting of the beta adrenergic receptor kinase by protein kinase C. *Proceedings of the National Academy of Science USA* **93**, 2105–2109.

Protein kinase B

Hajduch, E., Litherland, G.J., and Hundal, H.S. (2001) Protein kinase B (PKB/Akt)—a key regulator of glucose transport? *FEBS Letters* **492**, 199–203.

Scheid, M.P. and Woodgett, J.R. (2003) Unravelling the activation mechanisms of protein kinase B/Akt. *FEBS Letters* **546**, 108–112.

AMP-activated protein kinase

Dyck, J.R.B., Gao, G., Widmer, J., Stapleton, D., Fernandez, C.S., Kemp, B.E., and Witters, L.A. (1996) Regulation of 5′-AMP-activated protein kinase activity by the noncatalytic β and γ subunits. *Journal of Biological Chemistry* **271**, 17798–17803.

Gao, G., Fernandez, S., Stapleton, D., Auster, A.S., Widmer, J., Dyck, J.R.B., Kemp, B.E., and Witters, L.A. (1996) Non-catalytic β and γ-subunit isoforms of the 5′-AMP-activated protein kinase. *Journal of Biological Chemistry* **271**, 8675–8681.

Haem-regulated protein kinase

Kozak, M. (1992) Regulation of translation in eukaryotic systems. *Annual Review of Cell Biology* **8**, 197–225.

Proud, C.G. (1992) Protein phosphorylation in translational control. *Current Topics in Cellular Regulation* **32**, 243–369.

Plant-specific serine/threonine kinases

Coughlan, S.J. and Hind, G. (1986) Purification and characterisation of a membrane-bound protein kinase from spinach thylakoids. *Journal of Biological Chemistry* **261**, 11378–11385.

Stone, J.M. and Walker, J.C. (1995) Plant protein kinase families and signal transduction. *Plant Physiology* **108**, 451–457.

Tyrosine kinases

Bellot, F., Crumley, G., Kaplow, J. M., Schessinger, J., Jaye, M., and Dionne, C.A. (1991) Ligand-induced transphosphorylation between different FGF receptors. *EMBO Journal* **10**, 2849–2854.

Fantl, W.J., Escobedo, J.A., Martin, G.A., Turck, C.W., Del Rosario, M., McCormick, F., and Williams, L.T. (1992) Distinct phosphotyrosines on a growth factor receptor bind to specific molecules that mediate different signalling pathways. *Cell* **69**, 413–423.

Fantl, W.J., Johnson, D.E., and Williams, L.T. (1993) Signalling by receptor tyrosine kinases. *Annual Review of Biochemistry* **62**, 453–481.

Kawamura, M., McVicar, D.W., Johnston, J.A., Blake, T.B., Chen, Y.-Q, Lal, B.K., Lloyd, A.R., Kelvin, D.J., Staples, J.E., Ortaldo, J.R., and O'Shea, J.J. (1994) Molecular cloning of L-Jak, a Janus family protein-tyrosine kinase expressed in natrual killer cells and activated leukocytes. *Proceedings of the National Academy of Science USA* 91, 6374–6378 [L-Jak is now known as JAK3].

Schlessinger, J. and Ullrich, A. (1992) Growth factor signalling by receptor tyrosine kinases. *Neuron* 9, 383–391.

MAP kinases

Blenis, J. (1993) Signal transduction via the MAP kinases: Proceed at your own RSK. *Proceedings of the National Academy of Science USA* 90, 5889–5892.

Blumer, K.J. and Johnson, G.L. (1994) Diversity in function and regulation of MAP kinase pathways. *Trends in Biochemical Sciences* 19, 236–240.

Herskowitz, I. (1995) MAP kinase pathways in yeast: for mating and more. *Cell* 80, 187–197.

Irie, K., Takase, M., Lee, K.S., Levin, D.E., Araki, H., Matsumoto, K., and Oshima, Y. (1993) *MKK1* and *MKK2* which encode *Saccharomyces cerevisiae* mitogen-activated protein kinase-kinase homologs, function in the pathway mediated by protein kinase C. *Molecular and Cellular Biology* 13, 3076–3083.

Marshall, C.J. (1994) MAP kinase kinase kinase, MAP kinase kinase and MAP kinase. *Current Opinion in Genetic Development* 4, 82–89.

Nebreda, A.R. (1994) Inactivation of MAP kinases. *Trends in Biochemical Sciences* 19, 1–2.

Šamaj, J., Baluška, F., and Hirt, H. (2004) From signal to cell polarity: mitogen-activated protein kinases as sensors and effectors of cytoskeleton dynamicity. *Journal of Experimental Botany* 55, 189–198. Contains a good illustrated comparison of MAP kinases in mammals, yeast, and plants.

Roberts, R.L. and Fink, G.R. (1994) Elements of a single MAP kinase cascade in *Saccharomyces cerevisiae* mediate two developmental programs in the same cell type: mating and invasive growth. *Genes and Development* 8, 2974–2985.

Histidine phosphorylation

Alex, L.A. and Simon, M.L. (1994) Protein histidine kinases and signal transduction in prokaryotes and eukaryotes. *Trends in Genetics* 10, 133–138.

Besant, P.G., Tan, E., and Attwood, P.V. (2003) Mammalian protein histidine kinases. *International Journal of Biochemistry and Cell Biology* 35, 297–309.

Inouye, M. and Dutta, R. (2003) *Histidine Kinases in Signal Transduction*. Academic Press.

Kofoid, E.C. and Parkinson, J.S. (1992) Communication modules in bacterial signalling proteins. *Annual Review of Genetics* 26, 71–112.

Volz, K. and Matsumura, P. (1991) Crystal structure of *Escherichia coli* CheY refined at 1.7Å resolution. *Journal of Biological Chemistry* 266, 15511–15519.

Phosphatases

Barford, D. (2001) The mechanism of protein kinase regulation by protein phosphatases. *Biochemical Society Transactions* 29, 385–391.

Cohen, P.T.W., Brewis, N.D., Hughes, V., and Mann, D.J. (1990) Protein serine/threonine phosphatases: a expanding family. *FEBS Letters* 268, 355–359.

Gang, L. (2003) Protein tyrosine phosphatase 1B inhibition: opportunities and challenges. *Current Medicinal Chemistry* 10, 1407–1421.

Guan, K.L., Broyles, S.S., and Dixon, J.E. (1991) A Tyr/Ser protein phosphatase encoded by vaccinia virus. *Nature* 350, 359–362.

Klumpp, S. and Krieglstein, J. (eds) (2003) Protein phosphatases. *Methods in Enzymology* 266. A series of articles on the topic.

Tonks, N.K. (2003) PTP1B: from the sidelines to the front lines! *FEBS Letters* 546, 140–148.

Virshup, D.M. (2000) Protein phosphatase 2A: a panoply of enzymes. *Current Opinion in Cell Biology* 12, 180–185.

5 Cyclic nucleotides, cyclases, and G proteins

Once signalling molecules have been perceived by a cell, intracellular cascades of events usually ensue. One of the most common ways for the signal to be propagated in a cell is by the alteration of the concentration of other molecules. In this chapter alteration of a class of molecules called cyclic nucleotides is discussed. These are crucial intermediates between the receptor and phosphorylation events in many cases, and control many important cellular processes; for example, glycogen metabolism and hence the availability of glucose. They are also mediators of NO signalling, discussed further in **Chapter 8**, and importantly allow for a large amplification of a signal inside a cell.

In some cases, production of cyclic nucleotides is controlled by a class of protein known as G proteins. These have been dubbed 'molecular switches' which in many ways succinctly sums up their action: they can exist in an 'on' state or an 'off' state. However, they control far more than nucleotide metabolism, and are found in cascades between receptor activation and phosphorylation cascades, for example. Many isoforms of both classes of these proteins have been identified in a wide range of organisms, highlighting their integral signalling role. Furthermore, some forms of cancerous cell growth have been attributed to G proteins being stuck in the 'on' state.

5.1 Introduction

The reception of a signal at the cell surface is a crucial event in many cell signalling cascades, as discussed in previous chapters. However, once the cell has received an external signal, such as the binding of a ligand to a cell surface receptor, a response inside the cell needs to follow rapidly. Such responses often involve molecules that have been referred to as 'second messengers', as they are, or were thought to be, the second signalling component in the chain of action. However, the term second messenger, although widely used, is somewhat misleading. Many of the second messengers are not in fact the second component at all, and the term is rather loosely used to refer to small molecules that are formed inside the cell and diffuse to their point of action where they relay their message to the next part of the signalling cascade.

In this chapter the role of one group of messenger molecules, the cyclic nucleotides, is discussed, along with the mechanisms that are involved in the control of their production.

5.2 cAMP

One of the major groups of 'second messengers' is the cyclic nucleotides, and in particular cyclic adenosine monophosphate or adenosine 3',5'-cyclic monophosphate to give it its more chemical name. It is understandable that it is usually simply referred to as cyclic AMP, or more commonly cAMP. It is in fact the intracellular concentration of cAMP that constitutes the signal here. Some of the hormones which have their cellular effects mediated by cAMP are extremely important and widely found, and include epinephrine and glucagon along with those listed in Table 5.1.

cAMP is produced at the plasma membrane of the cell by the enzyme adenylyl cyclase, alternatively known as adenylate cyclase. The cAMP is produced on the inner side of the membrane and is released into the cytosol where it can diffuse and act on the next part of the signal transduction

Table 5.1 Some signalling molecules that act via cAMP.

Hormone	Major tissues affected
Adrenocorticotropic hormone (corticotropin)	Adrenal cortex, fat
Epinephrine (adrenalin)	Muscle, fat, heart
Glucagon	Liver, fat
Luteinizing hormone	Ovaries
Parathormone	Bone
Thyroid-stimulating hormone	Thyroid gland, fat
Vasopressin	Kidney

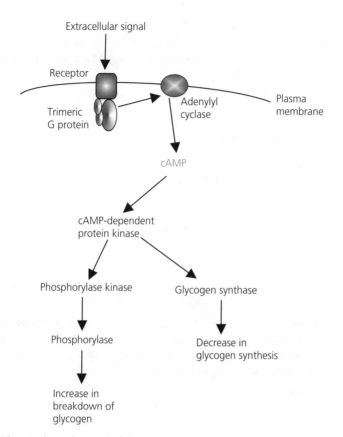

Fig. 5.1 The coordinated control of glycogen metabolism. Hormones binding to receptors activate a G protein (G$_s$), which activates adenylyl cyclase. The rise in cAMP activates cAMP-dependent protein kinase, leading to the phosphorylation of phosphorylase kinase and glycogen synthase, causing inhibition of the latter. Phosphorylase kinase's phosphorylation of phosphorylase causes an increase in glycogen breakdown. Therefore the overall result is an inhibition of glycogen synthesis and in increase in glycogen breakdown.

pathway. For example, the next response might be the binding of cAMP to cAMP-dependent protein kinase (PKA), so increasing the phosphorylation of certain proteins, with the concomitant alteration of their activity (for a discussion of PKA see Chapter 4; section 4.2). An excellent example of the role of cAMP in a signalling pathway is the control of glycogen metabolism, as illustrated in Fig. 5.1. Here, an extracellular signal such a epinephrine binds to a membrane receptor which in turn activates a G protein (the action and role of G proteins will be discussed below). The G protein activates adenylyl cyclase to produce cAMP. The increase in intracellular cAMP subsequently activates PKA which phosphorylates phosphorylase kinase. This latter enzyme phosphorylates, and in doing so activates, phosphorylase, which breaks down glycogen, releasing glucose for energy metabolism. This of course is the action within the cell which the arrival of the epinephrine at the cell surface was sent to provoke.

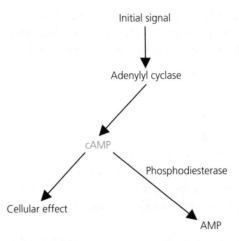

Fig. 5.2 A simplified scheme showing how cAMP fits into a signalling pathway.

> **Futile cycle**
>
> As discussed in **Chapter 4**, if the product of a reaction is simply converted back again to its original form—for example, glycogen being broken down to form glucose, and then the glucose being converted back to glycogen—then such cycling of chemicals is referred to as a futile cycle. Clearly an event like this is a waste of energy for the cell and needs to be avoided.

However, if the epinephrine was simply to result in the increase in the glucose level then this new glucose would be rapidly converted to glycogen again, and a futile cycle would be created. To prevent this, cAMP-dependent protein kinase also phosphorylates and inhibits glycogen synthase, reducing the synthesis of glycogen in the same cell, so allowing the glucose to be used for the purpose for which the epinephrine was sent in the first place.

As with all signals the cAMP message needs to be reversed. Cyclic nucleotides like cAMP are rapidly broken down by a family of enzymes called phosphodiesterases, and once removed from the cytosol of the cell no further signalling will take place. Therefore, an extremely simplified scheme of how cAMP and other related chemicals might signal is shown in Fig. 5.2.

With this brief discussion of the role of cAMP, it can be seen that it is ideal as a signalling molecule. cAMP is made rapidly by an enzyme, is small and readily diffusible, and it is readily and quickly broken down by another enzyme, phosphodiesterase. Therefore the signal can be rapid and reversed. Furthermore, the production and use of cAMP in cells, like most signalling routes, may lead to massive amplification of the signal.

The activation of one receptor can cause the activation of several adenylyl cyclase enzymes, each leading to the production of hundreds or thousands of cAMP molecules, each of which can go on to have an effect. A large amplification is therefore seen.

Although the majority of the effects of cAMP are mediated through the kinase PKA, as seen above, cAMP can also have more direct effects. For example, in *E. coli* it can act through cAMP receptor protein, or CRP, encoded by the *crp* gene. Alternatively this is known as catabolite gene-activator protein (CAP), a dimer of 22 kDa subunits. On an increase in intracellular cAMP concentration, cAMP binds to CRP to form a cAMP–CRP (or cAMP–CAP) complex which binds to DNA and alters the expression of several genes, including those of the lactose and arabinose operons.

In the majority of cases cAMP is seen as an extremely important intra-cellular signal. However, cAMP has also be seen as an extracellular signal, acting as a type of hormone in the slime mould *Dictyostelium*, where cAMP is used as a signal between cells, controlling cellular aggregation and differentiation.

5.3 Adenylyl cyclase

The enzyme that produces cAMP is adenylyl cyclase. It is also referred to as adenylate cyclase and in the past as adenyl cyclase. To understand how cAMP fits into signalling pathways it is necessary to discuss the structure, function, and control of this important enzyme.

In mammals adenylyl cyclase is a single polypeptide that resides in the plasma membranes of cells. It catalyses production of cAMP from the ever-present ATP, the majority of which is produced by the mito-chondria in many cells. The 3′-OH ribose group of the ATP attacks the α-phosphoryl group, resulting in a cyclization of the molecule. The by-product from the reaction is inorganic pyrophosphate which is itself broken down by the enzyme pyrophosphatase. The energy released from this latter reaction probably helps to drive the cyclization reaction (Fig. 5.3).

As with many of the proteins involved in signal transduction, there are various isoforms of the enzyme adenylyl cyclase. The cellular location of adenylyl cyclases is usually in the plasma membrane, as with the mam-malian ones which are integral in that membrane, but other forms have been found to be peripheral plasma membrane proteins. For example, peripheral membrane forms are found in *E. coli* and in the yeast *S. cerevisiae*.

Soluble forms of adenylyl cyclase have also been reported, originally in some bacteria. However, a soluble form of adenylyl cyclase also exists in

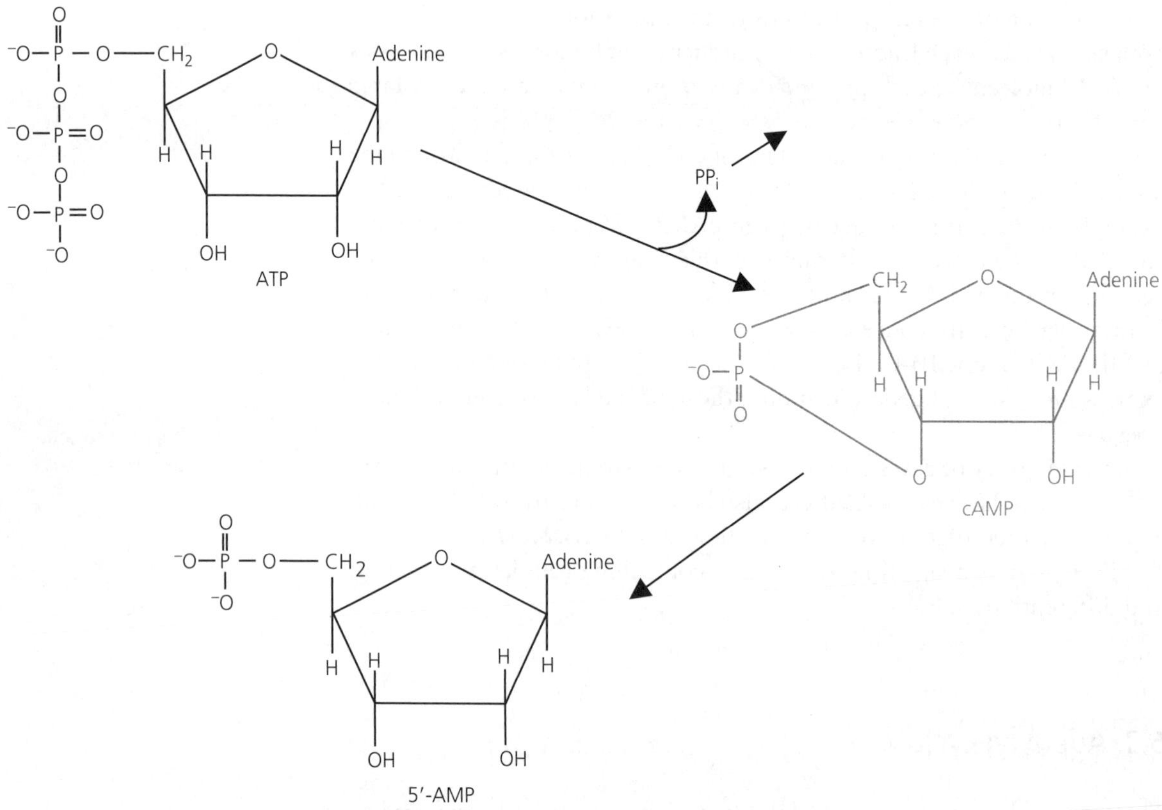

Fig. 5.3 The production and hydrolysis of cAMP.

higher organisms. This enzyme was originally found primarily in mammalian sperm and testis, but has a much wider tissue distribution. It has a molecular weight of around 48 kDa, and interestingly it is activated by bicarbonate, suggesting that this form of cyclase might be responding to the metabolic status of the cell. There is now a growing interest in this enzyme and how its activity fits into the control of intracellular cAMP levels.

At least eight isoforms of the plasma membrane adenylyl cyclase have been identified in mammals (types I–VIII), but they all share a similar structural topology. In general the proteins contain two clusters of six membrane-spanning, highly hydrophobic domains which separate two catalytic domains on the cytoplasmic side of the membrane (Fig. 5.4).

It makes sense to have the catalytic areas of adenylyl cyclase on the cytoplasmic side of the plasma membrane, as it is here that the product, cAMP, needs to be created to have its desired effect. Some areas of the two catalytic domains have been shown to be well conserved across the cyclases studied, and interestingly they also show homology to regions of membrane-bound guanylyl cyclases. In this latter enzyme the catalysis is very similar—it also produces a cyclic nucleotide—so a similar catalytic

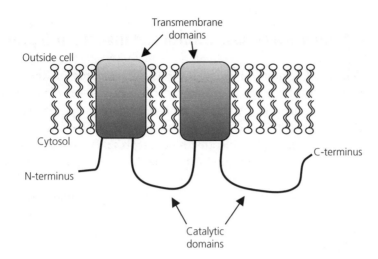

Fig. 5.4 The proposed structural topology of adenylyl cyclase.

mechanism and structure might be expected. It is probable that both these conserved regions in the catalytic domains of adenylyl cyclase, called C_{1a} and C_{2a}, are responsible for the activity of the enzyme, as it has been found that point mutations in either region are extremely inhibitory.

Studies have concluded that the adenylyl cyclase enzyme has a molecular weight of 200–250 kDa but the actual molecular weight is approximately 120 kDa, begging the question as to whether the enzyme exists as a dimer in the membrane. The extra apparent molecular weight may come from its association with other proteins, such as G proteins or even receptors.

All enzymes that are involved in signal transduction need to be controlled carefully, and adenylyl cyclase is no different. One of the most important controlling mechanisms is through the action of G proteins, and the next section will be dedicated to that. However, for experimental purposes, adenylyl cyclases can be non-competitively inhibited by Mg^{2+}/ATP analogues such as 3′-AMP, and it can be activated by a lipid-soluble diterpene, forskolin. It is presumed that the hydrophobic domains of the enzyme are the target for forskolin. In fact, isoforms without the six transmembrane domains seem to be insensitive to forskolin.

In *Dictyostelium*, the slime mould that uses cAMP as an extracellular signal, one form of adenylyl cyclase conforms to the mammalian model, although another form differs in only having one membrane-spanning region. A hypothesis has also been put forward that the hydrophobic domains might act as a channel across the membrane, and this is seen as important in cells that excrete cAMP. However, little evidence for the adenylyl cyclase acting as a transporter has been found and the heterogeneity of the sequences in these regions between different isoforms would suggest that this is not a role of the enzyme.

5.4 Adenylyl cyclase control and the role of G proteins

When a hormone binds to the relevant receptor on the cell surface, activation of the enzyme adenylyl cyclase is not a direct process, but rather requires the use of other proteins, and further it was discovered that breakdown of GTP was also involved. The other proteins here are in fact guanyl nucleotide-binding proteins, commonly referred to as G proteins. The receptors that use this mechanism are referred to as G protein-linked receptors. In this case, activation of the receptor causes the activation of the G protein which regulates the activity of adenylyl cyclase. Some extracellular signalling molecules which bind to G protein-linked receptors are discussed in Chapter 3 (see section 3.2 and Table 3.1).

An example of a signalling pathway involving the receptor/G protein/ adenylyl cyclase pathway was discussed above (see Fig. 5.1), in particular the control of glycogen metabolism. However, the action of a G protein on the activity of adenylyl cyclase can be either stimulatory or inhibitory, depending on the G protein involved. If the result of the G protein activation is a stimulation of adenylyl cyclase activity, the G protein is known as a stimulatory G protein or G_s, but if it is inhibitory the G protein is referred to as an inhibitory G protein or G_i.

G proteins, at their simplest, can be seen as molecular switches; that is, they can exist in an on state and an off state, and can toggle between the two forms. Here, with their role in controlling adenylyl cyclase, they associate with the receptor in their off, inactive, state and then conformational changes induced in the receptor by ligand binding induce changes in the G protein so that it adopts the on state. In this new state, it can therefore transmit its message on to the next component in the signalling cascade; here, the adenylyl cyclase.

G proteins can be grouped into two main classes, those that consist of a single polypeptide, the monomeric family, and those that have three subunits, the heterotrimeric G proteins. To consider the control of adenylyl cyclase, the heterotrimeric G proteins will be discussed. In the inactive state these heterotrimeric G protein exists as a complex of three polypeptides, α, β, and γ. The α subunit has a molecular weight of approximately 45 kDa, the β subunit approximately 35 kDa whereas the γ subunit has a molecular weight of only 7 kDa.

As discussed above, the G proteins are guanyl nucleotide-binding proteins and in the inactive state the G protein has guanosine diphosphate (GDP) bound to a single GDP-binding site on the $G\alpha$ subunit. On activation of the receptor, the GDP on the G protein is released in exchange for GTP. This causes the breakdown of the G protein complex into a free α subunit (which has bound to it a GTP) and a β/γ complex which does not

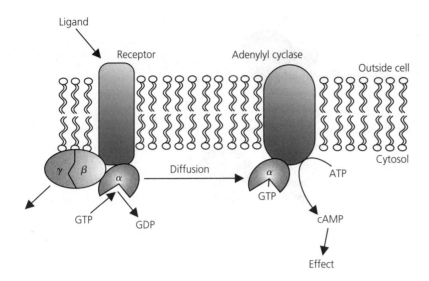

Fig. 5.5 The regulation of adenylyl cyclase by the activation of a trimeric G protein by a G protein-linked receptor.

dissociate further. The free Gα–GTP subunit diffuses to the adenylyl cyclase, where it binds and causes activation (in the case of G_s), with the subsequent release of cAMP (Fig. 5.5).

Any signal that is turned on needs to be turned off again. Here the deactivation of the G protein is achieved by the breakdown of GTP back to GDP on the α subunit. This breakdown of GTP is catalysed by an intrinsic GTPase activity which resides on the α subunit itself. This GTPase activity is activated by the complexing of the α subunit to adenylyl cyclase, and once it has converted the GTP back to GDP the α subunit subsequent releases from the adenylyl cyclase and reforms the original complex with the β/γ subunits. Therefore, it can be seen that the G protein can undergo a cycle of activation and deactivation, with the former involving the exchange of GTP and GDP along with dissociation of the complex, and the latter involving the catalytic cleavage of GTP back to GDP, and the re-association of the complex (see Fig. 5.6).

Certain oncogenes have been shown to code for G proteins which contain a defect in their intrinsic GTPase activity. Such proteins, once bound to GTP, are unable to return to the inactive state, causing the cell to continue to receive the on signal, even in the absence of receptor binding to a ligand. A heterotrimeric G protein that contains a lack of intrinsic GTPase activity would only be able to go half way around the G protein cycle, as shown in Fig. 5.7, and would be left permanently in the active state. A similar effect can be emulated *in vitro* by the addition of guanosine 5′-[γ-thio]triphosphate (GTPγ-S) (*you may also see this written as* GTP[S]). This analogue of GTP contains a sulphur in the bond that is usually broken down by the GTPase when converting GTP to GDP. Hence GTPγ-S is not able to be broken down by the GTPase and again the α subunit will

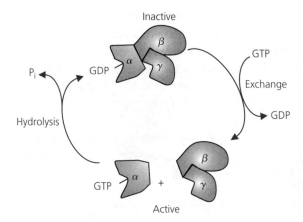

Fig. 5.6 The G protein cycle: activation by nucleotide exchange and subunit dissociation with deactivation by nucleotide hydrolysis and subunit re-association.

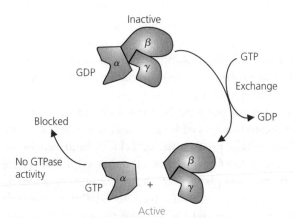

Fig. 5.7 Blockage of the G protein cycle causes permanent activation of the G protein.

remain in an active state. The importance of turning off the G protein signal is emphasized by the disease cholera. The toxin released from the bacterium *Vibrio cholerae* inactivates the α subunit of these G proteins, in particular G_s. Cholera toxin, otherwise known as choleragen, is in fact an oligomeric enzyme, consisting of a complex of an A_1 subunit of 25 kDa linked to an A_2 subunit of 5.5 kDa by a disulphide bond, along with five B subunits of 16 kDa. The A_1 subunit of the enzyme uses the oxidized form of nicotinamide adenine dinucleotide (NAD^+) as a substrate, and catalyses the transfer of ADP-ribose to a specific arginine residue of the α subunit of G_s, destroying the α subunit's intrinsic GTPase activity. The efficient modification of the $G_s α$ subunit requires the presence of another G protein called ADP-ribosylation factor (ARF) which is subgroup of the Ras family of proteins. Once modified, the $G_s α$ submit has no GTPase activity, cannot convert its GTP back to GDP, and is permanently active. Therefore it continues to activate adenylyl cyclase which is the G protein's target in the signalling cascade, and the subsequent loss of cAMP control in the gut epithelial cells leads to movement of Na^+ and water into the intestine, causing the resultant diarrhoea. It is worth noting here that the $G_s\alpha$ subunit can also be activated by the presence of aluminium tetrafluoride (AlF_4^-) if together with Mg^{2+}.

As mentioned above, G proteins are not only able to cause activation of adenylyl cyclase, as seen with Gs, but another form can lead to inhibition of the enzyme. This G protein, termed G_i, is analogous to G_s described above, except that on activation and diffusion to adenylyl cyclase the release of cAMP from adenylyl cyclase is depressed. The G protein contains the same β/γ subunit complex as the G_s, but the $G\alpha$ subunit differs. It has been suggested that there are two possible ways in which this G_i protein might be causing adenylyl cyclase inhibition. Firstly, once released the $G_i\alpha$ subunit may be able cause direct inhibition of adenylyl cyclase, in a similar manner to the way G_s causes activation. Secondly, the $G_i\alpha$ subunit appears to be less inhibitory than the β/γ subunit released by the G protein's activation and interestingly the β/γ subunit is much more inhibitory in the presence of G_s. This suggests that the β/γ subunit from the G_i is able to bind to the $G_s\alpha$ which emanates from the dissociation of G_s and stop activation of the adenylyl cyclase by the G_s route. As the β/γ complexes of both G_i and G_s are proposed to be the same this would be possible. This interaction of the β/γ complex with different $G\alpha$ subunits has been termed the subunit exchange (Fig. 5.8), and further suggests that the exact activation profile of adenylyl cyclase relies on a balance of the activation of the G_s and G_i subunits in some cases.

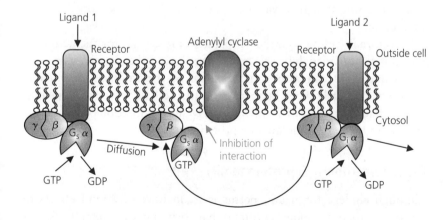

Fig. 5.8 A diagram of subunit exchange: activation of G_i prevents stimulation by G_s through the action of the β/γ subunits.

In a similar way to the permanent activation of the $G_s\alpha$ subunit by cholera toxin, the $G_i\alpha$ subunit is susceptible to a modification mediated by another toxin, this time the pertussis toxin. Pertussis toxin is released by the bacterium which causes pertussis or whooping cough. Here, this toxin causes an ADP-ribose moiety to be added to a specific cysteine residue at the C-terminal end of the $G_i\alpha$ polypeptide, destroying its interaction with the receptor, and this prevents $G_i\alpha$ activation. The lack of activation of the G_i protein will stop the inhibition of the adenylyl cyclase. As this is a double negative, the end result may be similar to the permanent activation of the G_s protein; that is, the adenylyl cyclase will be able to be turned off. However, this might depend on whether the adenylyl cyclase had been turned on by some other mechanism. Again, it is the balance of activation and inhibition that might be crucial here, and toxins will sway that balance one way or the other with profound effects.

The interplay of the G_s and G_i proteins is illustrated by the action of epinephrine on different receptors. On binding to the cell's β-adrenergic receptors the activity of adenylyl cyclase is increased through a route which uses G_s. On the other hand, epinephrine binding to α_2-adrenergic receptors causes a decrease in the activity of the cell's adenylyl cyclase through a route which uses G_i.

The exact nature of the binding of the adenylyl cyclase to the G proteins has been investigated by the use of chimeric G proteins; that is, proteins made with different parts of different G proteins. For example, part of the α chain may be derived from an α subunit which is known not to stimulate adenylyl cyclase, such as a chimera made of segments from G_{i2} and G_t. G_t is a G protein known as transducin and has a defined role in the photoreception of the rod cells of the eye (discussed in Chapter 10). By the use of various combinations in such chimeras, the active parts of the α subunits can be deciphered. It was found that if a protein contained 40% of the C-terminal end of the G_s subunit, adenylyl cyclase was activated, indicating that this was a crucial part of the polypeptide for this interaction.

Therefore it can be seen that G proteins are crucial for the activation and control of adenylyl cyclase but there are other mechanisms too. Ca^{2+}/calmodulin can also activate some of the isoforms of adenylyl cyclase, particularly type I but also types III and VIII as well as some nonmammalian forms. Types V and VI seem to show inhibition by Ca^{2+} while types II and VII are stimulated by protein kinase C.

The heterotrimeric G protein family

Although cholera toxin and pertussis toxin have profound effects on G proteins, there are some G proteins that are immune to their effects. This suggests that there are many isoforms of these important proteins, and that is exactly what has been found. For example, there is a subfamily known as G_q, the members of which are very important in the control of phospholipase C activity (also see Chapter 6; section 6.3) but are not effected by either toxin.

This is no surprise, for as with other signalling proteins heterotrimeric G proteins form a family of related proteins. For example, there is not a single G_s polypeptide, but rather a family of G protein subunits, with $G_s\alpha$ arising from alternative splicing of the same mRNA. This leads to at least four forms of the $G_s\alpha$ subunit. Since the discovery of heterotrimeric G proteins in the early 1980, numerous forms have been isolated, and the number of different receptors that have their effects through the action of G proteins must run into the hundreds. The different α subunits of the trimer are used in general to define the G protein, but as well as the different α subunits at least five isoforms of the β subunit exist and more than 10 γ subunits have been identified. If every type of α, β, and γ sub-unit could randomly mix in the trimer there could theoretically be around 1000 differing forms of heterotrimeric G proteins. However, reconstitution studies show that not all combinations are possible. For example, the isoform Gγ1 can complex with Gβ1 but not with Gβ2. It is probable, in fact, that the stability and therefore existence in the cell of some of the subunits may be dependent on the presence of and complexing with other relevant partners. If a Gβ polypeptide is formed in the absence of a Gγ polypeptide, which it is normally able to complex with, it appears to be aggregated and degraded. Therefore not all combinations will exist; some will be reliant on coordinated protein expression and many combinations expressed will be tissue-specific. It is the specific mixture of G proteins in a cell that will enable it to tailor its response to receptor activation and react appropriately to the initial signal.

Along with G_s, G_i, and G_q, other heterotrimeric G proteins that have been characterized include the G_o of brain neurons, and G_{olf} and G_t of the olfactory and light-sensitive cells of the eye respectively. These do not all of course have an affect on adenylyl cyclase; for example G_q regulates the action of phospholipase C which leads to the release of other intracellular messengers through the inositol phosphate pathway, further leading to the release of calcium and the activation of phosphorylation, while G_t controls the activity of cGMP phosphodiesterase. Some of the G protein α subunits that allow functional diversity are listed in Table 5.2 while the functions of G proteins will be considered again in Chapters 6, 9, and 10.

As well as being able to use the inhibitory effects of toxins to elucidate the role of such G proteins in pathways, the ability to stimulate heterotrimeric G proteins artificially in the laboratory will also help greatly. For example, a tetradecapeptide isolated from wasp venom, mastoparan, mimics the action of G protein-coupled receptors and causes the activation of G proteins, particularly G_i and G_o.

Table 5.2 The more common members of G protein α subunit families.

G_s					
	α_{olf}	α_s			
G_i					
	α_{i1}	α_{i2}	α_{i3}		
	α_{oA}	α_{oB}			
	α_{t1}	α_{t2}			
	α_g				
	α_z				
G_q					
	α_q	α_{11}	α_{14}	α_{15}	α_{16}
G_{12}					
	α_{12}	α_{13}			

Although not directly related to this discussion it is interesting to note that signalling is never obvious in cells. We may assume that G proteins control cAMP levels in many cells, but can it be the other way around, where the levels of cAMP are regulating the activation of the G protein? This is the mechanism of the slime mould *Dictyostelium*, where the level of extracellular cAMP is detected by receptors which lead to the activation of the G protein subunit $G\alpha2$.

The roles of the β/γ complex

When looking at the roles of G proteins in signal transduction the active part of the heterotrimeric G proteins has traditionally been thought to be the α subunit, with the β/γ complex only being involved in its association with the α subunit when the complex is inactive. However, the β/γ complex of heterotimeric G proteins has also been shown to be important in its own right. As many isoforms of the β and γ subunits exist, in studies similar to those carried out with the chimeric α subunits defined β/γ complexes with known β and γ subunits were formed by expressing the subunits in insect Sf9 cells. These could be subsequently purified by subunit affinity chromatography and used to study whether there was any interaction of the β/γ complexes with, for example, the adenylyl cyclase enzyme. It transpired that different isoforms of adenylyl cyclase are controlled differently by the subunits of the G protein trimer. The action of β/γ can be stimulatory, as with type II or IV adenylyl cyclases, or inhibitory, as seen with type I. Furthermore, some adenylyl cyclase isoforms do not appear to be affected by β/γ at all. It was also found that the interactions of the different isoforms were dependent on prenylation of the γ subunit of the G protein. The discussion of subunit exchange above suggests that the β/γ subunits have a role in the modulation of the action of the G proteins themselves. In fact, it was noted that the concentration of β/γ subunits needed to regulate adenylyl cyclase is greater than that of the $G_s\alpha$ subunit that is normally released. As the heterotrimeric G proteins undergo a stoichiometric dissociation—that is, one $G_s\alpha$ gives rise to one β/γ subunit when the G protein is activated—then the β/γ that can give rise to the effect probably does not arise solely from the breakdown of the G_s trimer. The β/γ complex may come from G_i or another G protein, for example G_o, which is more abundant in some tissues, such as in the brain. This sort of interaction may be commonplace, enabling different G protein trimers to have different effects on adenylyl cyclase at the same time, as well as effecting the activity of each other through subunit exchange.

However, the β/γ subunits may have far wider roles in signal transduction. Beside the possible regulation of adenylyl cyclase, other roles in which the β/γ subunits have been found to be involved include regulation of K^+ channels in myocytes and the pheromone signalling pathway of yeasts, here through the activation of MAP kinase cascades. β/γ subunits have also been implicated in the control of the kinase activity and phosphorylation of β-adrenergic receptors, as discussed in Chapter 4 (section 4.2); β/γ subunits activate GRK. A similar interaction is with a protein called phosducin which may be involved in the regulation of the concentration of β/γ subunits by forming complexes with them. In the photoreceptor cells of the eye, phosducin is found to be a phosphoprotein whose phosphorylated state is light-sensitive, mediated by light-induced changes in cyclic nucleotide concentrations. In the dark, the protein is phosphorylated by cAMP-dependent protein kinase, while it is dephosphorylated in the light by phosphatase 2A. The role of the phosducin seems to be to complex with the β/γ subunits thereby stopping their interaction with Gα subunits and preventing the Gα subunits from recycling (discussed again in Chapter 10; section 10.2). Similar proteins to phosducin have been reported to be present in the cytosol of bovine brain cells, and this type of control might be quite widespread.

The β/γ subunit has also been shown to activate one of the forms of phospholipase C, that is PLCβ, leading to the activation of the inositol phosphate pathway. The isoforms of PLC which appear to be most susceptible are β2 and β3 (see section 6.3). Therefore, depending on the PLC isoform, members of the family of this enzyme can be controlled by the α or β/γ subunits of heterotrimeric G proteins.

Other roles of the heterotrimeric G proteins

As can be seen from the discussion above, there are many roles of the heterotrimeric G proteins, not just the activation of adenylyl cyclase. As well as those mentioned, the α subunit released from the activation of G_i in muscle tissue, the $G_i\alpha$ subunit, is also able to activate K^+ membrane channels, allowing K^+ to exit the cells, thus inhibiting the contraction of the muscle. So the exact signalling process that transpires from the dissociation of any G protein trimer may be very complex, as subunits may have more than one effect, the exact nature of the response being likely to be dependent on the tissue in which the G protein resides. Similarly, $G_s\alpha$ subunits have been seen to inhibit cardiac Na^+ channels and an effect of $G_s\alpha$ on voltage-gated Ca^{2+} channels has also been reported.

In most cases, the G proteins are activated at the membrane, and have their effect there too. If these subunits from the G proteins are readily diffusible along the membrane from receptor to enzyme, what is it that keeps them attached to the membrane when dissociation takes place and stops them from diffusing out into the cytosol? For the γ subunit it has been proposed that the subunit is bound to an isoprenoid lipid, which being a hydrophobic molecule will attach the subunit to the membrane. For the α subunits the attachment is probably through another lipid, this time myristic acid (see section 4.7 for a discussion on protein modifications).

Although classically these G proteins have been thought of as activators of enzymes on the plasma membrane, particularly adenylyl cyclase, it is thought that their action may be far more widespread throughout the cell. For example, an isoform of the subunit $G_i\alpha$ has been localized to the Golgi apparatus where it is probably involved in the control of the packaging of proteins and in the regulation of vesicular activity.

It is likely that the roles of G proteins in disease and disorders will be far more widespread than those reported to date. Defects of the G_s and G_i G proteins have been implicated in the excessive proliferation of pituitary cells which leads to pituitary cancer, and it is reasonably certain that the role of these G proteins will be implicated in many disorders and be the target for many therapies. Of course, most of the discussion here is based on data obtained from mammalian tissue, but the existence of G proteins is not limited to these organisms.

5.5 Guanylyl cyclase

In an analogous way to the production of cAMP, cGMP is also used by cells as a signalling molecule. Such nucleotides were identified from urine in 1963 by Ashman and his colleagues, and much research has been targeted on these molecules since. cGMP is produced by the enzyme guanylyl cyclase, otherwise known as guanylate cyclase. However, like adenylyl cyclase, the cAMP-producing enzyme, guanylyl cyclase, is found in two forms, a soluble type which resides in the cytoplasm of the cell and

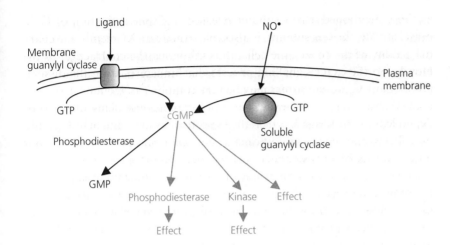

Fig. 5.9 A scheme to show how cGMP may fit into signalling pathways.

a membrane-bound form, located in the plasma membrane. These forms are distinct proteins which have their own modes of regulation, as discussed below. The catalysis of the enzymes is similar to that of adenylyl cyclase, however, as the cGMP is produced from GTP, as cAMP is produced from ATP, and so the chemistry shown in Fig. 5.3 would be pertinent (with guanine and not adenine as the base).

Therefore, cGMP can fit into pathways as a diffusible messenger, in a similar manner to cAMP (Fig. 5.9). One of the most discussed roles for cGMP is in the regulation of cGMP-gated ion channels, as in the photosensitive cells of the retina, as discussed further in Chapter 10 (section 10.3 and Fig. 10.4). However, cGMP has also been found to control cGMP-dependent phosphodiesterases and cGMP-dependent protein kinase (cGPK; also known as PKG).

Soluble guanylyl cyclase

A very important version of the guanylyl cyclase enzyme is the soluble form, which is characterized by the presence of haem at the catalytic site. The haem prosthetic group is an iron-containing protoporphyrin IX, as it is in most cytochrome molecules in cells. It is the haem group within the enzyme that is the target of nitric oxide (NO), one of the main activators of the enzyme, as discussed in Chapter 8 (section 8.2). In this signalling scheme, NO may emanate from another cell, cross the plasma membrane as it is non-charged, and activate soluble guanylyl cyclase with the concomitant rise in intracellular cGMP levels (Fig. 5.9).

The topology of the guanylyl cyclase enzyme shows it to be a heterodimer of α and β subunits of approximately 70 and 85 kDa. Each of these subunits contains a region that is homologous to the C_{1a} and

C_{2a} catalytic regions of adenylyl cyclase, and both are required for catalysis. The presence of Mn^{2+} ions as opposed to Mg^{2+} ions increases the activity of the enzyme, which also is slightly stimulated by Ca^{2+} ions. However, ATP acts as an inhibitor. Haem binding has been shown to involve a histidine residue which is on the β subunit of the enzyme.

However, the enzyme appears to exist in tissue-specific isoforms. For example, two isoforms have been reported to exist in bovine lung. The two forms contained one subunit in common but the second subunit differs; one is 85 kDa whereas the other is a little smaller, being 73 kDa. Interestingly as it is so important for function, the attachment of the haem group may vary, as seen by differences in the haem spectrum for the guanylyl cyclase enzymes. It has also been suggested that the haem group of some enzymes is penta-coordinate while that of others is hexa-coordinate, which suggests that the two enzymes may have subtle differences in their catalytic action and in their control. Such subtle differences, as discussed above, are not uncommon among families of proteins and enzymes involved in cell signalling.

Membrane-bound guanylyl cyclase

As well as soluble forms of guanylyl cyclase, there are also membrane-bound forms. However, except for relatively small regions, these show little topological similarity to adenylyl cyclase, or indeed to the soluble forms of guanylyl cyclase. The membrane guanylyl cyclase contains a single membrane-spanning domain, and therefore has domains on the outside and inside of the cell. On the cytoplasmic side, along with a protein kinase-homology domain, is an intracellular catalytic domain (Fig. 5.10). This makes sense, as any cGMP produced needs to be in the cytoplasm to enable it to pass its message down the signalling pathway. Some areas of homology to the mammalian adenylyl cyclases have been reported for this catalytic area, but as was noted above, this is no great surprise as the chemistry which needs to be catalysed is very similar, i.e. the triphosphate converted to cyclic monophosphate forms. On the outside of the membrane is the extracellular domain which is in fact a ligand-binding site, and so these enzymes are receptors. In the classification given in Chapter 3 (section 3.1, and see section 3.2) they would be referred to as receptors 'containing intrinsic enzymatic activity'. They act as receptors for two main classes of molecule: signalling peptides or the so-called naturietic peptides and the heat-stable enterotoxins including guanylins.

These membrane cyclase forms, like many receptor proteins, probably function as dimers or oligomers. ATP has been shown stimulate the activity of these cyclases but despite the presence of a protein kinase-homology domain non-hydrolysable forms of ATP will have the same effect and

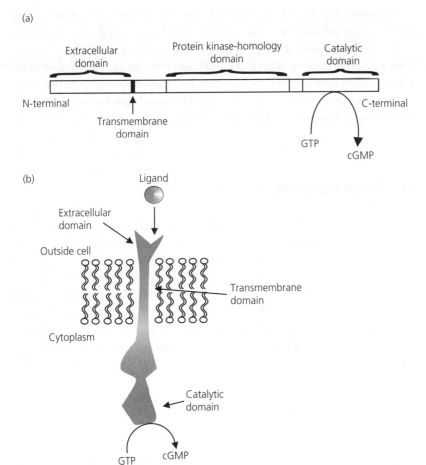

Fig. 5.10 The domain structure and schematic topology of membrane-associated guanylyl cyclase. (a) Depicted as a simple linear representation. (b) Schematic of how the protein will arrange in a membrane, with the ligand-binding site on one side and the catalytic site on the other.

neither ATPase activity nor phosphorylation activity is required for the functioning of these receptors. However, it appears that the cyclase normally resides in a highly phosphorylated state. Further, a novel protein phosphatase has been found associated with one of the forms of the receptor and it is thought that the state of phosphorylation of the cyclase may be crucial to its activity, dephosphorylation causing a desensitization of the receptor.

Calcium ion concentrations are also important in the regulation of the membrane forms of guanylyl cyclase. In the retina, Ca^{2+}-sensitive proteins have been found which interact with guanylyl cyclase. One such protein, of 26 kDa, is recoverin. Other such proteins have molecular weights of 20 and 24 kDa; these are known as p20 and p24 respectively. These proteins are cytoplasmic, where they respond to the changes in the Ca^{2+} concentration, but their exact interaction with the cyclase remains to be resolved. However, increases in Ca^{2+} will lead to a complex of the calcium-sensitive protein and the cyclase and the production of cGMP will be decreased. It is possible that Ca^{2+} might also have a direct interaction with the cyclase but this has never been demonstrated.

Interestingly, one particulate form of guanylyl cyclase, that from bovine rod outer segments, is reported to be activated up to 20-fold by nitric oxide, suggesting that it might not just be the soluble forms of guanylyl cyclase that are targets of this intercellular signal.

5.6 Phosphodiesterases

All signals need to have the potential to be reversed, and cAMP and cGMP are no different in this respect. However, despite the fact that they are produced by distinct and surprisingly different enzymes, the destruction of both of them is down to a single family of enzymes. The breakdown of cAMP and cGMP and hence the reduction of their intracellular concentrations, usually leading to a cessation of the message, are catalysed by the enzymes called cyclic nucleotide phosphodiesterases (PDEs), usually referred to simply as phosphodiesterases. This reaction can be written as follows (see also Fig. 5.3):

$$cAMP + H_2O \xrightarrow{Mg^{2+}} AMP + H^+$$

or

$$cGMP + H_2O \xrightarrow{Mg^{2+}} GMP + H^+$$

Although cAMP and cGMP are relatively stable in the cell, removal of the cyclic nature of the molecules and hence their breakdown overall is exergonic; that is, the change in Gibbs free energy (ΔG) is negative, the enzyme overcoming the activation energy barrier which stops the reaction becoming spontaneous.

In mammals at least 11 different types of phosphodiesterase have been distinguished (see Table 5.3), encoded for by at least 20 genes, with the existence of more classes suggested by some researchers. As would be expected, some of the types have more than one form; for example, type I has three forms, type III has at least two forms, and type IV has at least four forms. Phosphodiesterases all seem to be highly homologous in their C-terminal regions but show greater divergence towards the N-terminal end of the polypeptides. The expression of different forms of the same

Class of enzyme	Characteristics	Comments
I	Ca^{2+}/calmodulin-dependent	Two calmodulin domains
II	cGMP-stimulated	Two GAF domains, which are cGMP-binding domains
III	cGMP-inhibited	Transmembrane domains (six?)
IV	cAMP-specific	At least four members in this family
V	cGMP-specific	Two GAF domains
VI	Photoreceptor type	Associated with control proteins
VII	High-affinity, cAMP-specific	Appears simple, having only a catalytic domain
VIII	cAMP-specific	Ligand-binding domain?
IX	High-affinity, cGMP-specific	Appears simple, having only a catalytic domain
X	Hydrolyses cAMP and cGMP	Might act as a cAMP-inhibited cGMP enzyme
XI	Hydrolyses cAMP and cGMP	Contains a GAF domain

family of PDEs is also tissue-specific, suggesting subtle differences in their functionality.

Table 5.3 The classes of phosphodiesterase.

If the cDNAs for all the PDEs are compared, the 3′ ends show a great deal of homology, over a stretch of approximately 800 base pairs, as would be expected from the homology of the C-terminal ends of the polypeptides. These regions code for a common catalytic core of the molecules. As with the enzymes that produce the cyclic compounds, similarity here would be expected in the enzymes that break them down, as the chemistry is the same, regardless of whether the base is adenine or guanine. However, comparisons of the 5′ ends show low homology and this accounts for the differences which make up the seven family types. The N-terminal end of the polypeptides may also influence their subcellular location, containing signal peptides which either direct their post-translational route within the cell or allow their attachment to membranes.

The genes for at least five cAMP-specific phosphodiesterases have been located in both humans and mice. In humans, two of the genes are found on chromosome 19, with others on chromosomes 5, 1, and 8, whereas in mice phosphodiesterase genes were found on chromosomes 4, 8, 9, and 13. However multiple forms of a PDE within a family may also arise from the same gene. In some cases multiple promoters within one gene have been found, enabling transcription to start at different points and therefore resulting in polypeptides of varying length. Even if the same promoter is used, the gene transcript may undergo alternative splicing, and again varying polypeptides will result. Such differences in expression and post-transcriptional modification may be tissue-specific, each of the PDEs retaining the catalytic core but having alternative modules which alter their function and cellular location. However, some PDEs are particularly susceptible to proteolysis and therefore several reports in the literature of the presence of new isoenzyme forms may only be the result of breakdown products from larger PDEs. Despite this, it is clear that there are many forms of this enzyme, and subtle differences will no doubt be required in different tissues to maintain the balance of intracellular cyclic nucleotides used in signalling.

Although cAMP and cGMP are often broken down by different phosphodiesterases, the presence of one cyclic nucleotide may still influence the intracellular concentration of the other. Some forms of cAMP PDE are influenced by the presence of cGMP. For example, type III has been shown to be inhibited by the presence of cGMP. Such phosphodiesterases are therefore referred to as cGMP-inhibited phosphodiesterases, or cGI-PDEs. These PDEs have been shown to have a molecular weight of approximately 110 kDa but proteolytic products between 30 and 80 kDa have been reported also. Other cAMP phosphodiesterases have been reported to contain allosteric binding sites for cGMP which cause a stimulation. These PDEs also have an apparent molecular weight of approximately 105 kDa, but further analysis suggests that the native structure is as non-spherical dimers, although a tetrameric form has been reported to be have been isolated from rabbit brain tissue.

Other factors, besides the cyclic nucleotides, might influence the activity of these PDE enzymes. One family of PDEs (type I) is controlled by the intracellular concentration of Ca^{2+} in association with calmodulin, and a look at the structure of these enzymes shows that they have two calmodulin-binding domains in their N-terminal end. This family has been further subdivided into four groups, which have the molecular weights of 60, 63, 58, and 67 kDa.

One of the most well-characterized cGMP PDEs is that from the rod cells of the eye (type VI). Here, as discussed in section 10.2, cGMP acts as a regulator of ion channels. The membrane-bound PDE from bovine rods exists as a $\alpha\beta\gamma_2$ complex where the α subunit has a molecular weight of 88 kDa, the β subunit has a molecular weight of 84 kDa, and the γ subunit has an apparent weight of only 11 kDa. However, other complexes may exist such as $\alpha\alpha\gamma_2$ and $\beta\beta\gamma_2$. Here, as well as the membrane form, there is also a soluble form where the α, β, and γ subunits appear to be of the same sizes as the membrane PDE but they are joined by a further δ subunit of 15 kDa. The analogous PDE found in the cone cells, responsible for colour vision, has a large subunit of 94 kDa which is joined by three small subunits of 11, 13, and 15 kDa. The overall weight of this complex has been estimated to be approximately 230 kDa.

Of significant importance are the type V phosphodiesterases, the cGMP-specific members of the family. It has been known for many years that the blood supply though the vessels in mammals can be influenced by the signalling molecule nitric oxide. However, as we have seen above, NO targets the soluble form of guanylyl cyclase, causing its activation. Therefore NO causes a rise in intracellular cGMP. But, if type V phosphodiesterase is active, the rise in cGMP will be very transient, and in some individuals the rise is too transient. The obvious phenotype here is the lack of a sustained penile erection. Therefore, a drug, commonly known as Viagra™, has been developed to enhance the longevity of the raised cGMP levels. It does this by inhibiting type V phosphodiesterase. Therefore, despite some reports of Viagra™ causing NO release, it does not, but it is a phosphodiesterase inhibitor. However, as mentioned, the chemistry and catalytic domains of the phosphodiesterases is very similar across the family, and therefore it is no surprise to learn that the drug also has inhibitory effects on other members of the family in some individuals, in particular phosphodiesterase type VI. This manifests itself as an alteration in eyesight, in that a blue haze is reported, and this is because phosphodiesterase type VI is so important in the signalling that takes place in the rods and cones of the mammalian eye. This is also discussed in Chapters 8 (section 8.2) and 10 (section 10.2).

Viagra™

Chemically Viagra™ is sildenafil citrate or more fully 1-[[3-(6,7-dihydro-1-methyl-7-oxo-3-propyl-1*H*-pyrazolo[4,3-*d*]pyrimidin-5-yl)-4-ethoxyphenyl]sulphonyl]-4-methylpiperazine citrate.

It is also possible that binding of cyclic nucleotides to phosphodiesterases takes place in which there is no catalytic turnover, and this might serve to buffer the intracellular concentration of such signalling molecules, so that production of a cyclic nucleotide such as cGMP may not automatically mean that the molecule will be free in solution to have a further effect of other enzymes.

Some phosphodiesterases may in fact be exported from the cell, and these are then used to control the levels of extracellular cyclic nucleotides. This would be of importance to organisms such as *Dictyostelium*, which uses the concentrations of extracellular cAMP as a signal for aggregation.

The isolation of a phosphodiesterase which appears to be specific for cyclic cytidine monophosphate (cCMP) suggests that this third cyclic nucleotide may also be found to have a role in signalling and no doubt such data will lead to further research in this area.

Hot-spots

Many signalling molecules have been found to exist in high concentrations in particular areas of the cytoplasm, and the term hot-spots has been coined to describe this. Hot-spots of cAMP have been found, and it is thought that the exact location of specific isoforms of the synthesizing enzymes, cyclases, and breakdown enzymes, phophodiesterases, within the cell finely control the local concentrations of such signals, allowing greater subtly to their action.

5.7 The GTPase superfamily: functions of monomeric G proteins

The role of G proteins—that is, proteins which bind GDP and GTP, have an exchange of the two nucleotides leading to an activated state, and hydrolysis of the nucleotide to restore the protein back to an inactive state—is far more common than that outlined above for the control of adenylyl cyclase. As well as the plasma membrane-associated heterotrimeric G proteins discussed above there exists a group of small monomeric G proteins which control such diverse cellular functions as proliferation, gene expression and protein synthesis, differentiation, and regulation of movement of proteins through the cytoplasm. All these G proteins possess intrinsic GTPase activity and the phrase GTPase superfamily has been coined to encapsulate this group of proteins. In a manner which is very similar to that seen with the heterotrimeric G proteins, the monomeric G proteins can act as molecular switches, in a mechanism which allows for a large amount of amplification of the signal.

As with the heterotrimeric G proteins, the monomeric G proteins are inactive when bound to GDP, which on activation is exchanged for GTP causing a conformational change in the structure of the proteins and so allowing the signal to be transmitted along the pathway. The intrinsic GTPase activity converts the GTP back to GDP, turning off the signal and allowing the protein to await the next signal (Fig. 5.11).

It has also been postulated that a third state of the G protein exists in which GDP has been removed but GTP has yet to bind. This is a transient state in which either GTP can bind to turn the protein on, or GDP can bind to prevent the system turning on. Normally in the cell the GTP-binding

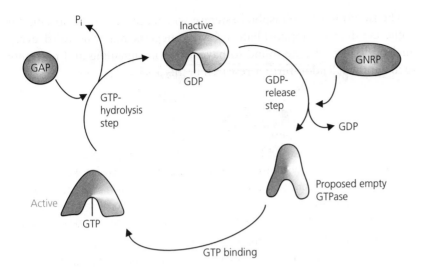

Fig. 5.11 The G protein cycle of the monomeric G proteins showing the actions of GTPase-activating protein (GAP) and guanine nucleotide-releasing protein (GNRP).

step would be favoured. An analogous empty state of the guanine nucleotide-binding site probably also exists in the cycle of the α subunit of the heterotrimeric G proteins too.

However, there are significant differences between the monomeric system and that of heterotrimeric G proteins described above. The G proteins here are monomeric, and in general act as monomers, but in no sense do they act alone, but rather they associate and are aided in their function by other polypeptides. Both the dissociation step resulting in the loss of the GDP, and the GTPase step converting the GTP back to GDP, require the assistance of other proteins. The first, GDP-releasing, step is aided by the presence of guanine nucleotide-releasing proteins (GNRPs), while the GTP hydrolysis is aided by the presence of GTPase-activating proteins (GAPs; Fig. 5.11).

It is probable that the α subunit of heterotrimeric G proteins, although acting as a monomer when associated with GTP, does not need the assistance of a GAP for activating the GTPase activity because it contains a region on the polypeptide that has GAP-like activity; that is, a GAP-like domain.

So, how do these G proteins fit into a signalling pathway? An overall but very simplified scheme is shown in Fig. 5.12. A receptor, often a tyrosine kinase receptor here, will bind its ligand, and an autophosphorylation event will lead to the formation of binding sites for an intracellular adaptor molecule. This will activate a GNRP, which will facilitate the exchange of the GDP for GTP on the monomeric G protein. Once active, the G protein can activate its downstream effector, often a kinase cascade, which will lead to the cellular effect required by the ligand perception. More detailed and characterized pathways are shown in Figs 4.8 and 9.7.

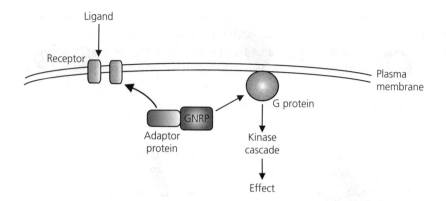

Fig. 5.12 An extremely simplified scheme to show how a monomeric G protein might fit into a signalling pathway.

The most well-characterized type of monomeric G protein is the product of the *ras* oncogene, p21^{ras}, or known simply as Ras. This is a plasma membrane-associated protein of molecular weight 21 kDa. Like many such proteins it was first identified as being encoded by a gene from acute transforming retroviruses, but now the Ras protein has been crystallized and a structure resolved.

The role of Ras was originally hinted at by the discovery that transforming genes were in fact alterations of normal genes, activation of which was due to the presence of a point mutation. Their involvement has been suggested in the formation of several forms of cancer, including formation of tumours induced by physical and some chemical reagents. Most of the assays for the presence of Ras depend on the ability of Ras to induce cell proliferation. Bioassays have been based on DNA synthesis, transformation of established cultured cells, and the formation of tumours. Other assays have exploited the ability of Ras to induce differentiation of cells. However, it is interesting that full transformation of cells *in vitro* often also depends on the presence of another oncogene, for example *myc*.

The *ras* gene has been well characterized. In humans and some other mammals there are three functional *ras* genes present in the genome. These are H-*ras*, K-*ras* and N-*ras*. H-*ras* and K-*ras* are related to the Harvey (Ha) and Kirsten (Ki) sarcoma viruses of mice. The structure of the *ras* gene is unusual in that the first 5′ exon appears to be non-coding. The actual coding region is split into four exons, although the K-*ras* gene contains two alternative fourth exons, which can give rise by alternative splicing to two separate gene products. The introns (non-coding regions within the genes) are very different between the various genes, leading to very different lengths of mRNA for each.

The *ras* products have been detected in all tissues explored, including fetal tissue. However, histochemical staining was more intense in cells that were proliferating, as opposed to those which were fully differentiated.

Northern blot analysis, to reveal the levels of gene expression, showed that the genes did show some tissue specificity. Studies of transcriptional control for these genes suggests that the first intron may contain an enhancer, with another small weak enhancer downstream of the gene, while a short alternative exon within another intron also appears to be important, although the exact nature of its role remains obscure. However, the presence of serum or some growth factors can enhance the expression of Ras.

With the gene relatively well characterized, what is known about the protein itself? Most Ras proteins contain 189 amino acids, except the K-*ras*B product which has 188 amino acids. A striking feature is that the first 164 amino acids are homologous across all the *ras* products, with the C-terminal region being referred to as the heterogeneous region, it being more divergent. However, sequence patterns are conserved even here between species, suggesting that certain C-terminal sequences confer functional specificity to the proteins. If a region of many proteins is so highly conserved it suggests that there is a function of great importance being harboured there, and indeed the first 164 amino acids contain the GTPase activity of the proteins. Also, in all Ras proteins a cysteine residue is conserved at a position four residues before the C-terminus, again suggesting an importance attached to this amino acid. The protein is made as a pro-p21^{ras} product and this C-terminal region also undergoes a great deal of post-translational modification, including farnesylation at the conserved cysteine amino acid, cleavage of the end amino acids, and methylation (see also Chapter 4; section 4.7). These modifications are involved in the increase in the hydrophobicity of the protein which encourages its association with the inner face of the plasma membrane. Mutant proteins without this conserved cysteine at the C-terminus are cytosolic and have no transforming activity.

The general structure of the p21^{ras} protein, Ras, shows that it has a hydrophobic core consisting of six β sheet strands connected by hydrophilic loops and α helices. Five regions of the protein, which have been designated G1–G5, lie on one side of the protein and are associated with the hydrophilic loops. G1 is probably involved in binding of the first two phosphate groups, α and β, of either GDP or GTP. The catalytic step, that is the GTPase activity, of the cycle requires the presence of Mg^{2+} and this is probably associated with regions G2 and G3. These regions, G2 and G3, have also been implicated as the sites of conformational change induced by GTP binding, and so activation of the protein. They have also been implicated in the association with GAP proteins. The other association needed for the functioning of the GTPase is an interaction with the protein next in line down the signalling pathway that the GTPase needs to act on; that is, the effector. Here again the GTPase region G2 has been implicated.

The sequences of Ras oncoproteins, which have the ability to promote tumour formation, can be used to determine the active residues in the protein. Commonly, substitutions of residues at positions 12, a glycine, and 61, a glutamine, reduces the intrinsic GTPase activity of the protein. It is now believed that the catalytic cycle involves a glutamine residue on the protein, which activates a water molecule which then attacks the γ-phosphoryl group of the GTP in a nucleophilic fashion. However different GTPases also certainly vary in the precise mechanism of GTP hydrolysis.

Some viral *ras* genes encode a threonine at position 59, with autophosphorylation taking place at this threonine, but the role of this phosphoryl addition is unclear. Another, rather fascinating, mutant form of Ras has no ability to bind to guanine nucleotides, and therefore its conformation appears to be permanently locked into an active form.

Another good model for the monomeric G protein family is EF-Tu, elongation factor Tu, which is involved in protein synthesis. However, this protein is much larger, with a molecular weight of approximately 43 kDa, but like Ras it has been crystallized. Other members of the GTPase family include the proteins Rab, Rho, and Rac. At least four Rho polypeptides have been identified along with two Rac polypeptides as well as other related proteins such as TC10 and CDC42H. Like Ras, such proteins also work in conjunction with GAP-like proteins, such as Rho-GAP, and GNRP-like proteins, such as Dbl and Rho-GDS.

Let us now turn to the proteins which interact with the monomeric G proteins, and which are so instrumental in the functioning of the monomeric G proteins. Several proteins have been found which contain GAP activity; that is, have a GTPase-activating function. One of these is a protein with an approximate molecular weight of 120 kDa called p120GAP. It consists of 1047 amino acids, and appears to be expressed everywhere. When in its active state, it has the ability to increase up to fivefold the GTPase activity of Ras, and this function is due to a catalytic domain in the C-terminal end of GAP. The polypeptide also contains several protein–protein-binding domains: two SH2 domains, a SH3 domain, and a PH domain (Pleckstrin homology domain), suggesting that it has a great ability to interact with other polypeptides. Two such polypeptides, which are themselves phosphorylated, include one of 190 kDa, p190, and one of 62 kDa, p62. If p120GAP forms a complex with p190 it has a reduced ability to stimulate GTPase activity of Ras. p120GAP is also phosphorylated on tyrosine residues by tyrosine kinase receptors which therefore probably have a measure of control on its activity. It is also inhibited by some lipids, including arachidonic acid.

A second group of GAP proteins is a family known as GAP1. Mammals contain at least two homologues, GAP1[m] and GAP1[IP4BP] (superscript [IP4BP] here denotes inositol 1,3,4,5-tetrakisphosphate binding). These proteins, of approximately 850 amino acids, contain a PH domain towards the

C-terminal end of the polypeptide, a GAP-related domain (GRD) in the middle, and towards the N-terminus two domains which are homologous to the C2 regulatory domains of protein kinase C. Interestingly, as the name suggests, these proteins have been shown to bind to inositol 1,3,4,5-tetrakisphosphate, one of the metabolites derived from InsP$_3$, which suggests exciting possibilities in its control. As discussed in Chapter 6 (section 6.4), several inositol derivatives have been thought to have signalling roles, and clearly here is an example. Like the other GAP these proteins also show inhibition by phospholipids. Expression of this protein is not so widespread as for p120GAP, with the highest expression seen in the placenta, brain, and kidneys.

GAP is not the only protein which has been found to stimulate GTPase activity in Ras. A rather large protein of approximately 250 kDa known as NF1, otherwise called neurofibromin, is responsible for Recklinghausen's neurofibromatosis (NF1 disease). Out of its 2818 amino acids a region of 350 amino acids found in the centre section of the polypeptide is analogous to GAP. The protein appears to be expressed primarily in the nervous system, particularly in neurons and Schwann cells, where it exists in at least three isoforms, the different forms arising from alternative splicing. The role of neurofibromin also certainly involves its interaction with other proteins in a complex and again, like other GAPs, its activity is inhibited by lipids.

The GAP1 family also shows GAP activity against other G proteins, such as Rap, although this activity is not affected by lipids or by the presence of inositol tetrakisphosphate (InsP$_4$). However, other members of the Ras superfamily are not necessarily acted upon by GAP itself, but have their own proteins which contain GAP-like activity.

The other G protein-regulatory proteins are the GNRPs, the guanine nucleotide-releasing proteins. One of the most important, and certainly the most well studied, is the protein known as Son of Sevenless or Sos. It was originally found to be involved in signalling in the Sevenless system in the fly *Drosophila*. It is often found associated with an adaptor protein, such as GRB2 in mammals, or Drk in *Drosophila*, the interaction being possibly through the adaptor's SH3 domains. As discussed above, on binding of a ligand to its respective receptor protein kinase the receptor can usually autophosphorylate and the formation of the new phosphotyrosine groups allows an interaction with adaptor protein through its SH2 domains. This ensures that the adaptor protein and its associated GNRP, for example Sos, are now in close association with the membrane where the latter leads to nucleotide exchange and activation of the monomeric G protein. Such a system is discussed further in Chapter 9 (Fig. 9.4) and previously in Chapter 4 (section 4.3 and Fig. 4.8).

Sos is not, as expected, the only GNRP to be found and characterized. Other GNRPs include CDC25 in *S. cerevisiae*, a 1545 amino-acid polypeptide. Mutants of *S. cerevisiae* lacking the CDC25 protein, *cdc25*⁻ mutants, have been useful in the identification of similar proteins. An analogous gene in mammals, $cdc25^{Mm}$, encodes a protein of 140 kDa, which is found in brain tissue. The gene, however, is quite complex as it can give rise to at least four distinct proteins of very disparate molecular weights. Other GNRPs include a protein of 35 kDa that appears to have the role of stimulating guanine nucleotide exchange, while another cytosolic protein, encoded by the gene *smgp21GDS*, has been found to be active only if Ras has undergone its post-translational modifications. To complicate the issue further, inhibitor proteins of guanine nucleotide exchange have been reported which act on several members of the Ras superfamily, particularly members of the Rab, Rho, and Rac families; for example, Rho-GDI.

Although schemes can be drawn to show how these monomeric G proteins fit into transduction pathways, it is not always clear exactly how the signalling mechanism works. In particular, the activity of Ras can be increased in one of two ways. It is usually assumed that the G protein is activated by an increase in GTP/GDP exchange, through an increase of GNRP activity, perhaps involving Sos (as shown in Figs 4.8 and 9.4). However, if the GTPase activity is decreased, for example through a decrease of GAP activity, then the G protein will remain longer in its active state, and signalling will continue, the result being similar to an increase in GNRP activity.

As for the pathways in which monomeric G proteins are involved, they are numerous. Ras and its family members are often seen to be involved in pathways leading from the activation of growth factor receptors for example, and in the pathways leading from insulin perception. It has also been shown that the transformation abilities of several oncogenes encoding protein tyrosine kinases, for example *src* and *fms*, require the activation of Ras. Ras itself is often found to activate the serine/threonine kinase Raf, also a proto-oncogene product, and activation of Raf often leads to the activation of the MAP kinase cascades. Ras can also activate other serine/threonine kinases, and it has also been reported to activate MAP kinase kinases as well as MAP kinases. Activation by Ras of ribosomal protein S6 kinase as well as protein kinase C has also been reported. As with the heterotrimeric G proteins, in *S. cerevisiae* Ras has also been shown to activate the enzyme adenylyl cyclase, and so have an influence on cAMP pathways too.

However, the effector role of Ras may not necessarily be directly related to changes in the Ras protein. If Ras is in interaction with its associated proteins, for example GAP, then conformational changes are possible in the GAPs too, and that in itself might constitute a signal. Binding of Ras to GAPs is thought to lead to exposure of protein-interacting domains, such as SH2 and SH3 domains. Once exposed, these domains would enable GAP proteins to interact with other proteins and so propagate a signal within the cell, without the direct involvement of Ras. In this case, this might mean that the binding of GAP and Ras allows the divergence of the signal, with two distinct signals being propagated into the cell.

Other Ras-related proteins

A subgroup of the Ras family of proteins is the family of ADP-ribosylation factors (ARFs). These were first identified as factors needed for the efficient modification of $G_s\alpha$ of the heterotrimeric G proteins by cholera toxin (see above, section 5.4). These proteins have been shown to be involved in the vesicle-mediated protein traffic of cells as well as in the control of some lipid signalling by the regulation of the activity of phospholipase D, an enzyme which yields phosphatidic acid and choline from membrane lipids (see section 6.7).

Lastly here, as mentioned above, an elongation factor involved in protein synthesis, EF-Tu, is also very similar to Ras in its mode of action. However, the GNRP involved in the GDP/GTP exchange is another elongation factor, EF-Ts. This protein contains constitutive activity, unlike the GDP/GTP exchange on many other G proteins which involves the activation of a receptor or receptor-associated proteins.

5.8 SUMMARY

- Cyclic nucleotides and G proteins are central signalling components in many transduction pathways, (Fig. 5.13).

- The main cyclic nucleotides used in signalling are cAMP and cGMP.

- cAMP is produced from ATP by the enzyme adenylyl cyclase.

- In mammals adenylyl cyclase is a single polypeptide integral to the plasma membrane, but several isoforms are seen in nature.

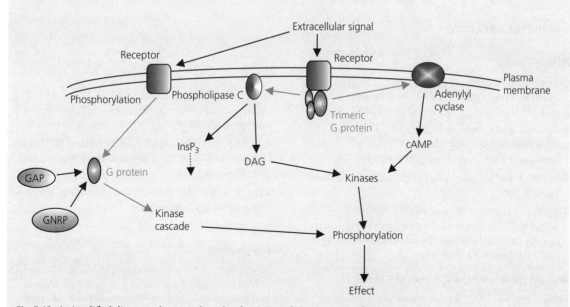

Fig. 5.13 A simplified diagram showing the role of cAMP and G proteins in the transduction of cellular signals.

- The main control of adenylyl cyclase activity is through an interaction with heterotrimeric G proteins, particularly G_s and G_i, which are stimulatory and inhibitory respectively.

- Heterotrimeric G proteins are constructed of an α subunit, which binds to GDP in its inactive state and GTP in its active state, along with a β/γ subunit complex.

- Although most of the signalling by heterotrimeric G proteins has been reported to be through the action of the α subunit, it has now been found that the β/γ complex has many signalling roles.

- Heterotrimeric G proteins are in fact a large family of signalling proteins and include members which interact with phosphodiesterases, G_t, and regulate phospholipase C, G_q.

- cGMP is produced by enzymes analogous to that producing cAMP. Guanylyl cyclase has two forms: soluble and membrane-bound.

- Cyclic nucleotide signalling is ended by the hydrolysis of the cyclic forms to the monophosphate forms by the enzymes phosphodiesterases.

- Eleven classes of phosphodiesterases have been reported in mammals, including ones which are controlled by Ca^{2+}/calmodulin, stimulated by cGMP, and inhibited by cGMP.

- cGMP-specific phosphodiesterases (type V) are the targets of drugs such as Viagra™.

- Monomeric G proteins are an extremely important group of signalling molecules which among their number include the oncogene products Ras and Rac as well as the protein synthesis elongation factor EF-Tu.

- Like the heterotrimeric forms, monomeric G proteins are active when bound to GTP, and intrinsic GTPase activity hydrolyses the GTP to GDP accompanied by a conformational change to the protein and its return to an inactive state.

- GTPase activity in monomeric proteins is enhanced by the protein GAP, while the GDP release is enhanced by the proteins called GNRPs, such as Sos.

- Monomeric G proteins have been shown to be instrumental in the transduction of the signal from many receptors, particularly those for growth factors, on the plasma membrane to kinase cascades such as the MAP kinases, and hence often on to alterations of activity in the nuclei of cells.

5.9 FURTHER READING

Adenylyl cyclase

Cooper, D.M.F., Mons, N., and Karpen, J.W. (1995) Adenylyl cyclases and the interaction between calcium and cAMP signalling. *Nature* **374**, 421–424.

Tang, W.-J. and Gilman, A.G. (1991) Type specific regulation of adenylyl cyclase by G protein beta/gamma subunits. *Science* **254**, 1500–1503.

Tang, W.-J. and Gilman, A.G. (1992) Adenylyl cyclases. *Cell* **70**, 869–872.

Zippin, J.H., Levin, L.R., and Buck, J. (2001) CO_2/HCO_3^- -responsive soluble adenylyl cyclase as a putative metabolic sensor. *Trends in Endocrinology and Metabolism* **12**, 366–370.

Adenylyl cyclase control and the role of G proteins

Birnbaumer, L. (1992) Receptor-to-effector signalling through G proteins: roles of βγ dimers as well as α subunits. *Cell* **71**, 1069–1072.

Hepler, J.R. and Gilman, A.G. (1992) G proteins. *Trends in Biochemical Sciences* **17**, 383–387.

Iñiguz-Lluhi, J., Kleuss, C., and Gilman A.G. (1993) The importance of G-protein β/γ subunits. *Trends in Cell Biology* **3**, 230–235.

Simon, M.I., Strathmann, M.P., and Gautam, N. (1991) Diversity of G proteins in signal transduction. *Science* **252**, 802–808.

Smrcka, A.V. (Ed.) (2004) *G Protein Signaling: Methods and Protocols*. Humana Press, Totowa.

Guanylyl cyclase

Ashman, D.F., Lipton, R., Melicow, M.M., and Price, T.D. (1963) Isolation of adenosine 3′,5′-monophosphate and guanosine 3′,5′-monophosphate from rat urine. *Biochemical and Biophysical Research Communication* **11**, 330–334.

Garbers, D.L. and Lowe, D.G. (1994) Gyanylyl cyclase receptors. *Journal of Biological Chemistry* **269**, 30741–30744.

Goraczniak, R.M., Duda, T., Sitaramayya, A., and Sharma, R.K. (1994) Structural and functional characterisation of the rod outer segment membrane guanylyl cyclase. *Biochemical Journal* **302**, 455–461.

Stone, J.R. and Marletta, M.A. (1993) Bovine lung contains multiple isoforms of soluble guanylate cyclase. *FASEB Journal* **7**, A1152.

Phosphodiesterase

Beavo, J. and Houslay, M.D. (eds) (1990) *Cyclic Nucleotide Phosphodiesterases: Structure, Regulation and Drug Design.* John Wiley & Sons, Chichester.

Gijsbers, R. (2003) *Domain Structure and Function of Nucleotide Pyrophosphatases/Phosphodiesterases (Npps).* Leuven University Press, Leuven.

Milatovich, A., Bolger, G., Michaeli, T., and Francke, U. (1994) Chromosome localizations of genes for five cAMP-specific phosphodiesterases in man and mouse. *Somatic Cell and Molecular Genetics* **20**, 75–86.

Soderling, S.H. and Beavo, J.A. (2000) Regulation of cAMP and cGMP signaling: new phosphodiesterases and new functions. *Current Opinion in Cell Biology* **12**, 174–179.

The GTPase superfamily

Bourne, H.R., Sanders, D.A., and McCormick, F. (1991) The GTPase superfamily: conserved structure and molecular mechanism. *Nature* **349**, 117–127.

Cullen, P.J., Hauan, J.J., Truong, O. Letcher, A.J., Jackson, T.R., Dawson, A.P., and Irvine, R.F. (1995) Identification of a specific Ins(1,3,4,5)P$_4$ binding protein as a member of the GAP1 family. *Nature* **376**, 527–530.

Fukuda, M. and Mikoshiba, K. (1996) Structure-function relationships of the Mouse Gap1^m: determination of the inositol 1,3,4,5-tetrakisphosphate binding domain. *Journal of Biological Chemistry* **271**, 18838–18842.

Lowy, D.R. and Willumsen, B.M. (1993) Function and regulation of Ras. *Annual Review of Biochemistry* **62**, 851–891.

Pai, E.F., Krengel, U., Petsko, G.A., Goody, R.S., Kabsch, W., and Wittinghofer, A. (1990) Refined crystal structure of the triphosphate conformation of H-Ras P21 at 1.35Å resolution: implications for the mechanism of GTP hydrolysis. *EMBO Journal* **9**, 2351–2359.

Schlessinger, J. (1993) How receptor tyrosine kinases activate Ras. *Trends in Biochemical Sciences* **18**, 273–275.

Schlichting, I., Almo, S.C., Rapp, G., Wilson, K., Petratos, K., Lentfer, A., Wittinghofer, A., Kabsch, W., Pai, E.F., Petsko, G.A., and Goody, R.S. (1990) Time resolved X-ray crystallographic study of the conformational change in Ha-Ras P21 protein on GTP hydrolysis. *Nature* **345**, 309–315.

Symons, M. and Settleman, J. (2000) Rho family GTPases: more than simple switches. *Trends in Cell Biology* **10**, 415–419.

Vojtek, A.B., Hollenberg, S.M., and Cooper, J.A. (1993) Mammalian Ras interacts directly with the serine threonine kinase Raf. *Cell* **74**, 205–214.

Other Ras-related proteins

Brown, H.A., Gutowski, S., Moomaw, C.R., Slaughter, C., and Sternweis, P.C. (1993) ADP-ribosylation factor, a small GTP-dependent regulatory protein, stimulates phospholipase D activity. *Cell* **75**, 1137–1144.

Inositol phosphate metabolism and roles of membrane lipids

Inositol lipids and compounds derived from these lipids are used by cells to transmit messages into their interior. One of the main pathways here can be controlled by G proteins (as discussed in **Chapter 5**), which transmit the signal from a cell-surface receptor and leads to the production of compounds which control phosphorylation events and calcium signalling. This leads to the role of calcium ions as discussed in the next chapter. As with other intracellular signals, mechanisms using lipids and their derivatives allow for amplification of the signal, and allow for divergence, potentially regulating more than one pathway. The inositol compounds used as signals can also be further modified, giving the potential for much more signalling. Hence, mechanisms discussed in this chapter are central in the control of many cellular events.

As with all signalling events, dysfunction of these pathways can also lead to disease, and such pathways are targets for manipulation by drugs.

6.1 Introduction

One of the key events in signal transduction in cells that takes place on the plasma membrane is the perception of the signal, facilitated by a receptor. But this is not the only role that the plasma membrane has in signalling. An important mechanism in transducing the signal into the cell involves the modification of some of the lipids which comprise the membrane. Such modifications might involve their phosphorylation, or their breakdown to yield further signalling molecules.

Between 2 and 8% of the lipids of eukaryotic membranes are lipids which contain inositol. The three main forms of these lipid structures are phosphatidylinositol (PtdIns; sometimes called PI), phosphatidylinositol 4-phosphate (PtdIns4P), and phosphatidylinositol 4,5-bisphosphate (PtdInsP$_2$; otherwise abbreviated to PIP$_2$), with the non-phosphorylated form, phosphatidylinositol, accounting for more than 80% of the total inositol lipid content. Although in relatively small concentrations, these lipids are nonetheless extremely important, not just as part of the membrane structure, but as molecules that can be used in signalling.

A major route in signalling can occur using inositol lipids, as key proteins on the membrane are responsible for cleavage of certain inositol lipid molecules into smaller, diffusible molecules which then convey the signal to other parts of the cell. Primarily, PtdInsP$_2$ is broken down by the enzyme phospholipase C (PLC) to give the primary products, inositol 1,4,5-trisphosphate (InsP$_3$) and diacylglycerol (DAG), both of which can diffuse and signal into the cell. This pathway, like others, gives a great deal of amplification to the signal, as one activated phospholipase C molecule will produce many InsP$_3$ and DAG molecules, but further to this, and importantly, the system leads to great divergence of the signal. InsP$_3$ is responsible for the release of Ca^{2+} ions from the intracellular stores, which will lead to the activation of the calcium arm of the signalling pathways, including turning on of calmodulin and its associated affects, as discussed in the next chapter, while DAG leads to the activation of PKC and the associated phosphorylation of a host of proteins along with modulation of their activity, as discussed in Chapter 4. A typical scheme is depicted in Fig. 6.1. Therefore, inositol phosphate metabolism can be seen as a keystone pathway linking events at the membrane with other transduction pathways deep within a cell.

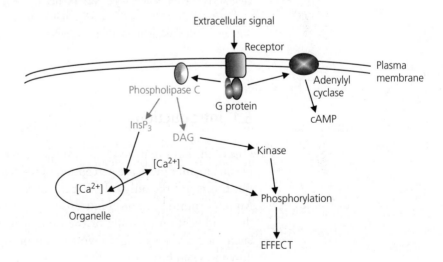

Fig. 6.1 The control and role of phospholipase C in inositol phosphate metabolism and its influence on kinase activity and intracellular calcium release. The role of phospholipase C is highlighted in blue.

Once this system was discovered, it was thought that this was probably the end to the story. It appeared to answer many questions. However, a plethora of reports since have indicated that it is just not that simple, perhaps as would be expected in cell signalling. For example, more than one type of membrane-associated lipid can be cleaved, while $InsP_3$ can lead to a host of secondary products, and gradually many of these have been assigned signalling roles within the cell.

At this point it is probably worth mentioning nomenclature. Here, inositol 1,4,5-trisphosphate is referred to as $InsP_3$, but many reports now find that a more full abbreviation (that is, $Ins(1,4,5)P_3$) is needed to clarify which trisphosphate is being referred to, and a similar story goes for other phosphorylated derivatives. For example, phosphatidylinositol bisphosphate ($PtdInsP_2$) could be phosphorylated at positions 3 and 4, or at positions 4 and 5. However, to try to simplify the situation, here I have used a minimalist approach and the abbreviation refers to the most important isomers recorded; that is, $InsP_3$ for inositol 1,4,5-trisphosphate, and $PtdInsP_2$ for phosphatidylinositol 4,5-bisphosphate, while other derivatives which have a less well-defined role within the cell are given a more full name, including the numbers, for clarity. It should also be noted that some researchers and journals use another type of nomenclature, the Chilton forms; that is, $InsP_3$ is known as IP_3 and $PtdInsP_2$ as PIP_2. Conventions have been published which give guidelines to which nomenclature to use, but many journals still accept publication of both forms, $InsP_3$ or IP_3.

> ■ The numbering for the inositols is derived from the positions of the carbon atoms around the inositol-ring to which the extra phosphates are attached, i.e. not the one used as part of the backbone of the structure. An example can be seen in Fig. 6.2. Thus, when $PtdIns(4,5)P_2$ is cleaved, the numbering of the derivative includes the phosphate on carbon number 1, and hence the product is a trisphosphate, $Ins(1,4,5)P_3$.

6.2 Events at the membrane

The main inositol-containing lipid cleaved in the membrane, which leads to the production of $InsP_3$ and DAG, is $PtdInsP_2$. The hydrophobic tails of the lipid comprise part of the inner leaflet of the membrane bilayer, while the inositol group and phosphates stick out to the cytoplasm of the cell (Fig. 6.2). Therefore, on cleavage, the lipid part—that is the DAG—is left in the membrane, while the inositol phosphate, which as we shall see needs to diffuse to the endoplasmic reticulum, is produced in the cytoplasm where it can have its action.

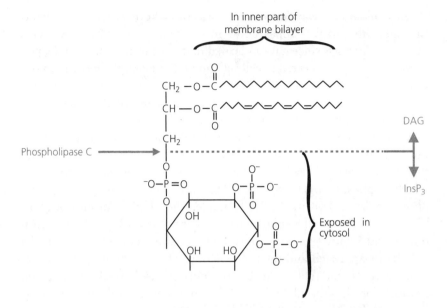

Fig. 6.2 The molecular structure of phosphatidylinositol 4,5-bisphosphate (PtdIns P_2) and its breakdown by phospholipase C to form inositol 1, 4, 5-trisphosphate (InsP$_3$) and DAG.

PtdInsP$_2$ is itself produced by the sequential phosphorylation of PtdIns. PtdIns is first phosphorylated to phosphatidylinositol 4-phosphate (PtdIns4P) by Phosphoinositide 4-kinase (PtdIns 4-kinase) and then further by the addition of a phosphoryl group to the 5 position of the inositol ring, catalysed by PtdIns4P 5-kinase, to produce PtdInsP$_2$. The former enzyme, PtdIns 4-kinase, has been purified from several sources, including liver and brain, and in most tissues has been found to be associated with the membranes. PtdIns4P 5-kinase on the other hand has been found to be in both the particulate and soluble fractions of cells. In erythrocytes, two immunologically distinct forms of the enzyme have been found, a 53-kDa form which is both soluble and membrane-associated and a second form which is only found in the membranes.

Therefore, it can be seen that PtdInsP$_2$ is extremely important for signalling in cell. However, as we shall see when discussing the inositol lipids and their derivatives, the story is never quite as simple as first thought. PtdInsP$_2$ is phosphorylated on positions 4 and 5 of the inositol ring, but several phosphatidylinositol derivatives that have also been phosphorylated on the 3 position of the inositol ring have been found to exist (Fig. 6.3). For example, phosphatidylinositol 3,4,5-trisphosphate (PtdIns(3,4,5)P$_3$). These were first identified in fibroblasts by Whitman and colleagues in 1988, and it was suggested that they were involved in the control of cell proliferation. However, their presence of compounds such as PtdIns(3,4,5)P$_3$ in cells which are differentiated and do not undergo proliferation suggests that other functions also need to be assigned to these lipids. More lately, such lipids have been shown to be

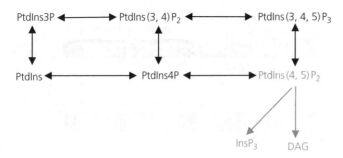

Fig. 6.3 The phosphorylated forms of phosphatidylinositol which may be formed and their interconversion. The substrate and products of phospholipase C activity are highlighted in blue.

involved in events that lead from the perception of insulin, for example, as discussed further in Chapter 9.

These inositol lipids are interconvertible by simple phosphorylation and dephosphorylation steps, as shown in Fig. 6.3, and this suggests that some of these lipids are themselves signalling molecules. For example, PtdIns(3,4,5)P_3, derived mainly from the phosphorylation of PtdInsP_2, may be involved in the activation of the small G protein Rac, and has been shown to activate a specific kinase known as 3-phosphoinositide-dependent kinase (PDK1; although there are other forms too). PDK1, as discussed for insulin signalling (Chapter 9), can activate PKB and possibly PKC for continuation of signal transduction.

The kinase involved in the production of these three phospho-derivatives of the inositol lipids, Phosphoinositide 3-kinase (PtdIns 3-kinase), is activated probably though several routes, depending on the type, including by tyrosine kinase activity such as receptor tyrosine kinases, by the action of monomeric G proteins such as Ras, or in the case of insulin signalling via the multiphosphorylated protein insulin receptor substrate 1 (IRS1). Other adaptor proteins might be instrumental in the control of PtdIns 3-kinase too. Research into the role of PtdIns 3-kinase has been greatly facilitated by the finding that it can be reasonably specifically inhibited by the fungal metabolite, wortmannin, as well as by other inhibitors, such as LY294002.

PtdIns 3-kinase is not a single entity, but a family of proteins, which can be grouped in classes: IA, IB, II, and III. They all contain a kinase domain as expected for their activity, and they all appear to contain a kinase accessory domain (PI3Ka domain; N-terminal to the kinase domain), and most contain towards the N-terminus a PKC homology domain 2 (C2 domain; class II has an extra C2 domain at the C-terminal end). Classes IA, IB, and II also contain Ras-binding sites, while class IB also has binding sites for βγ subunits from heterotrimeric G proteins. Therefore, with a close look at the domain structures here, the spectrum of interactions which may be used for the control of each class of PtdIns 3-kinase can be predicted (Fig. 6.4).

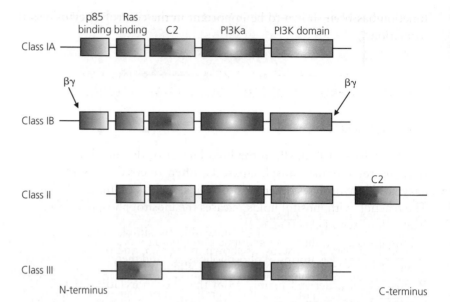

Fig. 6.4 The domain structures of PtdIns 3-kinase isoforms. C2, PKC homology domain 2; PI3Ka, PtdIns 3-kinase accessory domain; βγ, indicates binding site for β/γ subunits of trimeric G proteins.

As discussed in previous chapters (section 1.2 in particular), the reversal of signalling must take place, and the signalling through the PtdIns 3-kinase pathway is no exception. The PtdIns(3,4,5)P$_3$ produced has to be dephosphorylated to remove the signal, and this is catalysed by a phosphatase. The phosphatase that seems to be most involved here is the product encoded by the tumour-suppressor gene *PTEN*. The protein PTEN is expressed in all eukaryotic cells. The human PTEN enzyme structure shows that it has a phosphatase domain and C2 domain, although in yeast the C2 domain is missing. PTEN is not the only phosphatase that has this activity that is, the removal of phosphates from PtdIns(3,4,5)P$_3$; other phosphatases are also involved in the reversal of the phosphorylation of PtdIns(3,4,5)P$_3$, including ones called SHIP1 and SHIP2.

PTEN

The name PTEN comes from 'phosphatase and tensin homologue on chromosome 10', but it is also known as MMAC1 or TEP1.

The lipid PtdInsP$_2$, which is used as a starting point for making other phosphorylated forms, as well as being the main source of DAG and InsP$_3$, has been shown to have roles in its own right. For example, it has been found to be involved in interactions with proteins associated with the cytoskeleton, particularly with gelsolin and profilin, and such function has been shown to be important in the control of cytoskeletal formation.

6.3 The breakdown of the inositol phosphate lipids

Phospholipase C

The hydrolysis of PtdInsP$_2$ in the lipid bilayer of the membrane is catalysed by the enzyme phospholipase C, often referred to as PLC. This enzyme also hydrolyses other inositol lipids, such as PtdIns and PtdIns4P. The reaction with inositol lipids releases the inositol phosphates, such as InsP$_3$, along with DAG (Fig. 6.2).

Protein-based studies, and genomic and cloning work, has revealed the existence of many isoforms of phospholipase C, and therefore again we see that what was once thought of as a single protein entity is in fact a family of polypeptides, each with the same overall function, but with subtleties in their actions and control, which would account for their roles in signalling in different tissues. In mammals, for example, nine isoforms have been identified which can be classified into four main groups, α, β, δ and γ. At the N-terminal end of most, for example forms β, δ, and γ, there is a PH domain which probably aids in the enzymes' association with the lipids of the membrane. Next comes an EF-hand region (see Chapter 7; section 7.1), allowing calcium ion chelation and hence control by the intracellular calcium ion concentration. Then we find the catalytic core, which breaks down the lipids and creates the signalling molecules. At, or towards, the C-terminal end we see a C2 domain. The β class of phospholipases also have a G protein-interaction domain at their extreme C-terminal end, whereas the γ class have extra domains associated with protein–protein interactions, including two SH2 domains, an SH3 domain, and a split PH domain.

Therefore activation of PLCs can again be predicted from a study of the structures of the isoforms (as above with PtdIns 3-kinase). One of the major mechanisms for turning PLC on is through the interaction with components of the trimeric G proteins (which were discussed in more detail in Chapter 5; section 5.4). PLCβ1 is activated by the α subunit of the trimeric G protein G$_q$, that is G$_q\alpha$, (Fig. 6.5). As discussed above, the β class of phospholipases contain a G protein-interaction domain, and two sites, referred to as P and G, at the C-terminal end of PLCβ1 have been found to be important for this interaction. However, again as predicted from the structures, this G protein subunit has no stimulatory activity on any of the other forms of PLC. However, it should be noted that it is not only the α subunit of the trimeric G proteins that is important for the control of the β isoforms, as the β/γ subunit complex of the trimeric G protein family has be seen to activate the β2 and β3 isoforms

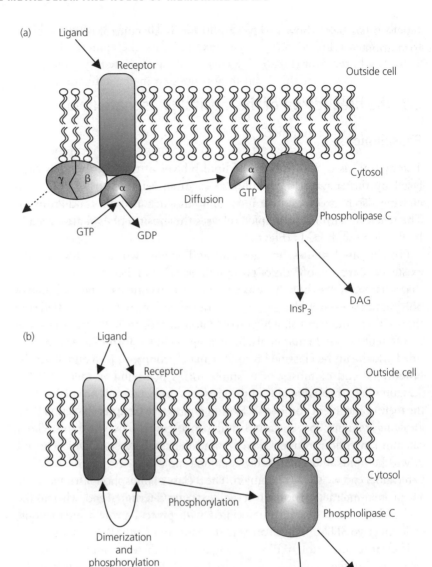

Fig. 6.5 The activation of phospholipase C can be by different mechansims depending on the isoform. (a) Activation may be through a receptor linked to a trimeric G protein, where the α subunit activates phospholipase, as with PLCβ1. Alternatively (b) activation may be through a tyrosine kinase-linked receptor, leading to phosphorylation of phospholipase C on tyrosine residues, as seen with PLCγ.

of phospholipase C in some cell types, again highlighting the role of more than the α subunits in the functioning of heterotrimeric G proteins.

Phosphorylation has also been seen to be crucial in the activation of other isoforms. PLCγ is phosphorylated on some tyrosine residues, usually three at positions 771, 783, and 1254. This may be catalysed by a tyrosine kinase-linked receptor, for example, by the epidermal growth factor (EGF) receptor (Fig. 6.5b). Treatments with protein tyrosine phosphatases suggest that this is an important step in the activation of this enzyme. When the cell is in the unstimulated state PLCγ is found mainly in the cytosol. However, a sequence of events similar to that seen with the activation of the monomeric G proteins (Chapter 5; section 5.7) may occur. On binding to its ligand, the tyrosine kinase-linked receptor firstly has to dimerize and phosphorylate itself. The phosphoryl groups added to the receptor itself create binding sites for the SH2 domains of PLCγ and so allow the translocation of PLCγ to the membrane and its association with the receptor, and subsequent phosphorylation, and activation, of the PLC enzyme. Therefore, such a mechanism not only activates this enzyme but brings the PLCγ to the membrane where its substrate resides. Hence it is only active when and where needed.

Interestingly, phosphorylation on a serine residue by both PKC and cAMP-dependent protein kinase (PKA) has been reported to be inhibitory, and therefore may reflect a negative control on the activity of PLC and a reduction in the formation subsequent intracellular signals.

It has also been reported that Ca^{2+} may well also be involved in PLC control. PLCδ contains one EF-hand motif which will confer Ca^{2+} binding potential to this protein and therefore PLCδ may represent a Ca^{2+}-controlled form of the PLC family. Therefore, taken together, it can be seen that various PLC isoforms can be controlled in a variety of ways, depending on the isoform. Ca^{2+}, G proteins, and kinases are all involved in controlling $InsP_3$ and DAG formation.

As mentioned earlier, inositol-containing lipids have been found to be associated with the cytoskeleton and it has been shown that if the lipid is associated with profilin, an actin-binding protein, then the activity of PLCγ is reduced severely, suggesting that this too might be a regulatory mechanism.

A PLC has also been found that hydrolyses phosphatidylcholine (PC). This enzyme is a heterodimer of 69 and 55 kDa and preferentially hydrolyses PC rather than $PtdInsP_2$ as a substrate.

6.4 Inositol 1,4,5-trisphosphate and its fate

Although originally the only inositol phosphate formed from the hydrolysis of the inositol lipids to be assigned a role in cell signalling was $InsP_3$, it is clear that the hydrolysis of the membrane lipids is only the first step in a very convoluted and complicated pathway. $PtdInsP_2$ is not the only

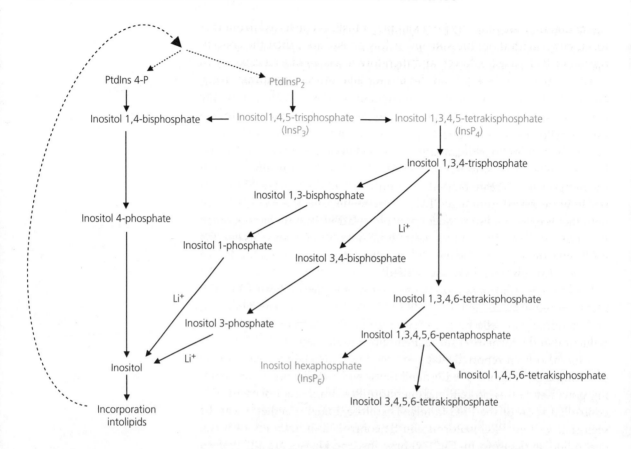

Fig. 6.6 Some of the most important inositol phosphate metabolism pathways, with the influence of lithium ions included. The most highly studied compounds are highlighted in blue.

lipid to be hydrolysed. The cleavage of PtdIns will lead to the formation of Ins1P, while the hydrolysis of PtdIns4P leads to the formation of Ins(1,4)P$_2$, with InsP$_3$ itself arising from the hydrolysis of PtdInsP$_2$. So, it can be seen that different lipids will give rise to different inositol phosphate products, each with a potential signalling role. Furthermore, as alluded to, inositol has the capacity to contain a variable number of phosphate groups, ranging from none, that is inositol, right through the spectrum of compounds up to the inositol hexaphosphate form, InsP$_6$(or IP$_6$). InsP$_3$ is poised in the middle of this jungle of compounds and therefore can lead to the formation of lower inositol phosphates, that is the inositol compounds containing one or two phosphate groups, or alternatively InsP$_3$ can lead to the formation of the higher phosphate forms, that is the four-, five-, or six-phosphate-containing compounds. Some, but not all, of the compounds that can be formed, along with their interconversions, are shown in Fig. 6.6. The addition of phosphate groups to the inositol ring is catalysed, as expected, by kinase enzymes, whereas the removal is catalysed by phosphatases.

So what function does this interconversion of the phosphate forms serve? Firstly, it deactivates inositol phosphates; that is, turning off the signal that they are relaying. For example, if InsP$_3$ is converted to another form, either with more phosphates or less phosphates, then the concentration of InsP$_3$ itself is lower, and therefore it has a reduced capacity to cause the release of Ca^{2+} from the intracellular stores (its normal role), and so the end result is in effect a turning off of that signal. Second, the new inositol phosphate formed might itself have a signalling role. InsP$_3$ is not the only inositol phosphate to have been shown to be a signal. More and more of these inositol phosphates are being assigned roles in the control of cellular functions, so the removal of one form, reducing the signal associated with the rise in its concentration, may well be leading to the rise in the concentration of another form and the formation of a new signal. It is beyond the scope of this book to attempt to look at all the isoforms involved and their postulated roles but some of the more prominent ones and the enzymes involved will be discussed.

As mentioned, InsP$_3$ lies in the middle of the inositol phosphate phosphorylation states, where phosphates can be added or removed to form other compounds. Addition of a phosphate can be carried out by InsP$_3$ 3-kinase, yielding a new inositol, inositol 1,3,4,5-tetrakisphosphate, commonly known as InsP$_4$ (or IP$_4$). The kinase enzyme involved is a soluble protein which like most kinases uses ATP as a substrate and requires Mg^{2+}. The K_m value for its substrate is lower than that of the phosphatases that can remove phosphate groups from InsP$_3$ and hence this might be the favoured reaction; that is, InsP$_4$ may be more likely to be made than InsP$_2$. The kinase contains two 53-kDa catalytic subunits and also contains calmodulin, allowing stimulation via the Ca^{2+} pathway, while additional regulation of this kinase is almost certainly via other kinases and phosphorylation, by either PKC or PKA.

As for the role of the InsP$_4$ that is formed from the phosphorylation of InsP$_3$, some debate has occurred. It is thought that it is involved, like InsP$_3$, in the immobilization of Ca^{2+} from internal stores, such as is seen with histamine receptor stimulation, but some papers refute the role of InsP$_4$ in Ca^{2+} release here. Some studies suggest that InsP$_4$ stimulates an ATP-independent uptake of Ca^{2+} into the sarcoplasmic reticulum, whereas other suggest that it may be involved in the control of Ca^{2+} back into the cell through the plasma membrane. Certainly proteins that are capable of binding InsP$_4$ have been reported. Putative plasma membrane-associated receptors for InsP$_4$ in human platelets have been reported. One protein isolated from pig cerebellar tissue, which has a molecular weight of 42 kDa, binds InsP$_4$ 100 times better than InsP$_3$. Other groups have reported two InsP$_4$-binding sites associated with the nuclear envelope. One of these is in the outer nuclear membrane, and is a 74-kDa protein that binds to InsP$_4$ with relatively high affinity and is involved in the uptake of nuclear Ca^{2+}. The other binding protein is on the inner membrane and has much lower InsP$_4$-binding affinity. Interestingly, binding of

$InsP_4$ and $InsP_3$ to putative receptor sites in rat cortical membranes appears to be enhanced by peptides derived from β-amyloid protein, which is a protein associated with Alzheimer's disease.

One of the key proteins seen to be able to bind to $InsP_4$ is a protein with GTPase-activating protein (GAP) activity which was discussed in Chapter 5 (section 5.7). This protein, GAP1[IP4BP], a member of the GAP1 family, affects the activity of monomeric G proteins and therefore $InsP_4$ may not only be involved in the control of intracellular Ca^{2+} but can control G protein function. Often, such G proteins are involved in the pathways which control gene expression, so $InsP_4$ may be involved in long-term as well as short-term effects in the cell.

As well as the tetrakis forms of the inositol phosphates (that is, containing four phosphate groups), further phosphate groups can be added to create the penta- and hexaphosphate derivatives. There are theoretically six possible forms of inositol pentaphosphate, but in animals the most favoured form is the one referred to as inositol 1,3,4,5,6-pentaphosphate. Such compounds are also found in plants and in *Dictyostelium*, suggesting that inositol pentaphosphate too might have a defined role. Inositol pentaphosphates, at least in animals, appear to be produced by the route outlined in Fig. 6.6. that is, $InsP_3$ is converted to inositol 1,3,4,5-tetrakisphosphate which, by the removal of a phosphate from position five, creates the trisphosphate, inositol 1,3,4-trisphosphate. Addition of a phosphate to position six will lead to the production of inositol 1,3,4,6-tetrakisphosphate and a further addition of a phosphate to position 5 will produce inositol 1,3,4,5,6-pentaphosphate. Once produced, this new inositol compound can lead to the production of lower inositols, such as inositol 3,4,5,6-tetrakisphosphate or inositol 1,4,5,6-tetrakisphosphate, or, by the further addition of a phosphate group, can lead to the production of $InsP_6$. Inositol 3,4,5,6-tetrakisphosphate has been shown to accumulate in a dose-response-related manner but its exact role remains to be determined, although it has been shown to selectively block epithelial calcium-activated chloride channels. In the meantime, such inositol compounds which have no defined function have been dubbed orphan signals.

Upon stimulation of some cells, for example the premyeloid cell line HL-60, the concentrations of the inositol penta- and hexaphosphates rise rapidly, certainly within minutes, although in other systems the peak of accumulation of these compounds may take hours if not days. However, the exact function of these higher inositol phosphates is unclear and it has been proposed that the pentaphosphate form may just be a precursor of the four-phosphate derivative inositol 3,4,5,6-tetrakisphosphate. It has even been proposed that the penta and hexa forms might have an extracellular role. However, using cerebellar membranes, an $InsP_6$-binding protein has been found, but a full understanding of the functions of such proteins has yet to be uncovered.

As well as the addition of phosphates to the inositol ring, removal of phosphates is also crucially important, not only to create even more derivatives of the inositol phosphates but also to liberate free inositol which will be used to make new inositol lipids. The removal of the phosphate from the 5 position of several inositol phosphates is carried out by inositol polyphosphate 5-phosphatase. This enzyme can use $InsP_3$ as a substrate, so creating inositol 1,4-bisphosphate. Inositol 1,4-bisphosphate is also formed by the hydrolysis of the lipid PtdIns4P. Removal of the phosphate at position 5 from $InsP_3$ will remove $InsP_3$ from the signalling pathway and so effectively turn off its message. However, this enzyme can also use inositol 1,3,4,5-tetrakisphosphate and cyclic inositol 1:2,4,5-trisphosphate as substrates. The phosphatase enzyme exists in many isoforms which have been found located both in the soluble and particulate fractions of cells. In human platelets, for example, two immunologically distinct forms have been isolated, a 45-kDa Mg^{2+}-requiring enzyme (type 1) and a 75-kDa enzyme (type II). The type I enzyme can be phosphorylated by PKC with a concomitant increase in its activity, suggesting that PKC may be instrumental in the reduction of $InsP_3$ levels in cells, and be a trigger for the termination of the $InsP_3$ signal. In other tissues other isoforms have been isolated, with diverse molecular weights, and in some cases the substrate specificities vary. The existence of a variety of forms of the phosphatases, and the potential reactions that such enzymes can catalyse, suggests that they are not there merely to remove $InsP_3$ from the cell, and suggests that their products have defined roles in their own right.

Removal of the phosphate at position 5 from inositol 1,3,4,5-tetrakisphosphate leads to the production of another trisphosphate; that is inositol 1,3,4-trisphosphate. This molecule itself can undergo dephosphorylation to produce inositol 3,4-bisphosphate. The enzyme involved here is inositol polyphosphate 1-phosphatase. Its only other substrate recorded is another bisphosphate, inositol 1,4-bisphosphate, and here inositol 4-phosphate is formed. The phosphatase enzyme is monomeric with a molecular weight of approximately 44 kDa. Again there is a requirement for Mg^{2+} ions, but the phosphatase is inhibited by the presence of Ca^{2+}. Of particular significance here is the fact that this reaction is also inhibited by lithium ions. Lithium ions have been used for years as a treatment for manic depression, and experimentally in the laboratory lithium ions have been used to stop inositol metabolism. They inhibit the activity of this enzyme and another phosphatase, inositol monophosphatase, and hence prevent the recycling of inositol, ultimately preventing its reincorporation back into PtdIns lipids. Lithium ion concentrations found in patients undergoing the treatment are consistent with the K_i of the inhibition of these enzymes by these ions, adding weight to the hypothesis that it is through this action that lithium is having its therapeutic affect. However, lithium ions have effects on other signalling systems too, so care needs to be exercised here in the exact interpretation of data obtained following lithium ion addition.

The recycling of the inositol may also proceed via a different route. As well as the conversion of inositol 1,3,4-trisphosphate to the bisphosphate, inositol 3,4-bisphosphate, an enzyme called inositol polyphosphate 4-phosphatase can convert the 1,3,4-trisphosphate form to a different bisphosphate, inositol 1,3-bisphosphate. It is also responsible for the conversion of inositol 3,4-bisphosphate to inositol 3-phosphate. This relatively large enzyme, of molecular weight 110 kDa, is monomeric and does not have a requirement for metal ions.

Inositol 1,3-bisphosphate formed is converted to inositol 1-phosphate by the enzyme inositol polyphosphate 3-phosphatase. This phosphatase enzyme exists in more than one isoform; for example, there are two in rat brain. Data from electrophoresis studies suggested that both isoforms exist as dimers. Type I appears as a dimer of 65-kDa subunits while the type II is a heterodimer of 65 and 78-kDa, with the 65-kDa subunits in the two cases probably being the same. The 78-kDa subunit of the type II isoform probably has a regulatory role as it seems to decrease the efficiency of catalysis. Interestingly, the substrate specificity of this enzyme is not restricted to the soluble forms of the inositols, and it has even been found that this enzyme can catalyse the removal of a phosphate group from the lipid PtdIns3P to produce PtdIns. Therefore this enzyme has the potential to be involved in two very disparate arms of the inositol signalling pathways.

Once the inositol phosphate has undergone dephosphorylation by the above array of enzymes and has been reduced to the monophosphate form, the final dephosphorylation can be carried out by the enzyme inositol monophosphatase. This monophosphatase enzyme will remove phosphates from all the positions around the inositol ring except those attached at the 2 position. Again Mg^{2+} is required for activity. The enzyme has an apparent molecular weight of 55 kDa and exists as a dimer of identical subunits. This monophosphatase enzyme is the second one in the inositol phosphate pathway which is inhibited by lithium ions, and here again the lack of production of free, non-phosphorylated inositol will prevent its reincorporation back into phosphatidylinositol and hence prevent further production of InsP$_3$ and continual cycling, and therefore signalling.

With the complete removal of all the phosphate groups, the resultant inositol is reused in the formation of PtdIns lipids in the endoplasmic reticulum, which will then be taken by vesicular transport to, and reincorporated back into, the plasma membrane, ready for another round of signalling. Such reincorporation into the plasma membrane will of course also help to restore the integrity of the membrane.

As well as the normal phosphorylated forms of the inositols described above, and shown in Fig. 6.6, there are also cyclic forms, as shown in

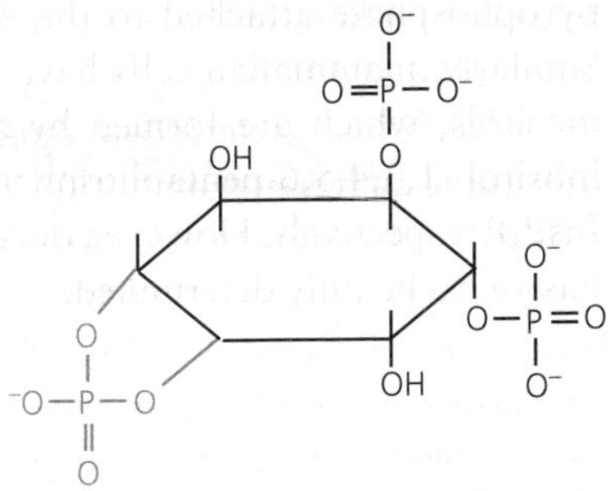

Fig. 6.7 The molecular structure of a cyclic inositol compound.

Fig. 6.7. Here, one of the phosphates bridges back across to a second carbon in the inositol ring. If some cells are stimulated for long periods of time the cyclic forms of the inositol phosphates are found to increase. Like their non-cyclic counterparts, they are produced by the action of phospholipase C but they are in general not substrates for the enzymes that add and remove phosphates from the non-cyclic forms of inositol phosphates. In the normal phospholipase C reaction, the hydrolysis of the bond between the phosphate of the inositol ring and the glycerol backbone of the lipid involves the supply of a hydroxyl group from water. However, this hydroxyl group can also be supplied from the inositol ring itself, and if this happens then the phosphate is cyclized (Fig. 6.7).

The enzyme that breaks the cyclic phosphate bonds is inositol 1:2-phosphate 2-phosphohydrolase or cyclic hydrolase. However, the only substrate for this enzyme appears to be cyclic inositol 1:2-phosphate. Therefore, presumably all of cyclic inositol phosphates are finally metabolized through this route and hence their concentrations in the cell are probably controlled by this enzyme, along with the rate of their production by PLC. The activity of cyclic hydrolase is inhibited by the metal ion Zn^{2+} but activated by Mn^{2+}. Inhibition also occurs by inositol 2-phosphate, suggesting that the levels of cyclic inositols can be controlled by the non-cyclic forms. Interestingly, when cyclic hydrolase was isolated from human placenta it was found to be the same as another protein, lipocortin III. This protein, belonging to a group of eight related polypeptides, was known to be able to bind lipids and also calcium ions. By the study of cDNAs and the use of overexpression techniques, it has been shown that the activity of cyclic hydrolase is correlated to the levels of proliferation of cell cultures and it was suggested that this enzyme is antiproliferative and could be an example of an anti-oncogene product, but cyclic hydrolase was not shown to prevent transformation of cells.

Other forms of inositol phosphates are the inositol pyrophosphates, where two phosphates are joined together and attached to the inositol ring. Such compounds have been mainly found in *Dictyostelium*. The pyrophosphate groups are usually at the 1 or 3 positions of the ring but pyrophosphate attached to the 4 and 6 positions have also been seen. Similarly, mammalian cells have been shown to contain pyrophosphate inositols, which are formed by an ATP-dependent phosphorylation of inositol 1,3,4,5,6-pentaphosphate and InsP6, designated $InsP_5P$ and $InsP_6P$ respectively. However, once again, the significance of these findings has yet to be fully determined.

6.5 The role of diacylglycerol

Any discussion of the role of DAG must emphasize that DAG is in fact not a single chemical but a family of related compounds, the structures of which are determined by the acyl groups which were present in the original lipid that was hydrolysed by the phospholipase (refer to Fig. 6.2). Therefore, it is likely that different DAGs will have subtly different roles in the cell. The main role of DAG has always been seen as an activator of PKC and there is clear evidence that activation of the phospholipase C leads to production of InsP3 and DAG and that this is accompanied by a rise in PKC activity. Such increases in PKC activity can also be induced by the use of DAG emulators such as phorbol esters.

Phorbol esters

Phorbol esters are tumour-promoting compounds that are used for the experimental activation of PKC. The most commonly used phorbol esters is phorbol 12-myristate 13-acetate (PMA), otherwise known as 12-O-tetradecanoyl-phorbol-12-acetate (TPA). Like many compounds used in biochemistry, for one's personal safety care needs to be taken during its use.

However, DAG may be further metabolized to other compounds, which themselves may have a signalling role. One such compound is phosphatidic acid (PA). PA has been seen to stimulate inositol 4,5-bisphosphate formation, activate phospholipase C, and act as a cell mitogen. PA production involves the phosphorylation of DAG by diacylglycerol kinase (DGK). Isoforms of this kinase have been found in both the membrane and cytoplasmic fractions of mammalian cells, with molecular weights of reported enzymes varying enormously. For example, the enzyme from porcine brain tissues appears to be 80 kDa, those from porcine thymus tissue are 80 and 150 kDa, whereas human platelets seem to have isoforms at 58, 75, and 152 kDa. Interestingly, when two of the isoforms which have a mass of approximately 80 kDa were cloned they were found to contain protein-folding regions known as zinc finger domains, suggesting a DNA-binding ability and possible control of gene expression. It has also been seen that receptor activation can lead to regulation of DGK activity and some forms of DGK are phosphorylated by PKC or PKA.

The breakdown of DAGs, and hence the termination of the signal they carry, is an open question, with the exact route for their metabolism depending on the acyl groups involved. However, like all signals their removal from transduction pathways is important, otherwise the signals would perpetuate beyond the period needed, with potentially poor consequences.

6.6 Inositol phosphate metabolism at the nucleus

When the inositol pathways were first unravelled, metabolism of inositol lipids was thought of as an event of the plasma membrane. However, the same biochemical pathways may also take place at the nucleus; that is, $PtdInsP_2$, and possibility PtdIns4P, are broken down, yielding $InsP_3$ and DAG, which can then carry the signal to their relevant effectors. Interestingly, many of the studies involving the nucleus have been carried out following the removal of the nuclear membranes, suggesting that it is the lipids and enzymes that are associated with the skeletal structures within the nucleus which are involved here and not the lipids that constitute the nuclear membranes. This has analogies to the association of the inositol lipids with the cytoplasmic cytoskeleton. Activation of this nuclear pathway is probably involved in the control of the cell cycle, with a likely involvement of PKC activated by the released DAG. DNA synthesis may also be under the control of inositol derivatives such as PtdIns4P or inositol 1,4-bisphosphate. Clearly, such involvement of inositol metabolism will be of interest to many cell biologists.

6.7 Other lipids involved in signalling

Most of the work which studies the involvement of lipids and lipid-derived products in cell signalling has over the years been centred on the role of the inositols, especially as it can be seen from the discussion above that there is such a plethora of these compounds, and no doubt a plethora of effects and roles. However, it is not just these inositol-based lipids that are active in cell signalling. Here, the roles of some of the others are briefly discussed.

Phosphatidylcholine and arachidonic acid metabolism

Lipids other than phosphatidylinositols are used for signalling from the plasma membrane to the inside of a cell. For example, it is well documented that phosphatidylcholine (PC) is also the precursor of signalling molecules. PC can be hydrolysed by several phospholipase enzymes, including phospholipase A2 (PLA$_2$), phospholipase C (PLC), and phospholipase D (PLD, Fig. 6.8).

As a general rule, agonists that lead to the increase in hydrolysis of PC also cause an increase in the breakdown of PtdIns in the same cell. This can lead to the rise in DAG being biphasic. Firstly, the breakdown of PtdIns leads to a rapid increase in the concentration of DAG, but slightly later the hydrolysis of PC leads to a more prolonged accumulation of DAG, which is accompanied by production of choline and phosphorylcholine.

Importantly however, PC, along with the breakdown of PtdIns and phosphatidylethanolamine (PE), is the major source of intracellular arachidonic acid (AA), which itself has been implicated as a major signalling molecule in its own right, as well as leading to the production of eicosanoids. As AA is most commonly found in the second (or middle) position of the glycerol backbone of the phospholipid, its release is catalysed by the enzyme PLA$_2$ (Fig. 6.9). PLA$_2$ will also hydrolyse PtdIns and PE at the second carbon of the glycerol backbone, and although arachidonic acid is the usual fatty acid there, PLA$_2$ is not specific for this substrate.

PLA$_2$ is commonly found associated with the plasma membrane of cells, but it can also reside in the cytoplasm, or even be associated with intracellular membranes. Purification has shown that PLA$_2$ enzymes

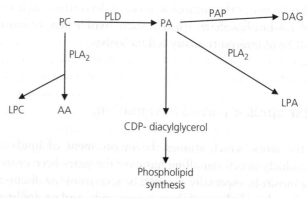

Fig. 6.8 The breakdown of phosphatidylcholine (PC) by phospholipase A$_2$ (PLA$_2$) and phospholipase D (PLD). AA, arachidonic acid; DAG, diacylglycerol, LPA, lysophosphatidic acid; LPC, lysophosphatidylcholine; PA, phosphatidic acid; PAP, phosphatidate phosphohydrolase.

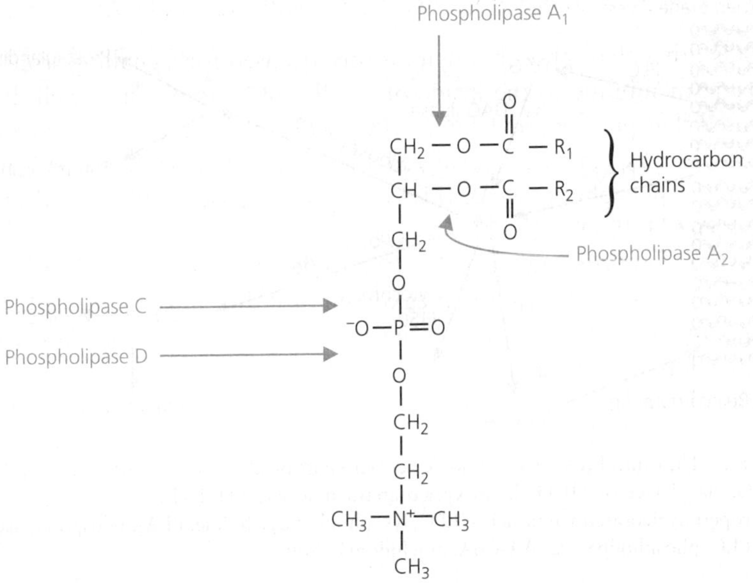

Phospholipase A_1

$$CH_2-O-\overset{\overset{O}{\|}}{C}-R_1$$
$$CH\ -O-\overset{\underset{\|}{O}}{C}-R_2$$
$$CH_2$$

Hydrocarbon chains

Phospholipase A_2

Phospholipase C

Phospholipase D

$$-O-\overset{O}{\underset{O}{P}}=O$$
$$CH_2$$
$$CH_2$$
$$CH_3-\overset{+}{N}-CH_3$$
$$CH_3$$

Fig. 6.9 The molecular structure of phosphatidylcholine and its breakdown by phospholipase A_2, phospholipase A_1, phospholipase D and phospholipase C. R denotes the fatty acid chains where R_2 is likely to be arachidonic acid.

are usually approximately 80–100 kDa in size and show activation by Ca^{2+}. The enzyme contains in its N-terminal region a Ca^{2+}-dependent phospholipid-binding sequence which mediates its interaction with the membrane and activation in this way is associated with a concomitant translocation of the enzyme to the membrane, and therefore it is activated and moved at the same time, ensuring that it is only active when and where needed. This type of activation/translocation is not uncommon and has been discussed above for other enzymes involved in cell signalling, such as PLCγ.

The activation of PLA_2 activity is certain to be complex. Besides its activation by Ca^{2+}, it has also been reported that PLA_2 can be activated by phosphorylation by MAP kinases as well as by specific isoforms of PKC. PKC can also lead to the activation of the MAP kinases which may subsequently cause the phosphorylation and activation of PLA_2. As well as phosphorylation of PLA_2, it is thought that G proteins also have a role in PLA_2 regulation. Although effects have been seen on the addition of GTPγ-S, a non-hydrolysable form of GTP used to lock G proteins in their active state, a direct interaction between a G protein and the PLA_2 enzyme has to be demonstrated and it is possible that the G proteins are themselves having an indirect affect through a kinase.

So what do the breakdown products created by PLA_2 actually do? AA, produced by hydrolysis of PC, is a 20-carbon unsaturated fatty acid containing four double bonds. AA can act as a signalling molecule, but it can also lead to the production of prostaglandins, thromboxanes, leukotrienes, and other eicosanoids. Some of the enzymes which act on

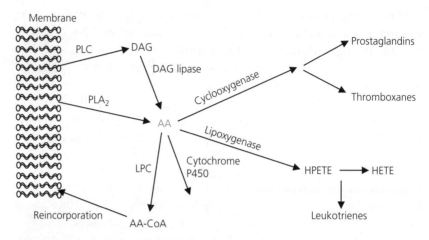

Fig. 6.10 The central role of arachidonic acid (AA): its production and further metabolism. DAG, diacylglycerol; HETE, hydroxyeicosatetraenoic acids; HPETE, hydroperoxyeicosatetraenoic acids; LPC, lysophophatidylcholine; PLA$_2$, phospholipase A$_2$; PLC, phospholipase C; AA-CoA, arachidonyl-CoA.

arachidonic acid or are involved in its further metabolism include cyclooxygenase, cytochrome P450, and lipoxygenases (Fig. 6.10). Cyclooxygenase, otherwise known as prostaglandin G/H synthase, or PGHS, leads to the production of prostaglandins G$_2$ and H$_2$ (PGG$_2$ and PGH$_2$) which further leads to the formation of other prostaglandins, prostacyclins, and thromboxanes. Cytochrome P450 leads to the production of epoxyeicosatrienoic acids (EET) which can be acted on by epoxide hydrolase to release diols, while lipoxygenases convert AA to hydroperoxyeicosatetraenoic acids (HPETE) leading to the formation of hydroxyeicosatetraenoic acids (HETE). Alternatively HPETE can lead to the production of leukotrienes and epoxyhydroxides. Added confusion arises when it is realized that some of these enzymes, for example lipoxygenases, can utilize other polyunsaturated fatty acids as well as arachidonic acid, for example linoleic acid, released by PLA$_2$, and this may also be leading to the production of even more molecules with signalling potential.

AA can also arise from the hydrolysis of DAG via the action of DAG lipase. DAG, as discussed, comes from the hydrolysis of phospholipids by PLC, which normally would be used in the activation of PKC. However, it can through DAG lipase also lead to AA and its further metabolism.

■ The enzyme system cytochrome P450, although discussed here as having a signalling role, is more usually associated with the removal of xenobiotics (foreign molecules) such as toxins from an organism.

Many of the products of the metabolism of AA have regulatory roles within the cell. Ion channels, Na^+/K^+-ATPase activity, and cell proliferation are among a whole host of functions under such control. Receptors for some of the eicosanoid family have recently been cloned. These, not surprisingly, contain seven putative membrane-spanning regions and are thought to act through heterotrimeric G proteins in the cell. As well as further products of AA, AA itself can lead to the activation of some isoforms of PKC and to the activation of PLD. No doubt, more and more regulatory functions for AA and the eicosanoids will come to light.

AA can also be used in the reformation of membrane lipids. This is through a route which includes firstly its conversion to arachidonoyl-CoA by the enzyme arachidonoyl-CoA synthetase and then esterification is catalysed by the enzyme arachidonoyl-lysophospholipid transferase. This route will obviously reduce the AA concentration in the cell available for the production of the host of signalling molecules previously mentioned, as well as reduce the signal which arises directly from the presence of AA.

The formation of AA is also modulated by the presence of a 37-kDa protein called lipocortin, otherwise known as annexin, possibly through the inhibition of PLA_2. Its identity was first discovered by studies of the suppression of eicosanoid production by glucocorticoid treatment. However, not surprisingly, lipocortin is in fact a family of at least 12 proteins. Different members of the family are characterized by their variable N-terminal regions, but all lipocortins contain between four and eight highly conserved repeating domains. These domains are capable of binding to phospholipids in the presence of Ca^{2+} ions. However, studies including the use of peptides designed from the known sequences of lipocortins have revealed that the proteins may have far wider roles in cellular regulation, and their presence has been implicated in the control of neutrophil migration, differentiation and growth of cells, and in the protection of neuronal tissues.

Hydrolysis of PC by PLA_2 not only produces AA but leaves behind a by product, lysophosphatidylcholine (LPC). This too is thought to have a signalling role, and although its exact action has not been fully defined it probably acts through PKC.

Some PLA_2 enzymes have also been found to be extracellular, for example in snake venom, synovial fluid, and pancreatic secretions. These enzymes are of very low molecular weight, less than 20 kDa, and fall into one of two groups, PLA_2-I or PLA_2-II. Although generally not thought to be involved in cell signalling cascades, their products might have a signalling role on cells in the vicinity of their production.

Phospholipase D

Another enzyme which has an important role in the hydrolysis of PC is phospholipase D (PLD; Fig. 6.8). Like all phospholipase enzymes it has a specificity for its cleavage site within the lipid, and here this cleavage of the phospholipid yields phosphatidic acid (PA) as a product, and it is thought that a significant amount of DAG in a cell arises from the further break-down of PA by the enzyme phosphatidate phosphohydrolase (PAP). This is contrary to the production of PA from DAG by DAG kinase as discussed earlier. The breakdown of PA to DAG, however, has been shown to be increased by several agonists in a wide range of cells. PAP exists in at least two forms, which differ in their subcellular locations as well as their enzymatic properties. PLD has a molecular weight of approximately 200 kDa and appears to exist in several isoforms with varying subcellular locations.

Early work was based on the observation that PLD activity was modu-lated by the addition of stable, non-hydrolysable GTP analogues such as GTPγ-S. The factor which was shown to be the activator of PLD activity is a low-molecular-weight GTP-binding protein known as ADP-ribosylation factor (ARF; also see Chapter 5, section 5.7). This protein was originally found to be required for the ADP-ribosylation of the G protein subunit $G_s\alpha$ in the presence of cholera toxin. ARF has also been shown to have a role in the control of vesicle protein trafficking. Other research groups have also suggested involvement of a monomeric G protein, Rho, in reg-ulation of PLD, while further control of the activities of PLD in the hydrolysis of the PC also comes from PKC. As well as serine/threonine phosphorylation, tyrosine phosphorylation has been shown to be involved here. Interestingly, hydrogen peroxide causes an increase in PLD activity, and this effect could be mediated by a tyrosine phosphorylation step, but evidence of a direct tyrosine phosphorylation of the PLD polypeptide leaves it open to debate what the exact mechanism of activation is.

The main role of PLD catalysed breakdown of lipids may be to result in the increase of DAG concentrations in the cell, but the product PA itself also may have signalling properties. It is thought that PA may have regu-latory roles on other signalling molecules such as adenylyl cyclase (with the production of cAMP), kinases, and phosphatases. PA has also been implicated in the control of superoxide release of neutrophils, which can signal itself or lead to the generation of other reactive oxygen species (ROS) signals, and in thrombin-induced actin polymerization of fibroblasts. Furthermore, the action of PLA_2 on PA leads to the production of LPA, itself an extracellular signal. Receptors for LPA have been found on some cells by the use of photoreactive analogues of LPA, although once again much work is needed to elucidate fully the exact mechanisms involved.

As mentioned above the concentration of DAG in the nucleus has been shown to be of crucial importance for the regulation of many nuclear events, and a substantial proportion of this DAG may be a result of the breakdown of PC.

An important point that should be borne in mind when studying the hydrolysis of membrane lipids, be it PtdIns, PC, or PE, is that the lipids are being removed from the membrane and therefore the structure and integrity of that membrane may be altered. Such changes in the membrane may well be acting as a control mechanism directly, besides the obvious effects of the products being made. The alteration of lipid-membrane fluidity caused by the removal of certain lipids may be involved in the regulation of secretion, neurotransmission, and the control of the activity of some membrane proteins. For example, detergents have been shown to activate the superoxide ion release of neutrophils, the enzyme involved probably responding to the alteration of the lipid structures of the membrane.

Sphingolipid pathways

Complement

The complement cascade is a series of proteins involved in the immune system of animals.

Sphingolipid pathways, which are further signalling pathways involving the use of lipids, independent of the PLC, PtdIns, and PC pathways, have also come to light in recent years. Sphingosine was reported to inhibit PKC activity, and this has led to a flurry of research which has revealed what is referred to as the sphingomyelin cycle. Essentially, extracellular signals such as γ-interferon, interleukin-1 and complement can lead to the activation of an enzyme, sphingomyelinase, which hydrolyses sphingomyelin (SM), releasing ceramide. Unlike other phospholipids, sphingomyelin does not have a backbone of glycerol but rather its backbone is sphingosine (Fig. 6.11). Sphingosine is an amino alcohol which contains a long unsaturated hydrocarbon chain. To produce SM, a fatty acid is added to sphingosine by an amide bond, creating ceramide, while its primary hydroxyl group is esterified to phosphoryl choline. In fact, in the biosynthesis, SM is made by an exchange of the phosphoryl choline head group from PC to ceramide, catalysed by the enzyme phosphatidylcholine:ceramide choline phosphotransferase. Breakdown by sphingomyelinase will once again yield ceramide which can enter a catabolic cascade.

Ceramide has been shown to be an important regulator in protein trafficking, cell growth, differentiation, and cell viability, having an important role in the orchestration of apoptosis in some cases. At a molecular level, ceramide has been shown to activate a nuclear transcription factor, nuclear factor κB (NF-κB), in permeabilized cells and to be important, in the control of transcription of several genes. Ceramide also has an effect through phosphorylation, with an activation of MAP kinase cascades, again which can lead to alterations in gene expression. However, ceramide

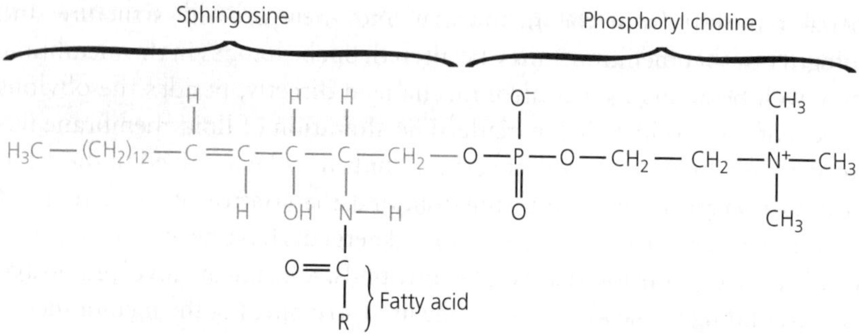

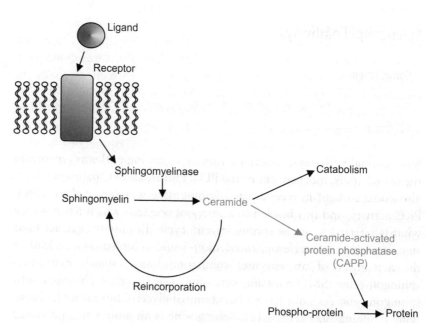

Fig. 6.11 The molecular structure of sphingomyelin (SM) showing its backbone of sphingosine (in blue).

Fig. 6.12 The production of ceramide and one of its possible roles within the cell.

has been shown to have a major control route through the use of a phosphatase. A serine/threonine phosphatase has been identified that is specifically and directly activated by the presence of ceramide and this enzyme has been called ceramide-activated protein phosphatase (CAPP; see Fig. 6.12). This phosphatase belongs to the 2A group of serine/threonine protein phosphatases, and therefore has a trimeric structure. This seems to be an elegant signalling cascade with ceramide as a central player, but the overall result is opposite to the promotion of protein phosphorylation seen with the activation of PKC by DAG.

Sphingosine itself has been shown to have a signalling role, and can stimulate the hydrolysis of PtdIns, resulting in the production of $InsP_3$ and the subsequent release of Ca^{2+} ions. Furthermore another metabolite, sphingosine 1-phosphate, has been shown to have second-messenger activity, where it is involved in inositol-independent release of calcium ions from intracellular stores. Such calcium signalling which does not involve inositols is creating much interest. This is further discussed in Chapter 7 (sections 7.7–7.9).

6.8 Related lipid-derived signalling molecules

It is now evident that many of the developmental stages of plant growth, as well as many cellular functions of plants, are under the control of complex compounds which can be derived from lipids. Some of these compounds were first shown to be of interest as they could induce tuber formation in plants, such as potatoes and yams. One of the major compounds is jasmonic acid (JA). JA is produced by a short metabolic pathway that starts with the action of lipoxygenase, which catalyses the incorporation of O_2 into linolenic acid. Most of the linolenic acid in the cell is found associated with lipids of the plasma membrane and can be released by the action of PLA_2. The 13-hydroperoxylinolenic acid produced by the action of PLA_2 is further acted upon sequentially by hydroperoxide dehydrase, allene oxide cyclase, and a reductase. Three further β-oxidation steps finally yield the 12-carbon JA (Fig. 6.13).

Other derivatives of this compound include methyl jasmonic acid, curcurbic acid, and tuberonic acid, the latter named after its tuber-inducing capacity. As with the inositol pathway there are further derivatives of these compounds which have been isolated and, because of the presence of chiral carbon atoms, several stereoisomers can exist for these molecules although there is probably a favoured conformation that occurs naturally.

Chiral

Chiral compounds are ones where there are two forms that are non-superimposable mirror images of each other, and therefore the two forms often have different biological activity.

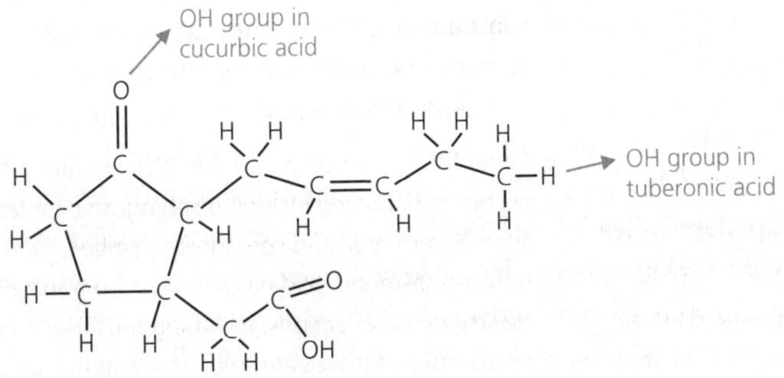

Fig. 6.13 The molecular structure of jasmonic acid, with derivatives shown.

JA and methyl jasmonic acid have been shown to induce senescence in leaves as well as the promotion of tuber formation. If leaves were floated on a solution of JA for 4 days, it was found that the polypeptide profile of the leaves changed as analysed by SDS/PAGE. These new proteins were termed jasmonate-induced proteins or JIPS. Further analysis of the regulation of the synthesis of these proteins indicated that the control was probably at the levels of both transcription and translation. JA has also been shown to induce the expression of other proteins with a known specific function, including storage proteins of leaves, known as vegetative storage protein (VSP). These proteins accumulate in vacuoles of the mesophyllic cells of the leaf and the bundle sheath cells before flowering. The mRNA coding for VSP is increased on the addition of JA. Cell division has also been suggested as a target for the control by JA or its related compounds.

6.9 SUMMARY

- A central part of many cell signalling pathways is the breakdown of lipids in the plasma membrane, with the subsequent formation of messenger molecules which can transmit the signal(s) into the cell.

- The formation of new inositol lipids by PtdIns 3-kinase has also been shown to be important in many transduction events, including insulin signalling (see **Chapter 9**).

- Products of PtdIns 3-kinase can activate kinases such as PKB.

- Two important signalling molecules are produced by breakdown of PtdInsP$_2$ by PLC: InsP$_3$ and DAG.

- Alternative abbreviations are used commonly for inositol compounds. Inositol 1,4,5-trisphosphate is shortened to InsP$_3$ or IP$_3$, whereas phosphatidylinositol 4,5-bisphosphate is shortened to PtdInsP$_2$ or PIP$_2$.

- DAG is known to act as an activator of PKC.

- InsP$_3$ releases Ca^{2+} from intracellular stores (Fig. 6.1, and further discussions in **Chapter 7, section 7.4**).

- PLC may be activated by the trimeric G protein G$_q$ (see **Chapter 5, section 5.4**), as well as by other means.

- InsP$_3$ may be the precursor of both higher and lower inositol phosphates; that is, it can be phosphorylated or dephosphorylated to create new metabolites, each of which may have a signalling role within the cell.

- Several inositol phosphates are thought to have defined signalling functions, including inositol tetrakisphosphate (InsP$_4$), and binding proteins for many of these compounds have now been reported.

- Several of the enzymes involved in this metabolism are the target for lithium ion inhibition, which might account for its therapeutic action in treating manic depression.

- Other inositols such as cyclic phosphate isomers and pyrophosphates have also been reported.

- Lipid-hydrolysing enzymes such as PLD and PLA$_2$ can generate other lipid-derived signalling molecules.

- AA, generated from lipid breakdown, can be a substrate for cyclooxygenase, cytochrome P450, and lipoxygenase, leading to the formation of prostaglandins, leukotrienes, and thromboxanes.

- Other lipid metabolism also has a clear influence on cellular functioning including sphingosine metabolites such as ceramide and sphingosine 1-phosphate, while in plants jasmonic acid has been shown to have a diverse range of actions, including the induction of senescence of leaves and tuber formation.

6.10 FURTHER READING

Berridge, M.J. (1993) Inositol trisphosphate and calcium signaling. *Nature* 361, 315–326.

Events at the membrane.

Agranoff, B.W., Bradley, R.M., and Brady, R.O. (1958) The enzymatic synthesis of inositol phosphatide. *Journal of Biological Chemistry* 233, 1077–1083.

Anderson, K.E. and Jackson, S.P. (2003) Class I phosphoinositides 3-kinases. *International Journal of Biochemistry and Cell Biology* 35, 1028–1033.

Divecha, N. and Irvine, R.F. (1995) Phospholipid signalling. *Cell* 80, 269–278.

Kodaki, T., Worchoski, R., Hallberg, R., Rodriguez-Viciana, P., Downward, J., and Parker, P.J. (1994) The activation of phosphatidylinositol 3-kinase by Ras. *Current Biology* 4, 798–806.

Koyasu, S. (2003) The role of PI3K in immune cells. *Nature Immunology* 4, 313–319.

Nishibe, S., Wahl, M.I., Rhee, S.G., and Carpenter, G. (1989) Tyrosine phosphorylation of phopholipase C-II *in vitro* by the epidermal growth factor receptor. *Journal of Biological Chemistry* 264, 10335–10338.

Rodriguez-Viciana, P., Warne, P.H., Dhand, R., Vanhaesebroeck, B., Gout, I., Fry, M.J., Waterfield, M.D., and Downward, J. (1994) Phosphatidylinositol-3-OH kinase as a direct target of Ras. *Nature*, 370, 527–532.

Smrcka, A.V., Hepler, J.R., Brown, K.O., and Sternweis, P.C. (1991) Regulation of polyphosphoinositide specific phospholipase C activity by purified G_q. *Science* 251, 804–807.

Stein, R.C. and Waterfield, M.D. (2000) PI3-kinase inhibition: A target for drug development? *Molecular Medicine Today* 6, 347–357.

Sulis, M.L. and Parsons, R. (2003) PTEN: from pathology to biology. *Trends in Cell Biology* 13, 478–483.

Whitman, M., Downes, C.P., Keeler, M., Keller, T., and Cantley, L. (1988) Type I phophatidylinositol kinase makes a novel inositol phospholipid, phosphadylinositol 3-phosphate. *Nature* 332, 644–646.

Wymann, M.P., Zvelebil, M., and Laffargue, M. (2003) Phosphoinositide 3-kinase signalling: Which way to target? *Trends in Pharmacological Sciences* 24, 366–376.

InsP₃ and its fate

Bansal, V.S., Caldwell, K.K., and Majerus, P.W. (1990) The isolation and characterisation of inositol polyphosphate 4-phosphatases. *Journal of Biological Chemistry* 265, 1806–1811.

Caldwell, K.K., Lips, D.L., Bansal, V.S., and Majerus, P.W. (1991) Isolation and characterisation of 2,3,-phosphatases that hydrolyse the phosphatidylinositol 3-phosphate and inositol 1,3-bisphosphate. *Journal of Biological Chemistry* 266, 18378–18386.

Cowburn, R.F., Wiehager, B., and Sundstrom, E. (1995) Beta-amyloid peptides enhance binding of the calcium mobilising 2nd messengers, inositol (1,4,5)trisphosphate and inositol (1,3,4,5)tetrakisphosphate to their receptor sites in rat cortical membranes. *Neuroscience Letters* 191, 31–34.

Cullen, P.J., Patel, Y., Kakkar, V.V., Irvine, R.F., and Authi, K.S. (1994) Specific binding-sites for inositol 1,3,4,5-tetrakisphosphate are located predominantly in the plasma membranes of human platelets. *Biochemical Journal* 298, 739–742.

Donie, F. and Reiser, G. (1991) Purification of a high affinity inositol 1,3,4,5, tetrakisphosphate receptor from brain. *Biochemical Journal* 267, 453–457.

Malviya, A.N. (1994) The nuclear inositol 1,4,5-trisphosphate and inositol 1,3,4,5-tetrakisphosphate receptors. *Cell Calcium* 16, 301–313.

Menniti, F.S., Oliver, K.G., Putney, J.W., and Shears, S.B. (1993) Inositol phosphates and cell signalling: new views of InsP₅ and InsP₆. *Trends in Biochemical Sciences* 18, 53–56.

Stephens, L.R., Hawkins, P.T., Stanley, A.F., Moore, T., Poyner, D.R., Morris, P.J., Hanley, M.R., Kay, R.R., and Irvine, R.F. (1991) Myoinositol pentakisphosphate: structure, biological occurance and phosphorylation to myoinositol hexakisphosphate. *Biochemical Journal* 275, 485–499.

The role of diacylglycerol

Florin-Christersen, J., Florin-Christersen, M., Delfino, J.M., and Rasmusssen, H. (1993) New pattern of diacylglycerol metabolism in intact cells. *Biochemical Journal* 289, 783–788.

Fujikawa, K., Imai, S., Sakane, F., and Kanoh., H. (1993) Isolation and characterisation of the human diacylglycerol kinase gene. *Biochemical Journal* 294, 443–449.

Inositol phosphate metabolism at the nucleus

Banfic, H., Zizak, M., Divecha, N., and Irvine, R. (1993) Nuclear diacylglycerol is increased during cell proliferation *in vivo*. *Biochemical Journal* **290**, 633–636.

Divecha, N., Banfic, H., and Irvine, R.F. (1993) Inositides and the nucleus and inositides in the nucleus. *Cell* **74** 405–407.

York, J.D., Saffritz, J.E., and Majerus, P.W. (1994) Inositol phosphate 1-phosphatase is present in the nucleus and inhibits DNA synthesis. *Journal of Biological Chemistry* **259**, 19992–19999.

Phosphatidylcholine and arachidonic acid metabolism

Brown, H.A., Gutowski, S., Moonaw, C.R., Slaughter, C., and Sternweis, P.C. (1993) ADP-ribosylation factor, a small GTP-dependent regulating protein stimulates phospholipase D activity. *Cell* **75**, 1137–1144.

Cockcroft, S., Thomas, G.M.H., Fensome, A., Geny, B., Cunningham, E., Gout, I., Hiles, I., Totty, N.F., Truong, O., and Hsuan, J.J. (1994) Phospholipiase D: a downstream effector of ARF in granulocytes. *Science* **263**, 523–526.

Durieux, M.E. and Lynch, K.R. (1993) Signalling properties of lysophosphatidic acid. *Trends in Pharmacological Sciences* **14**, 249–254.

Exton, J.H. (1994) Phosphatidylcholine breakdown and signal transduction. *Biochimica et Biophysica Acta* **1212**, 26–42.

Moolenaar, W.H., Jalink, K., and van Corven, E.J. (1992) Lysophosphatidic acid: a bioactive phospholipid with growth factor-like properties. *Reviews of Physiology, Biochemistry and Pharmacology* **199**, 47–65.

Piomelli, D. (1993) Arachiodonic acid in cell signalling. *Current Opinion in Cell Biology* **5**, 274–280.

Sphingolipid pathways

Dobrowsky, R.T. and Hannun, Y.A. (1992) Ceramide stimulates a cytosolic protein phospholipase. *Journal of Biological Chemistry* **267**, 5048–5051.

Hannun, Y.A. (1994) The sphingomyelin cycle and the second messenger function of ceramide. *Journal of Biological Chemistry* **269**, 3125–3128.

Spiegel, S. and Milstein, S. (1995) Sphingolipid metabolites: members of a new class of lipid second messengers. *Journal of Membrane Biology* **146**, 225–237.

Related lipid-derived signalling molecules

Anderson, J.M. (1991) Jasmonic acid dependent increase in vegetative storage protein in soybean tissue cultures. *Journal of Plant Growth and Regulation* **10**, 5–10.

Koda, Y., Kikuta, Y., Tazaki, H., Tsujino, Y., Sakamura, S., and Yoshihara, T. (1991) Potato tuber inducing activities of jasmonic acid and related compounds. *Phytochemistry* **30**, 1435–1438.

7 Intracellular calcium: its control and role as an intracellular signal

Calcium ions have widespread use as signals in cells. Cells expend large amounts of energy expelling calcium ions from the cytoplasm, either to the outside or into organelles. Such ions are subsequently allowed to return to the cytoplasm, constituting a signal. The movement of calcium ions may be controlled through the production of inositol phosphate compounds as discussed in **Chapter 6**, but many other mechanisms also exist—this is discussed in this chapter.

Calcium ions themselves control a myriad of functions in cells, their action often being mediated by a protein known as calmodulin. The production of other signals, phosphorylation events, and even calcium ion movements themselves, may be controlled.

The movement of calcium ions is so central to cell signalling events that it will have been mentioned in previous chapters, and will appear again in later chapters. Calcium ions control many metabolic and physiological functions in cells, and are even involved in the process of apoptosis; that is, cell suicide. Therefore, the discussion in this chapter is pivotal in gaining an understanding of cell signalling.

7.1 Introduction

Signalling in which calcium ions are involved is both extremely important and central to cellular communication. Numerous events and activities within the cell, as well as the release of products, perhaps even other signals, from the cell are under direct or indirect control by calcium ions. As will be discussed in this chapter, there are many mechanisms for the control of intracellular calcium, and many effectors that are regulated by the fluctuations that take place in calcium ion concentration within the cell.

However, despite their ubiquity, and importance, one aspect of the use of calcium ions as a signal is reasonably unusual. Most intracellular signalling molecules are produced by the cell, they have an effect, and then they are destroyed; for example, with the production of cAMP, its role in the activation of a kinase, and its destruction by phosphodiesterase (see Chapter 5, section 5.2). However, with calcium ions this rationale is not followed. The calcium ions are not made when needed, or destroyed after they have served their role. It is the variation in the concentration of calcium ions inside the cell, often denoted $[Ca^{2+}]_i$, which constitutes the signal. Calcium ions are simply moved around, not produced or destroyed, but it is the concentration of the ion that is experienced by the effector molecule in a particular compartment in which that effector acts as the message or signal. However, calcium signalling is ubiquitous, being found in bacteria, and right across the diversity of higher organisms, including plants and mammals: it is especially important in the neuronal pathways of higher animals.

The intracellular concentration of Ca^{2+} ions can be controlled in many ways, each of which have to work in a coordinated manner to allow the amount of calcium to be an effective signal. The steady-state level of calcium in the cytoplasm of a cell is usually less than 10^{-7} M. If the cell is activated by an influx of calcium, this level might rise to approximately 10^{-5} M, this higher concentration being the signal to trigger further events in the cell. Putting these concentrations into perspective, the extracellular Ca^{2+} concentration may be in the order of 2×10^{-3} M. Therefore, to maintain this very low steady-state level, with a relatively high extracellular Ca^{2+} concentration, requires calcium ions to be actively pumped out of the cell, a process which uses proteins in the plasma membrane, and a source of energy. Cells also sequester Ca^{2+} in organelles within the cell. For example, Ca^{2+} is sequestered in the endoplasmic reticulum (ER), or sarcoplasmic reticulum (SR) in muscle tissue, or in the mitochondria. Again this is an active process, requiring energy. Therefore, to drive these various processes, a considerable amount of a cell's energy is devoted to control of the $[Ca^{2+}]_i$. A signal may be propagated in the cytoplasm for example, when previously excluded calcium is released back into the cytoplasm, so rapidly increasing the Ca^{2+} concentration in the desired area. The calcium may be released from the organelles, or a sub-population of the organelles, or it might be released back into the cell from the outside. Such release of calcium ions is usually under the control of other signalling molecules, and in many cases several signalling molecules might be attempting to control calcium movements at the same time, some with antagonistic actions perhaps.

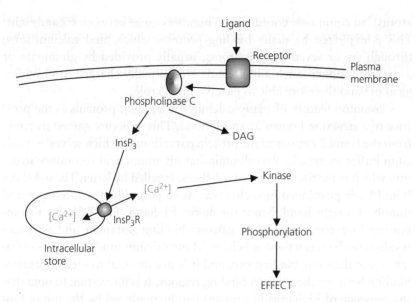

Fig. 7.1 An example of how ligand binding can lead to release of Ca^{2+} from intracellular stores. The activation of phospholipase C gives rise to the messenger $InsP_3$ which binds to $InsP_3$ receptors ($InsP_3Rs$) of the endoplasmic reticulum membranes. Such receptors are also channels and open to release calcium ions.

One of the earliest examples of a signalling molecule involved in the control of calcium ion movements, and perhaps one which has been best characterized, is the use of inositol trisphosphate ($InsP_3$), the production of which was discussed in Chapter 6 (sections 6.3 and 6.4). $InsP_3$ is released from the plasma membrane by the breakdown of phosphatidyl-inositol bisphosphate ($PtdInsP_2$) and once produced $InsP_3$ will bind to $InsP_3$ receptors ($InsP_3Rs$) on the membranes of the intracellular stores. These receptors are also Ca^{2+} channels which will allow the rapid release of Ca^{2+} back into cytoplasm of the cell. An example of such a signal trans-duction pathway is shown in Fig. 7.1.

The third, and important, way in which calcium concentrations are controlled in cells is by the presence of calcium-binding proteins. Such proteins actually have two roles. Firstly they remove free calcium ions from solution, and therefore effectively reduce $[Ca^{2+}]_i$, and second the binding of the calcium ions to such proteins is usually a mechanism by which the signal is propagated to the next component of the signal trans-duction cascade. For example, the action of calcium on an effector mole-cule, such as a kinase, may be through such a binding protein. One of the most important, and widespread, calcium-binding proteins is a protein called calmodulin, which is discussed in more detail below (section 7.2).

For calcium ions to bind to proteins there has to be a recognition of the ion and some form of physical attraction. For calcium ion binding, oxygen atoms are often involved. Calcium ions can coordinate up to 12 oxygen

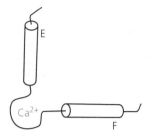

Fig. 7.2 The structure of an EF-hand. Two α-helices are connected by a loop structure that provides a calcium ion-binding site.

atoms, but commonly coordination numbers range between six and eight. This is exploited by many binding proteins which bind calcium ions, through six or seven oxygen atoms, usually provided by glutamate or aspartate residues. Such amino acids are normally charged at physiological pH and therefore able to partake in this role.

A common feature of many calcium ion-binding proteins is the presence of a structure known as an EF-hand. This structure gained its name from the E and F regions of the protein parvalbumin which serves as a calcium buffer in muscle. Parvalbumin has an amino acid secondary structure which comprises a series of α-helices, labelled by letter. The α-helices E and F are positioned in such a way as to point like the forefinger and thumb of a right hand, hence the name, EF-hand (Fig. 7.2). A calcium-binding loop containing active calcium-binding glutamate and aspartate residues lies between these α-helices. Many calcium-binding proteins contain more than one binding site, and it is not unusual to see cooperative binding between the calcium-binding regions. It is interesting to note that the presence of EF-hands in proteins can be predicted by the use of bioinformatics, and therefore the control by calcium of such proteins activity might be assumed and therefore tested for. An example of this will be met in Chapter 8 (section 8.3); when isoforms of the NADPH oxidase large subunit were identified from their cDNA sequences, bioinformatic analysis suggested that some variants had EF-hands towards their N-terminal ends. Calcium control of the oxidase had already been suggested, and further experimental studies will now enable researchers to confirm which isoforms are involved.

As discussed above, Ca^{2+} ion-binding proteins can potentially be fulfilling two roles in the formation of a calcium signal. They might simply be acting as buffers, so when the calcium concentration alters free calcium ions are not left in solution to have an effect. Alternatively, the proteins might be acting as the trigger, mediating the rise in free calcium and turning it into the signal, allowing the message to be transmitted down the transduction chain, or in many cases being the last part of the signal transduction chain and acting on the signal when perceived. On binding calcium ions, either at EF-hands or elsewhere, Ca^{2+} ion-binding proteins usually undergo a major alteration in their conformation, and this conformational change will undoubtedly expose active sites on the protein and so allow further reactions and propagation of the signal.

One of the most well-studied and best-characterized examples of a Ca^{2+}-binding protein with a clearly defined role is troponin. Troponin is, in fact, a complex of three polypeptides, TnC, TnI, and TnT, of 18, 24, and 37 kDa respectively. Along with the protein tropomyosin, troponin lies along the thin filaments of muscle fibres, and it is the TnC subunit that binds Ca^{2+} ions. TnC has two domains which are homologous, and each has two EF-hand binding sites for Ca^{2+}, those in the C-terminal domain being of high affinity while those in the N-terminal domain being of lower affinity. The domains are held together by an α-helix, the whole molecule showing high structural similarity to calmodulin (see below; section 7.2). On binding of Ca^{2+} to the low-affinity sites, that is when Ca^{2+} has been released from the sarcoplasmic reticulum, a conformational change takes place in the TnC polypeptide, which is transmitted through the other polypeptides in the complex and to tropomyosin. A forced change in the orientation of tropomyosin relieves the steric hindrance it caused over the actin–myosin interaction, allowing muscle function.

Calcium control of muscle

It is interesting to note here that calcium is also involved in the control of glycogen break-down, so calcium controls the movement of muscle and also the availability of glucose for the formation of ATP to drive the muscle movement.

There are many other examples of calcium-binding proteins, and a few will be highlighted here. Other calcium-binding proteins include a protein called calbindin-D_{28k}. This is a 28-kDa polypeptide which is related to vitamin D and has a wide but heterogeneous distribution in the brain. Another important, and rather interesting, example is calsequestrin. Calsequestrin is a 44-kDa protein which is incredibly acidic, having over one-third of its amino acids as either aspartic acid or glutamic acid, and amazingly each calsequestrin molecule can bind to 43 Ca^{2+} ions. Calsequestrin is found in the sarcoplasmic reticulum where it controls the concentration of free Ca^{2+} and so helps to keep down the calcium gradient across the membrane. Also present here is a polypeptide known as high-affinity Ca^{2+}-binding protein, which has a similar role to calsequestrin.

But what is it that calcium is controlling in the cell? To answer that question would take a long time as calcium seems to have a plethora of actions, some of which will be discussed, almost in passing, in the other chapters of this book. Perhaps, before answering it briefly, a quick discussion of some of the experimental approaches used to elucidate the role of calcium in cells would be useful. Many of the actions of calcium ions have been unravelled by the use of calcium ionophores, such as A23187 or ionomycin. Such chemicals effectively punch holes in the plasma membrane and let calcium ions flood into the cell, so mimicking the sudden rise of Ca^{2+} concentration which the cell could achieve by the release of intracellular stores or by allowing calcium ions in from outside by a signal-mediated response. Therefore, a researcher would measure the baseline rate of an activity, add A23817, and see what happens. If rates are modulated, it suggests that calcium is involved. If not, perhaps calcium has no bearing on the functioning being studied. Alternatively, calcium ions may be removed from the extracellular medium by the addition of chelating agents, such as EDTA, and therefore this will prevent the cell from using the plasma membrane channels as a means to increase cytoplasmic calcium. Cell-permeable analogues of chelators are also available to modulate intracellular Ca^{2+} ions concentrations. Radioisotopes can be used to measure calcium concentrations in cells, and with the use of fluorescent probes in conjunction with confocal microsopy (see later in this chapter; section 7.10) the concentrations of calcium over time, and the exact location in cells of the ions, can be determined. By the use of such methods the diversity of the roles of calcium ions in the signalling of cells can be discerned.

Therefore, by studying calcium ions *per se*, and by the use of binding studies and bioinformatic analysis, the answer to our question is revealed. Some, but by no means all, of the proteins which are known to have their activity modulated by the Ca^{2+}-signalling systems are listed in Table 7.1.

Certainly calcium is involved in the activation of Ca^{2+}/calmodulin-dependent protein kinase, PKC, Ca^{2+}-dependent phosphatases

Table 7.1 Some of the proteins which are controlled by calcium ion concentrations.

Ca^{2+}-dependent protein kinase
Ca^{2+}-dependent protein phosphatase (for example PP2B)
Phospholipase C (PLC)
Pyruvate dehydrogenase
Nitric oxide synthase (NOS)
Protein kinase C (PKC)
Adenylyl cyclase (particularly type 1)
$InsP_3$ receptors
Ryanodine receptors
Ca^{2+}/Mg^{2+}-dependent endonuclease
Phosphorylase kinase
Ca^{2+}-ATPase
Troponin

(see Chapter 4, sections 4.2 and 4.6), and in the activation of nitric oxide synthase (NOS) (see Chapter 8, section 8.2), leading to the propagation of intracellular and intercellular signals. It is also involved in the direct activation of many enzymes, such as pyruvate dehydrogenase, as well as being involved in more complex patterns of activity such as the control of the cell cycle. Spikes of calcium have been shown to be essential for triggering the completion of meiosis and the initiation of mitosis in some cells, while depletion of calcium stores can arrest cells in the G_0–G_1 phases. Gene expression can be under the control of calcium concentrations, and here again calcium-binding proteins such as calreticulin may mediate the changes in concentration into the final signal.

7.2 Calmodulin

Calmodulin is an extremely important protein which is involved in the mediation of the Ca^{2+} signal in a vast number of regulatory pathways, including the regulation of metabolic activity, the regulation of other signal pathways, such as the nitric oxide generation, and the regulation of gene expression.

It is sometimes found as an integral part of an enzyme complex, as in phosphorylase kinase, where it is known as the δ-subunit. Often it even appears to be involved in the contradictory regulation of an activity where it is responsible for an increase in the signal and a decrease at the same time. One example of this can be seen with its role in the control of cAMP levels, where it can increase the production of cAMP via the activation of adenylyl cyclase while at the same time activating its breakdown by the stimulation of Ca^{2+}/calmodulin (CaM)-dependent phosphodiesterase. However, with tissue-specific expression of enzyme isoforms, such situations are almost certainly not as simple and clear as they might appear. Interestingly, calmodulin can also lead to the activation of the plasma membrane and endoplasmic-reticular Ca^{2+} pumps, which will cause the cessation of the calcium signal. Calcium, through calmodulin, therefore appears to control its own destiny.

Calmodulin is a small, relatively acidic protein which has the capacity to bind four molecules of Ca^{2+}. Calmodulin has an affinity for Ca^{2+} of approximately 10^{-6} M, which lies between the resting Ca^{2+} concentration in the cell, usually less than 10^{-7} M, and the activated Ca^{2+} concentration which might reach as high as 10^{-5} M. The shape of the calmodulin protein molecule resembles a dumbbell, with two globular regions connected by a long, flexible, and very mobile α-helix (Fig. 7.3).

Each of the globular domains of calmodulin contains two EF-hand regions, with their characteristic helix-loop-helix topology (Fig. 7.2), each

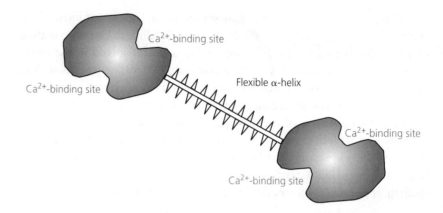

Fig. 7.3 The domain structure of calmodulin, showing the calcium-binding sites.

of which can bind to one molecule of Ca^{2+}. The two EF-hands within each globular region are connected by a short antiparallel β-sheet region. Interestingly, despite the similarity in the two regions their affinity for Ca^{2+} differs. The C-terminal domain has the higher affinity and will bind to Ca^{2+} first, leading to major conformational changes within the molecule. These structural changes reveal two hydrophobic patches, one in each half of the protein. It is these patches that can interact with the molecule, another protein, that calmodulin will control.

> ■ Small peptides (synthetic peptides) which have the same sequence as potential interaction regions in proteins can be used to bind to potential binding sites and therefore stop native binding. By the use of a range of such peptides, usually based on areas of a protein of interest, the exact nature of protein–protein interactions can be determined. Their use as therapeutics has also been patented, although getting such peptides into cells is a challenge.

With the use of synthetic peptides to block binding, proteolysis experiments to remove potential binding sites from proteins and the expression of cDNAs, the mapping of the areas on proteins which are controlled by calmodulin has been attempted. Such proteins bind to the hydrophobic patches revealed in calmodulin after it has bound to Ca^{2+}. Most of these calmodulin-binding sites in other proteins involve an α-helix of 16–35 amino acids within the interacting protein. If such an α-helix is viewed from one end, it can been seen that it possesses a polarity, with the majority of the polar, mainly basic, amino acids lying down one side, and the majority of the hydrophobic amino acids lying on the opposite side (Fig. 7.4). The binding regions also usually contain a potential phosphorylation site.

Besides the regulation of calmodulin by binding of Ca^{2+} to the four binding sites, it has also been shown that calmodulin is itself phosphorylated

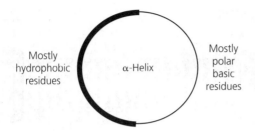

Fig. 7.4 Looking end-on down an α-helix which may be involved in the binding of a protein to activated calmodulin.

in vivo. The central helix is phosphorylated at two sites with a third site located in the third Ca^{2+}-binding region. It is thought that the interaction of the calmodulin with its target protein is reduced on phosphorylation and this might be caused by an inhibition of the structural changes which occur in the calmodulin molecule on Ca^{2+} binding, with the hydrophobic patches not being revealed properly.

7.3 The plasma membrane and its role in calcium concentration maintenance

The plasma membrane is crucial to the maintenance of the $[Ca^{2+}]_i$. Not only does it remove Ca^{2+} from the cell, but it also lets it back in again as required to create a signal. Therefore, these two activities need to be discussed.

Firstly, let us turn to the role of the plasma membrane in the removal of Ca^{2+} from the cell. The plasma membrane contains two types of pump which actively discharge Ca^{2+} from the cell, helping to maintain the low cytoplasmic Ca^{2+} concentration. One of these uses ATP as an energy source while the other uses the electrochemical gradient which occurs across the membrane. The one that uses ATP is the plasma membrane Ca^{2+}-ATPase (PMCA) pump (Fig. 7.5). It is a P-type ATPase, using the energy of one or two molecules of ATP to drive the transport of one molecule of Ca^{2+} out of the cell. The pumping cycle uses a phosphorylated intermediate (hence its classification as a P-type pump), involving the transient transfer of a phosphoryl group to an aspartate residue on the polypeptide. This phosphorylation of the protein causes a conformational change needed to transport the Ca^{2+} ions across the membrane, and dephosphorylation would restore the protein back to its former state ready for another round of Ca^{2+} pumping. The net result is the breaking and loss of the high-energy phosphate bond of ATP, the energy being used to drive the Ca^{2+}-pumping activity of the ATPase.

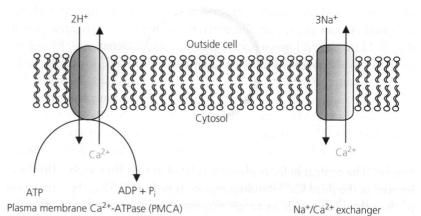

Fig. 7.5 Plasma membrane calcium pumps: a Ca^{2+}-ATPase (PMCA) and a Na^+/Ca^{2+} exchanger. P_i, inorganic phosphate.

The structure of the ATPase protein (the PMCA) is thought to have 10 membrane-spanning domains. Like many proteins involved in signalling, several tissue-specific isoforms exist, arising from four genes, as well as there being alternative splicing of the gene transcripts, so several variants exist; however, all will have a similar structure.

The pump that uses the electrochemical gradient which occurs across the plasma membrane is the Na^+/Ca^{2+} exchanger, and this is also active in the expulsion of Ca^{2+} ions from the cell (Fig. 7.5).

The Ca^{2+}-ATPase has a higher affinity for Ca^{2+} ions than the Na^+/Ca^{2+} exchanger, but the latter can pump Ca^{2+} ions at a much higher rate. A typical Ca^{2+}-ATPase has been shown to pump Ca^{2+} ions at a rate of approximately 30 s^{-1} in an *in vitro* situation whereas the Na^+/Ca^{2+} exchanger may pump Ca^{2+} ions at a rate as high as 2000 s^{-1}. Therefore, after an influx of calcium, which perhaps constituted a signal, large amounts of Ca^{2+} can be expelled from the cell quite quickly by the Na^+/Ca^{2+} exchanger, and when the Ca^{2+} ion concentration is down to micromolar levels the slower but higher-affinity ATPase can continue to decrease the cytoplasmic Ca^{2+} ion concentration further. The ability to reduce the $[Ca^{2+}]_i$ very quickly may be important for the maintenance of calcium spikes (see below).

The plasma membrane is also a major site of re-entry of Ca^{2+} back into the cell. In most cells the re-entry is controlled by polarization of the membrane as well as the presence of other second-messenger molecules. The opening of potassium (K^+) channels alters the membrane potential across the plasma membrane, and this drives the Ca^{2+} ions across into the cytoplasm through voltage-independent Ca^{2+} channels. The structure of the channels is such that the re-entry is very selective for Ca^{2+} ions, not allowing the entry of other ions.

The depletion of the intracellular stores can also serve as a signal for re-entry of Ca^{2+} through the plasma membrane, referred to as capacitative Ca^{2+} entry (CCE). Work has intensified to find the channels that are involved and mechanisms proposed, referred to as the Ca^{2+}-release-activated current (I_{CRAC}) or alternatively the depletion-activated current (I_{DAC}). Many second-messenger molecules and mechanisms have been suggested to be signals for opening of these channels, including $InsP_3$, $InsP_4$, tyrosine phosphorylation, cGMP, and G proteins, both the heterotrimeric and smaller monomeric ones, but the exact mechanism has yet to be discovered. Messenger molecules, such as the one named Ca^{2+}-influx factor (CIF) has been suggested as a candidate for this role. CIF has been characterized as being a stable non-protein molecule, perhaps a negatively charged phosphorylated anion of less than 500 Da, but more studies will be required to fully elucidate its mode of action.

As well as the signalling mechanisms involved, the proteins that act as the channels for Ca^{2+} ions have been sought. For example, proteins encoded by the *Trp* genes (trp, transient receptor potential) are thought to be instrumental here. Six trp homologues have been identified in mammals, but not all are likely to be instrumental in Ca^{2+} ion re-entry in all tissues.

In plants, such plasma membrane Ca^{2+} ion channels have also been sought. Candidates here include products of the *AtTPC1*, *AtGORK*, and *AtSKOR* genes in *Arabidopsis thaliana*, and members of the annexin family are likely to be involved.

7.4 Intracellular stores

Endoplasmic reticulum stores

Crucial aspects of calcium ion signalling are the maintenance of a very low $[Ca^{2+}]_i$, the ability to allow it to rise suddenly, and then its subsequent restoration. All these aspects of the signalling process are facilitated by the presence of intracellular stores of Ca^{2+} ions, in which Ca^{2+} can be sequestered, and from which Ca^{2+} can be rapidly released, and then subsequently sequestered again, in a never-ending cycle of potential signalling.

One of the major stores for Ca^{2+} ions is the lumen of the ER, or in muscle tissues the SR. Specialized areas of ER used for calcium storage have been observed in some cells and these have been called calciosomes.

As with the plasma membrane, Ca^{2+} needs to be pumped through the ER membrane, and then released back the other way at the appropriate time. In the membrane of the ER calcium ion pumps pump calcium ions from the cytoplasm to the inside of the membranous network (Fig. 7.6). These are commonly referred to as smooth endoplasmic reticulum calcium ATPase, or SERCA, pumps. In the specialized type of ER found in muscle, the SR, they are in fact a major protein component of the membrane, being approximately 80% of the integral proteins found in that

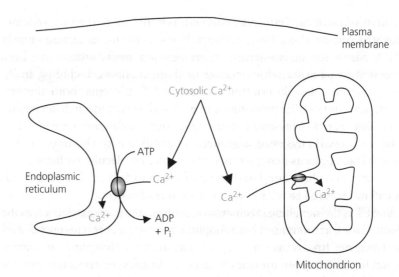

Fig. 7.6 Intracellular calcium stores: the endoplasmic reticulum and mitochondria both sequester calcium ions.

membrane. The density of these pumps has been estimated to be as high as 25 000/μm^2, and such data highlight the importance of this type of activity to the overall functioning of the cell. As the name suggests, being called an ATPase, this is an active process, pumping calcium from a region of very low concentration, the cytoplasm, to a region of relatively high concentration, the inner part (the lumen) of the ER. Therefore, an input of energy is required, and here the SERCA pumps use the breakdown of ATP. Again, like the plasma membrane equivalent, a phosphoryl intermediate is involved in the pumping steps (hence they are also P-type ATPases); that is, a phosphate group is added to an aspartate residue on the polypeptide causing a conformational change, resulting in the Ca^{2+} ions moving across the membrane where they can be released. Removal of the phosphate group again will restore the enzyme to its first conformation, allowing further rounds of ATP hydrolysis and Ca^{2+} pumping. It is thought that two Ca^{2+} ions are transported for every ATP hydrolysed. Therefore, as would be expected, like the PMCA, the SERCA is not only classified as a P-type ATPase, but also has 10 transmembrane regions defined in the polypeptide. However, despite this topological and functional similarity between the PMCA and the SERCA, very little sequence homology exists between the two types of Ca^{2+} pump. The endoplasmic reticular protein appears to contain two main domains, with a smaller one on the lumen side of the membrane. One main domain is buried in the membrane and acts as a Ca^{2+} channel, the Ca^{2+} ions binding near the centre line of the membrane. The other major domain contains the ATPase activity on the cytoplasmic side of the membrane and the site of phosphorylation.

When a functional Ca^{2+}-ATPase was reconstituted into a phospholipid membrane, it was seen to pump Ca^{2+} ions at a rate of approximately $30 \ s^{-1}$, but interestingly if a calcium gradient was artificially set up across the membrane, but the wrong way around, the ATPase could be forced to run in reverse and ATP formation was seen if ADP was present. Such potential for reversibility is often seen with ATPases, but it is hard to imagine this happening *in vivo*, and probably remains a curiosity which can be used to unravel the molecular mechanisms involved.

The SERCA pumps, at least in some mammals, arise from three genes which appear to be expressed in a tissue-specific manner. The gene products SERCA1 and SERCA2 are expressed in different types of muscle, while SERCA3 is expressed in non-muscle tissue. Thapsigargin, a lactone which also contains tumour-promoting activity, has been identified as a highly specific inhibitor of SERCA activity, trapping the pump in a Ca^{2+}-free state, and has been useful as a biochemical tool for the study of these activities as it does not inhibit plasma membrane pumps. Alternatively, orthovanadate can be used as a more general inhibitor of P-type ATPases.

The activity of Ca^{2+}-ATPases is probably not just continuous but is controlled. Such regulation would allow the calcium to be removed quickly, or perhaps more slowly if the response was required to be elongated. An increase in cytoplasmic Ca^{2+} accelerates the activity of the ER pump while an increase of free Ca^{2+} inside the ER inhibits the activity of the pump, half the activity being lost if Ca^{2+} ion concentrations reach approximately $300 \ \mu M$. Control of the pumps may also come from a receptor-mediated mechanism.

Once inside the ER, Ca^{2+} is sequestered and buffered, keeping the concentration of free ions down. As discussed above, one such sequestering protein is calsequestrin, which has an amazing affinity for calcium. No role for the sequestering of the calcium ions has been assigned other than as a store which can release calcium back to the cytoplasm at the appropriate moment. However, as one of the major roles of the ER is in the sorting and trafficking of proteins, it has been suggested that the calcium ions might have a dual role here, although there is little evidence of this. Certainly, if free Ca^{2+} was allowed to be high, then the normal functioning of the ER would be disrupted, and that of course would be detrimental to the cell. Therefore, the ER has a crucial role in calcium signalling, but this role cannot be to the hindrance of its other functions.

Therefore, calcium ion uptake by the ER is an energy-dependent pumping mechanism, but there also has to be mechanism to allow its quick release when required as a signal. As discussed in the Chapter 6 (section 6.3), one of the products of $PtdInsP_2$ breakdown is $InsP_3$, the major role of which is the release of calcium ions from intracellular stores, in particular the ER. To be able to do this the machinery of the ER has to recognize and respond to InsP$_3$. The ER, in fact, contains InsP$_3$Rs which facilitate the release of the calcium ions. The receptors are tetrameric in conformation with a C-terminal domain which spans the membrane of the ER. Each subunit has an approximate molecular weight of 310 kDa and contains a cationic pore, which shows relatively little selectivity, but allows the movement of Ca^{2+} through the membrane. Each of the four subunits also contains an arginine- and lysine-rich region at the N-terminal end which can bind one InsP$_3$ molecule, acting as a ligand-binding site for the protein. Therefore, analogies can be drawn here with the receptors of the plasma membrane (Chapter 3, sections 3.1–3.3). The InsP$_3$ acts as the ligand, which binds to the receptor (InsP$_3$R), and so opening a channel for the movement of Ca^{2+}. The N-terminal region, which is on the cytoplasmic side of the membrane, also has binding sites for two ATP molecules and at least one Ca^{2+} ion.

As is regularly seen with other signalling proteins, the InsP$_3$R can exist in several isoforms. In mammals, at least four genes encode InsP$_3$Rs with significant homology between them. Expression of the receptors, as might be expected, differs between different tissues.

However, InsP$_3$R activity is not only controlled by the binding of InsP$_3$ but very importantly the passage of Ca^{2+} through these proteins is modulated by Ca^{2+} itself, Ca^{2+} being referred to as a co-agonist. If InsP$_3$ concentrations are low, relatively low cytoplasmic Ca^{2+} concentrations accelerate Ca^{2+} release, but at higher cytoplasmic Ca^{2+} concentrations the release of Ca^{2+} through InsP$_3$R is inhibited. Therefore, if plotted, the release response to Ca^{2+} concentrations would give a bell-shaped curve, with a maximum around 0.2–0.3 μM (Fig. 7.7). This a very important phenomenon because, as discussed below, the passage through InsP$_3$R is not the only way to increase cytoplasmic calcium ion concentrations. If Ca^{2+} is released through other channels, perhaps from non-ER-like stores, then the arrival of Ca^{2+} at the InsP$_3$R can influence whether more

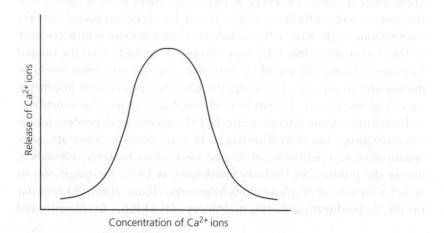

Fig. 7.7 Ca^{2+}-induced Ca^{2+} release from InsP$_3$R in the presence of lower concentrations of InsP$_3$.

Release of Ca^{2+} ions

Concentration of Ca^{2+} ions

Ca^{2+} is released through them from the ER, or SR. Therefore, by relatively low calcium rises having a control on Ca^{2+} release, small changes can very rapidly be converted to large changes, in what could be referred to as feed-forward-type control. Alternatively, inhibition at high Ca^{2+} might also be important, giving a level of moderation.

Control of $InsP_3Rs$ is not only through rises in $[Ca^{2+}]_i$ but may also be via phosphorylation. For example, control may be through cAMP-dependent protein kinase. $InsP_3R$ activity is also modulated by the presence of Mg^{2+}. Of particular interest to those who drink lots of strong coffee, caffeine has been shown to inhibit Ca^{2+} release from $InsP_3Rs$ in response to $InsP_3$, while heparin, better known for its anti-coagulant properties, competitively inhibits $InsP_3Rs$ by binding to the $InsP_3$-binding sites. Interestingly, ethanol is also a potent inhibitor of $InsP_3Rs$.

It is also interesting to note that $InsP_3R$-like receptors have been found in the plasma membrane of some cells, but their role there has not been fully elucidated. As discussed above, re-entry of Ca^{2+} into the cell through the plasma membrane, having been expelled previously, is very important and the use of $InsP_3$ might give one mechanism by which this might happen.

It is worth a quick note to point out that as well as $InsP_3$, other inositol phosphate metabolites might be involved in the release and control of Ca^{2+}. For example, inositol 1,3,4,5-tetrakisphosphate ($InsP_4$) has been shown to affect the Ca^{2+} influx in cells and also to be an inhibitor of Ca^{2+}-ATPase pumps. Such inositol compounds are further discussed in Chapter 6 (section 6.4).

Leaving the inositol metabolites for a while, the release of Ca^{2+} from intracellular stores can also involve other proteins other than $InsP_3R$. The other calcium ion-release channel found in the ER is the ryanodine receptor (RyR). Like the $InsP_3R$, the ryanodine receptor is also a tetramer, but this time the subunits are much bigger, having molecular weights of up to 560 kDa, and the receptor has been purified as a 30 S complex, which highlights its size. Three genes (*ryr-1, ryr-2, and ryr-3*) code for separate isoforms of the ryanodine receptor, with expression of the three being very tissue-specific. The gene *ryr-1* is expressed in skeletal muscle, while *ryr-2* is found expressed in cardiac muscle. The gene *ryr-3* represents a non-muscle form. Of the proteins encoded by these genes, RyR1 and RyR2 show a large amount of sequence homology but the third form, RyR3, is much smaller, with less than 650 amino acids. The C-terminal ends of the RyR subunits create a Ca^{2+} channel through the membrane, having between four and ten putative membrane-spanning regions, while the N-terminal ends of the RyR subunits protrude into the cytosol. As the name suggests, these channels are sensitive to ryanodine, which is a plant alkaloid. Low concentrations of ryanodine cause opening of the RyR channels but at higher, micromolar, concentrations of ryanodine the channels are once again closed. Opening of RyR is also seen in the presence of caffeine, opposite to that seen with $InsP_3R$.

As ryanodine is a plant alkaloid, and unlikely to be an endogenous signalling molecule in mammals, what causes the RyR to open as a physiological response? In some tissues, for example in striated muscle, the SR comes into close contact with the plasma membrane and protein complexes exist which connect the two membranes. These protein complexes are known as T-tubule feet, and contain RyRs. The other part of the foot structure is a plasma membrane-associated, voltage-sensitive receptor, such as the dihydropyridine receptor. This receptor will sense changes in the voltage across the membrane, and undergo a conformational change. This change in conformation of the plasma membrane-associated receptor is sensed by RyR, as it is part of the same protein complex. A conformational change is induced in RyR which results in channel opening and the release of calcium ions (Fig. 7.8). This is a good example of conformational changes in one protein being transmitted across to another, having the desired effect.

Alternatively, like $InsP_3R$, RyRs are sensitive to low concentrations of Ca^{2+}, having a similar bell-shaped dependence on Ca^{2+} (as depicted in Fig. 7.7 for $InsP_3Rs$). Like $InsP_3Rs$, low concentrations of Ca^{2+} will induce opening of RyRs while higher, millimolar concentrations of Ca^{2+} will inhibit opening. Therefore, a voltage change across the plasma membrane may induce a small influx of Ca^{2+} from the outside, through a voltage-gated Ca^{2+} channel. This would be sufficient to induce the opening of RyR and the release of intracellular Ca^{2+} stores as illustrated in Fig. 7.8.

RyRs are also sensitive to, and open in, the presence of cyclic ADP-ribose (cADPr; discussed further below in section 7.8). RyR2 seems to be the target for cADPr as RyR1 is insensitive to this messenger molecule. It is possible that RyR3 is also controlled in this way. Control of RyR by calmodulin and phosphorylation has also been reported, although the exact influence of these has to be established.

Interestingly, many of the Ca^{2+} channels so far studied contain the amino acid sequence -Thr-X-Cys-Phe-Ile-Cys-Gly-. This is found in the C-terminal ends of the polypeptides and may play a role in the opening of the channel. It also seems to bestow thiol sensitivity to the channels, the SH groups of the cysteine residues being open to possible oxidation. It is conceivable, therefore, that such sequences are targets for reactive oxygen species (see discussion in Chapter 8, section 8.4).

Mitochondrial calcium metabolism

As well as sequestration of Ca^{2+} by the ER, other organelles also actively take up calcium ions. One of the most significant organelles which

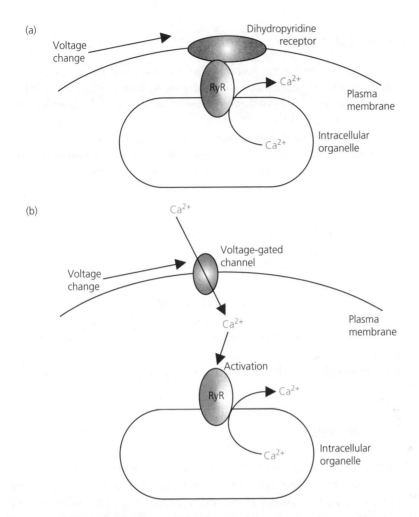

Fig. 7.8 The mechanisms for release of Ca^{2+} through ryanodine receptors (RyRs). (a) Voltage changes are sensed by a plasma membrane-associated receptor and protein conformational changes are relayed to the ryanodine receptor, causing its opening. (b) Voltage-gated Ca^{2+} channels allow small amounts of Ca^{2+} to enter the cell which trigger the ryanodine receptors to open.

partake in such activity are the mitochondria, which can acquire calcium ions through the use of a Ca^{2+} uniporter (Fig. 7.6), driven by the proton gradient across the inner mitochondrial membrane. This proton gradient is maintained by the respiratory electron transport chain. The affinity for Ca^{2+} of the pump is less than that of the ER pumps, but levels of Ca^{2+} ions inside the mitochondria can rise to approximately 0.5 mM. Once pumped into the mitochondria the calcium ions are not free, but appear to form a gel with phosphate, so allowing the calcium ions to be somewhat immobile, but in a state from which they can be rapidly released if required. Efflux from the mitochondria is via a Ca^{2+}/Na^+ exchanger, although the exact mechanism is uncertain, perhaps being 2 Na^+ for each Ca^{2+} (with no net movement of charge), or 3 Na^+ for each Ca^{2+} (with net movement of charge).

The proton gradient

The mitochondrial electron transport chain uses the energy from the transfer of electrons to pump protons across the inner mitochondrial membrane, and although this H^+ gradient is there extensively to drive ATP synthesis, it can also be utilized by many other proteins. For example, ADP and ATP are transported using its energy, as is phosphate, while it can also be utilized for the transport of calcium ions. In this latter case, the protein is referred to as a uniporter as it has movement of ions in one direction only, as opposed to an antiporter, where movement of one molecule in one direction is accompanied by the movement of another in the opposite direction.

The mitochondria sequester Ca^{2+} from the cytoplasm, and release it when necessary, and therefore are a useful organelle for the control of cytoplasmic calcium, like the ER, but the Ca^{2+} in the mitochondria is also used to control the activity of enzymes such as pyruvate dehydrogenase. This enzyme acts as a crucial link between glycolysis and the citric acid cycle and hence is involved in the production of ATP. Calcium ion concentrations in the mitochondria have also been linked to the onset of degenerative diseases, where a lack of control can lead to a depletion of ATP in neuronal tissue, leading to neuronal death.

7.5 Nerve cells

The plasma membrane re-entry of Ca^{2+} ions in neuronal tissue differs from that of other cells. Nerve cells contain voltage-gated calcium ion channels. Depolarization of the resting membrane potential causes a conformational change in Ca^{2+}-selective ion channels. The channel proteins contain special regions, S4 regions, which act as voltage-sensing regions. The membrane potential then supplies the driving force for the movement of the Ca^{2+} ions through the open channels. The process to a certain extent is self-controlling, as the channels appear to close in a time-dependent manner, and if the membrane is once again depolarized then the driving force—that is, the membrane potential—is depressed, not supplying the energy for Ca^{2+} movement.

The internal stores of calcium in neurons are released mainly through RyRs as discussed above for Ca^{2+} release from the ER.

7.6 Gradients, waves, and oscillations

An interesting and challenging aspect of the calcium-signalling story is the fact that the concentration of calcium is not uniform throughout the cytoplasm, or indeed in the compartment in which it might be sequestered. Intuitively, it would be expected that if calcium ions rush in from the outside of the cell through the plasma membrane then the concentration would rapidly dissipate throughout the cytoplasm by diffusion. However, this is clearly not the case. It has been estimated that calcium ions only migrate for approximately 50 μs, covering a distance of only 0.1–0.5 μm. By that time the ions would have been picked up by local binding proteins. However, such estimations will be altered by the saturation of the binding proteins and by the uneven distribution of such proteins. Some buffering proteins might be free in the cytoplasm, while others might be bound and immobile, again altering the equation. Therefore, a gradient of Ca^{2+} concentration may extend away from the site of Ca^{2+} release. Such gradients are called micro-gradients as they occur within a space of 1–10 μm. The highest concentrations of Ca^{2+} may reach up to 1 μM. As well as these micro-gradients very high local concentrations will be found around an open channel. Mobilization through a channel may well be more efficient than the diffusion of the Ca^{2+} ions through the cytosol, resulting in a very steep gradient extending away from that channel. These gradients will decay within tens of nanometres of the channel opening and are therefore referred to as nano-gradients. Typically such nano-gradients will only last for less than a millisecond but they could have Ca^{2+} concentrations as high as 100 μM. Gradients of Ca^{2+} have been proposed to be important in the control of cell migration, exocytosis, ion transport, and gap-junction regulation. Such uneven distribution of $[Ca^{2+}]_i$ is often visualized using fluorescent probes in conjunction with confocal microscopy (see discussion below). Hot-spots of Ca^{2+} are seen, often near the point of perception of a signal or stress. Therefore, these calcium signals are local, not pertinent to the whole cell, and global measures of both $[Ca^{2+}]_i$ and resulting enzyme activities are often erroneous: such data have to be interpreted with caution.

Other phenomena which remain as puzzles are the observations that the calcium concentrations can be measured as oscillations or waves. That is, the calcium ion concentration in the cytoplasm, for example, appears to rise steeply, drop steeply, and then rise again, in a series of oscillations. Therefore, it might not be the overall concentration of the calcium that actually constitutes the signal but either the amplitude or frequency of the concentration oscillation. Such oscillations, if created locally in the cell, perhaps on a particular part of the ER, will not necessarily be static either, and may move through the cell, like a wave of Ca^{2+}. Again, this gives new parameters to what the signal might be: above a threshold of Ca^{2+} or the frequency of the wave.

So how could a wave of Ca^{2+} be created in the cell? The creation of a wave might use the following as a typical mechanism, with such a scheme that could propagate a wave of calcium across the membrane illustrated

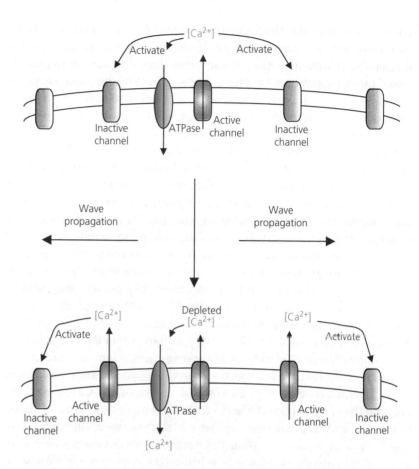

Fig. 7.9 A proposed scheme to explain the propagation of waves of calcium ions along the membranes of cells.

in Fig. 7.9. It is probable that local concentrations of Ca^{2+} are created around the opening of a channel in the endoplasmic reticulum, which in turn may cause the activation of channels in the locality, so releasing Ca^{2+} further along the membrane. Simultaneously, such local concentrations will inhibit the first channel from releasing any more calcium while the local Ca^{2+} ATPases will start to actively remove the ions, once again lowering the Ca^{2+} concentration in the location of the first channel to open. The new release further along the membrane can then initiate the same response with the channels and ATPases even further along the membrane and so the wave propagates out from the site of initiation. Interestingly, it has been reported that overexpression of the Ca^{2+}-ATPase pump of ER increases the frequency of the concentration waves recorded. Such waves may move through the cytosol at speeds of 5–100 μm/s and interestingly wave speed may depend on the concentration of agonist applied to the cell.

Importantly, calcium ion waves have been reported to travel from cell to cell through gap junctions which, as we saw in Chapter 1 (section 1.4), allow the passage of small signalling molecules from cell to cell, and therefore such a mechanism may constitute a longer-range signalling mechanism than first envisaged. Such signalling is important to consider when we discuss apoptosis. Here, if one cell is programmed to die, the organism may wish to prevent the spread of the death signal, and potentially detrimental Ca^{2+} movement through gap junctions will be prevented if the gap junctions close.

Having considered how waves might be created and propagated, let us turn back to oscillations. These were first demonstrated in blow fly salivary glands where it was observed that the Ca^{2+} concentration could be measured as a series of spikes and that the frequency of the spikes increased as the agonist concentration increased. Subsequent research has revealed that the patterns of the oscillation in fact depend on many factors, including agonist concentration, the type of receptor activated, and Ca^{2+}-buffering capacity of the cell. So how do such oscillations occur? Several models have been put forward to try to explain this phenomenon. One such model is known as the $InsP_3$–Ca^{2+} cross-coupling model (ICC model). Here, an extracellular signal would lead to the production of $InsP_3$ as outlined above (Fig. 7.1) and this would lead to the subsequent rise in cytoplasmic Ca^{2+} concentration. This increase in $[Ca^{2+}]_i$ may then stimulate the activity of phospholipase C leading to more $InsP_3$ and hence a greater release of Ca^{2+} from intracellular stores. However, $InsP_3R$ is inhibited by high concentrations of Ca^{2+}, and therefore once the cytoplasmic Ca^{2+} reaches a critical level the $InsP_3R$ channels are closed and no further Ca^{2+} release takes place. The cell now actively removes the calcium ions from the cytoplasm, as the pumps would be activated, and so restores the cell back to the non-stimulated situation. The reduction in cytoplasmic Ca^{2+} would also lead to a reduction of phospholipase C activity and a reduction of $InsP_3$ in the cytoplasm. The presence of agonist can once again initiate a spike in calcium release. The release of Ca^{2+} from the intracellular stores in this model would be a consequence of the spiking in phospholipase C activity. However, research has shown that even if the cytoplasmic $InsP_3$ concentration is kept constant, the calcium ion concentration can still be seen to oscillate. Therefore further models are needed.

Two further models are based on the proposed presence of a single pool of Ca^{2+} which is able to be released, the one-pool model, or the presence of two independent pools of Ca^{2+} which could be released, the two-pool model. One pool implies that all the calcium is sequestered in organelles which are equal; that is, the kinetics of activation, uptake, and release are the same. Two pools implies that there is more than one type of calcium-sequestering organelle, where the kinetics of activation, uptake, and release are different. This allows for the release from one pool before the other, or the re-uptake into one before the other. For example, early release from one

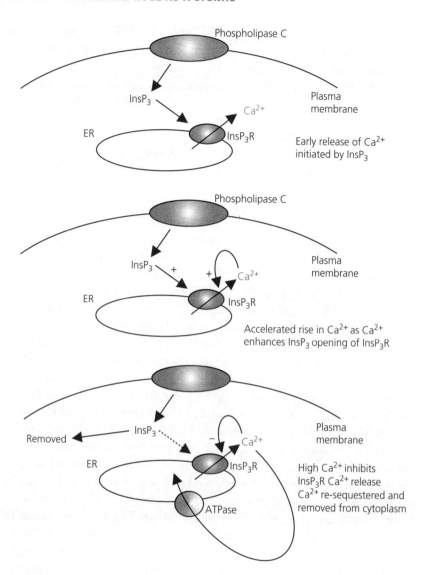

Fig. 7.10 A simplified scheme showing the control and role of calcium ion concentrations within the cell.

pool can cause rises in $[Ca^{2+}]_i$ which induces a later release in another. Of course, the true situation in some cells might be the presence of multiple pools, each of which has different kinetics, and which can be activated or inhibited by different arrays of molecules.

In the one-pool model (Fig. 7.10), the initial release of Ca^{2+} is relatively slow. However, because opening of $InsP_3Rs$ by $InsP_3$ is enhanced by the presence of Ca^{2+}, as the cytoplasmic Ca^{2+} rises, so the release of more Ca^{2+} through $InsP_3Rs$ is accelerated. This gives the rapid rise of Ca^{2+} seen as the early part of the spike. However, higher concentrations of Ca^{2+} inhibit $InsP_3Rs$ and the channels close, allowing no further rise in cytoplasmic calcium. A drop is seen as calcium ions are re-sequestered

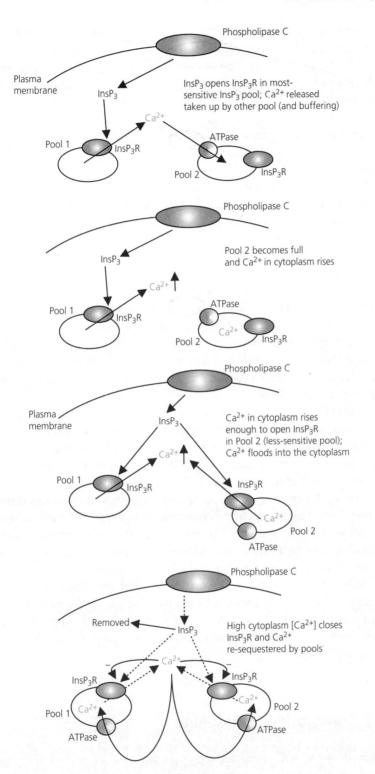

Fig. 7.11 Two-pool model of calcium ion oscillations. The two pools here have InsP$_3$Rs of differing sensitivities, but other models using two pools of calcium ions have also been proposed.

and removed from the cytoplasm by the various calcium ion pumps, perhaps activated by the high $[Ca^{2+}]_i$. It has also been suggested that the rise in Ca^{2+} causes the activation of phospholipase A_2, with the concomitant release of arachidonic acid. This could also potentially inhibit $InsP_3R$ and so play a role in the formation of the calcium spike.

As discussed above, the two-pool model suggests that Ca^{2+} may be sequestered into different intracellular stores which have receptors which differ in their sensitivity to $InsP_3$. In this model (Fig. 7.11), only the most sensitive $InsP_3Rs$ will cause a release, just from those independent stores. Other stores that are not releasing Ca^{2+} can still be actively taking in calcium and so their internal Ca^{2+} concentrations will rise, while cytoplasmic calcium ion buffers will also become saturated. Once the concentrations inside the second stores reach a critical level, their membrane pumps are shut off, and the cytoplasmic calcium level rises, as no more Ca^{2+} can be buffered and no more can be sequestered. Above a threshold cytoplasmic Ca^{2+} concentration all the $InsP_3Rs$ open and all the stores release calcium. Hence the cytoplasmic calcium ion concentration rises sharply, seen as the early phase of the spike. Again, as in the one-pool model, a high cytoplasmic Ca^{2+} concentration will result in the inhibition of $InsP_3Rs$ and the cessation of Ca^{2+} release. Pumps in the membranes will now actively re-sequester and remove Ca^{2+} from the cytoplasm, giving the down part of the spike.

A modification of the two-pool model may be that one store contains $InsP_3$ receptors while the other contains ryanodine receptors. It must also be remembered that it is not only the release and pumping of calcium at the intracellular store membranes which is important, but the plasma membrane will play a vital role in these events too.

Clearly, the control of $[Ca^{2+}]_i$ is complicated, and no doubt differs in subtle if not major ways in different cells. It is likely that a true consensus model will not be drawn up to encompass the full plethora of calcium signalling, but underlying principles will remain, and $[Ca^{2+}]_i$ will remain a central key to many signal transduction pathways.

Early calcium signalling research centred on $InsP_3$ as the main compound which was produced and able to release sequestered Ca^{2+}, but more recently several other molecules from apparently disparate sources have been found to be involved in calcium signalling. These will be briefly discussed below.

7.7 Sphingosine 1-phosphate

As well as the well-characterized route for calcium release through the production of $InsP_3$, another lipid metabolic pathway can also lead to the release of Ca^{2+} from intracellular stores, this being the sphingosine pathway as discussed in Chapter 6 (section 6.7). Here, one of the main active second messengers appears to be sphingosine 1-phosphate, which can mobilize Ca^{2+} from internal cell stores in an inositol-independent manner, and in a way that is independent of extracellular calcium. Sphingosine 1-phosphate can be produced from sphingosine by sphingosine kinase, a kinase which as expected is dependent on ATP as a source of phosphoryl group and which can be competitively inhibited by DL-threo-dihydrosphingosine. It is likely that the kinase is located within the ER membrane where sphingosine 1-phosphate will have its action.

Another metabolite which has been reported to be important in calcium ion oscillations in the cell is sphingosylphosphorylcholine (SPC), and although it seemed to be the same Ca^{2+} stores that were released by SPC as were released by $InsP_3$, the SPC response was not mediated by $InsP_3Rs$.

7.8 Cyclic ADP-ribose

As well as lipid derivatives such as $InsP_3$ and sphingosine 1-phosphate which can be used by cells to mobilize calcium to increase $[Ca^{2+}]_i$, there are other small signalling molecules which can be used too. One of the most studied is cyclic ADP-ribose (cADPr). cADPr is synthesized by a family of enzymes, ADP-ribosyl cyclases, which use oxidized nicotinamide adenine dinucleotide (NAD^+) as a substrate. These cyclase enzymes have been found in both invertebrate and mammalian cells. cADP-ribose has been shown to release Ca^{2+} in sea urchin eggs and, as mentioned above, may be an agonist for type 2 ryanodine receptors (RyR2) and maybe type 3 ryanodine receptors (RyR3). Therefore it has been proposed that cADPr has an important second-messenger role in many cells. The effect of cADPr on ryanodine receptors and therefore Ca^{2+} release can be inhibited by the presence of either Ruthenium Red or 8-amino-cADPr.

Levels of cADPr might well be modulated by cGMP in some instances, giving a control mechanism which will be physiologically important. cGMP is the product of guanylyl cyclases (see section 5.5) which may be under the control of the presence of the gas nitric oxide (NO). NO might be produced by the same cell or may be produced by a neighbouring cell. The production and roles of NO are discussed further in Chapter 8 (section 8.2). Therefore, a transduction pathway can be drawn in which NO causes increases in cGMP, which leads to increases in cADPr, causing a rise in intracellular calcium, which brings about the physiological response. An example of this is seen in plants, where such a pathway has been suggested to control, for example, stomatal movements, which themselves control gas exchange and water loss.

7.9 Nicotinate adenine-dinucleotide phosphate

Recently, nicotinate adenine dinucleotide phosphate ($NAADP^+$), a deaminated derivative of oxidized nicotinamide adenine dinucleotide phosphate ($NADP^+$), has been shown to have the capacity to release Ca^{2+} from intracellular stores. It is synthesized by the same family of enzymes that produce cADPr; that is, the ADP-ribosyl cyclases. Significantly, the ability of $NAADP^+$ to release calcium from stores is independent of the release triggered by $InsP_3$ and cADPr. For example, the release triggered by $NAADP^+$ was not inhibited by heparin, which would interfere with $InsP_3$ responses, and was not effected by the presence of Ruthenium Red, used to block the cADPr response, and therefore such results suggest a separate pathway which a cell could use for signalling the release of calcium ions. The $NAADP^+$ response was selectively inhibited by the presence of thionicotinamide-$NADP^+$ which will no doubt be of great use in the elucidation of the exact role of $NAADP^+$ in calcium signalling in many cells. Very recent work has shown that $NAADP^+$ concentrations are raised in cells following stimulation, and revealed that the $NAADP^+$ effects were centred on lysosomal-like vesicles, separate from the membranes targeted by $InsP_3$ and cADPr. Further, it was shown that Ca^{2+} released from $NAADP^+$-dependent stores could further influence the release of Ca^{2+} from $InsP_3$- and cADPr-sensitive stores. However, the identity of the $NAADP^+$ receptor, and the mechanisms used to control $NAADP^+$ levels in cells, at the moment remain elusive.

7.10 Fluorescence detection and confocal microscopy

Methods of detection of many molecules involved in biological functions have often relied on the presence of reporter molecules, and often these can be visualized because they are fluorescent. Sometimes the fluorescence is constitutive, and the research aims to find it or not. However, with many signalling molecules the fluorescent reporter probes have altered fluorescence characteristics depending on whether the reporter has bound or reacted with the signal or not. This might be seen as an increase or decrease in fluorescence, depending on the reporter used. For example, such fluorescent probes are widely used to detect reactive oxygen species and reactive nitrogen species, as discussed in Chapter 8 (section 8.5), and are also widely used to detect the presence, or rather the amount, of Ca^{2+} that resides in a particular location in the cell.

There is a plethora of fluorescent probes now available to detect Ca^{2+}, such as Fura-1 and Fluo-3, and it is certainly beyond the scope of the discussion here to cover them all. However, these chemicals all possess the ability to bind to free Ca^{2+} ions, and on binding their fluorescent spectral characteristics change. These changes can then be readily detected by several techniques. The light-emission change may be simply detected on a gross scale by the use of a fluorimeter, which will monitor the change in emitted light if the sample is excited with light of a preset wavelength, either fixed or scanning, but this technique is limited in that a population of cells is required to obtain a signal which is detectable. Fluorescence microscopes can overcome this problem, as single cells can be focused on, but interestingly such studies often show that two cells from the same tissue treated with the same stimulant may not necessarily respond in the same way.

An even more powerful technique is the use of a confocal scanning laser microscope. Here, the light source is from a laser as opposed to a normal white-light source, and the laser is driven to scan in the x–y plane. The laser is focused on to the sample, enabling a single image to be obtained from a very small area, often a single cell, but more exciting than that the fluorescence emission is also focused back to the detector, usually a photomultiplier tube, which sends a signal to a computer which itself creates the image on a VDU screen. The inherently grey image can then be enhanced by the use of pseudo-colour imaging, enabling the users to get an easy-to-see scale of fluorescence in the field of view. The power of this type of system comes from the fact that the laser is not only scanned on to the horizontal plane but the signal is re-focused on its way back from the sample, and the image is also extremely finely focused in the vertical plane. Therefore, the microscope can be used to scan in the z plane (the vertical plane) too. Hence, single cells, or indeed whole tissues, can be optically sectioned allowing a three-dimensional image to be created of the sample. These techniques have led to great advances in the study of the role of intracellular calcium as a signal. The concentration of Ca^{2+} ions inside organelles can now be visualized, giving an insight into the changes in these concentrations in, for example, the nucleus. New roles of calcium signalling, such as in the control of gene expression, will undoubtedly come to light as the use of such equipment expands. However, one of the most puzzling findings, discussed above in more detail, is the fact that the concentration of calcium ions in the cytoplasm is not even. Ca^{2+} is not uniformly distributed throughout the cytoplasm, but seen as hot-spots. Such imaging has revealed that other signalling molecules also reside temporarily in hot-spots too, and therefore our view of the cytoplasm being the "watery bit of the cell", where all things are even and able to diffuse, needs to be revised. The notion of hot-spots was discussed in section 5.6 and will be further considered in Chapter 12.

Another useful Ca^{2+} probe is the photoprotein aequorin. This protein emits light on binding to Ca^{2+}. Aequorin can be purified from the luminous jellyfish *Aequorea forskalea*, and can be used to sense calcium concentrations in the nanomolar to micromolar range.

7.11 SUMMARY

- Calcium ions are unusual signalling molecules in that they are neither created nor destroyed, but the signal comes from a rapid change in their distribution (Fig. 7.12).

- Cells usually expend large amounts of energy in actively removing Ca^{2+} ions, using pumps in their plasma membrane, which maintains a gradient where the intracellular Ca^{2+} concentration is maintained at less than 10^{-7} M, with the extracellular Ca^{2+} concentration in the region of 2×10^{-3} M.

- Pumps in the endoplasmic reticulum, sarcoplasmic reticulum, and mitochondria are also used to sequester Ca^{2+} into intracellular stores.

- A signal arises when the cytoplasmic Ca^{2+} concentration is allowed to rise, often by the release of Ca^{2+} from the intracellular stores,

commonly mediated by the presence of $InsP_3$ derived from the breakdown of lipids on the plasma membrane, an event which itself is tightly controlled.

- Although $InsP_3$ has been shown to have a major role in calcium release from intracellular stores, it is not the only means by which this can result. As well as $InsP_3$ receptors, ryanodine receptors are involved in calcium release from intracellular stores, and other signalling molecules such as $NAADP^+$, cADPr, and sphinosine 1-phosphate have been found to be important in many systems.

- Along with the presence of Ca^{2+} ions, the signal commonly relies on the presence of calcium ion-binding proteins.

- Calcium ion-binding proteins quite commonly contain polypeptide domains known as EF-hands.

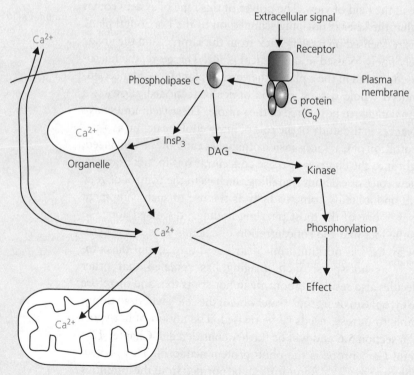

Fig. 7.12 Simplified scheme summarizing the control and role of calcium ion concentrations within a cell.

- The most common and certainly ubiquitous calcium ion-binding protein is calmodulin. This protein has two Ca^{2+}-binding domains (each capable of binding to two Ca^{2+} ions) connected by a flexible α-helix.

- Research into the movements of Ca^{2+} in cells has revealed the presence of hot-spots of high $[Ca^{2+}]$, oscillations in $[Ca^{2+}]$, and waves of high $[Ca^{2+}]$ moving through cells.

- Calcium is known to control many of the cells functions, including regulation of PKC, Ca^{2+}/calmodulin-dependent protein kinase, cell cycles, metabolic enzymes, and even other signalling routes such as that of nitric oxide synthase.

- The ubiquity of calcium signalling and its central importance in transduction cascades suggests that much more work will be carried out before the role of Ca^{2+} in cells is fully understood.

7.12 FURTHER READING

Berridge, M.J. (1993) Inositol trisphosphate and calcium signalling. *Nature* **361**, 315–325.

Berridge, M.J., Lipp, P., and Bootman, M.D. (2000) The versatility and universality of calcium signalling. *Nature Reviews Molecular Cell Biology* **1**, 11–21.

Clapham, D.E. (1995) Calcium signaling. *Cell* **80**, 259–268.

Clapham, D.E. (1995) Replenishing stores. *Nature* **375**, 634–635.

Clapham, D. E. and Sneyd, J. (1995) Intracellular calcium waves. In *Advances in Second Messengers and Phosphoprotein Research* (A.R. Means, ed.), pp. 1–24, Raven Press, New York.

Means, A.R. (1994) Calcium, calmodulin and cell-cycle regulation. *FEBS Letters* **347**, 1–4.

Nayal, M. and Di Cera, E. (1994) Predicting Ca^{2+} binding sites in proteins. *Proceedings of the National Academy of Sciences USA* **91**, 817–821.

Nicholls, D.G. and Ferguson, S.J. (2002) *Bioenergetics 3*. Academic Press. In particular sections 8.2.3 and 8.2.4 on calcium accumulation by mitochondria, and section 9.3.2 on measuring calcium concentrations.

Rutter, G.A., Burnett, P., Rizzuto, R., Brini, M., Murgia, M., Pozzan, T., Tavaré, J.M., and Denton, R.M. (1996) Subcellular imaging of intramitochondrial Ca^{2+} with recombinant targeted aequorin: significance for the regulation of pyruvate dehydrogenase activity. *Proceedings of the National Academy of Sciences USA* **93**, 5489–5494. An example of the use of aequorin.

Tran, Q.-K., Ohashi, K., and Watanabe, H. (2000) Calcium signalling in endothelial cells. *Cardiovascular Research* **48**, 13–22.

Calmodulin

Babu, Y., Sack, J.S., Greenhough, T.J., Bugg, C.E., Means, A.R., and Cook, W.J. (1985) 3-Dimensional structure of calmodulin. *Nature* **315**, 37–40.

Finn, B.E. and Forsen, S. (1995) The evolving model of calmodulin structure, function and activation. *Structure* **3**, 7–11.

James, P., Vorherr, T., and Carafoli, E. (1995) Calmodulin-binding domains: just two faced or multi-faceted. *Trends in Biochemical Sciences* **20**, 38–42.

Takuwa, N., Zhou, W., and Takuwa, Y. (1995) Calcium, calmodulin and cell-cycle progression. *Cellular Signalling* **7**, 93–104.

The plasma membrane and its role

Kiselyov, K. and Muallem, S. (1999) Fatty acids, diacylglycerol, Ins(1,4,5)P$_3$ and Ca^{2+} influx. *Trends in Neurosciences* **22**, 334–337.

Putney, Jr, J.W. and McKay, R.R. (1999) Capacitative calcium entry channels. *BioEssays* **21**, 38–46.

White, P.J., Bowen, H.C., Demidchik, V., Nichols, C., and Davies, J.M. (2002) *Biochemica et Biophysica Acta* **1564**, 299–309.

Intracellular stores

Finch, E.A. and Goldin, S.M. (1994) Calcium and inositol 1,4,5-trisphosphate-induced Ca^{2+} release. *Science* **265**, 813–815.

Meissner, G. (1994) Ryandodine receptor/Ca^{2+} release channels and their regulation by endogenous effectors. *Annual Review of Physiology* 56, 485–508.

Toyoshima, C., Sassabe, H., and Stokes, D.L. (1993) Three-dimensional cryo-electron microscopy of the calcium ion pump in the sarcoplasmic reticulum membrane. *Nature* 362, 469–471.

Gradients, waves, and oscillations

Allbritton, N.L. and Meyer, T. (1993) Localised calcium spikes and propagating calcium waves. *Cell Calcium* 14, 691–697.

Tsunoda, Y. (1991) Oscillatory Ca^{2+} signalling and its cellular function. *New Biology* 3, 3–17.

Sphingosine 1-phosphate

Ghosh, T.K., Bian, J., and Gill, D.L. (1994) Sphingosine 1-phosphate generated in the endoplasmic reticulum membrane activates release of stored calcium. *Journal of Biological Chemistry* 269, 22628–22635.

Mattie, M., Brooker, G., and Spiegel, S. (1994) Sphingosine 1-phosphate, a putative 2nd messenger, mobilizes calcium from internal stores via an inositol trisphosphate-independent pathway. *Journal of Biological Chemistry* 269, 3181–3188.

cADP-ribose

Gallione, A. (1993) Cyclic ADP-ribose: a new way to control calcium. *Science* 259, 325–326.

Lee, H.C. (1997) Mechanisms of calcium signalling by cyclic ADP-ribose and NAADP. *Physiological Reviews* 77, 1133–1164.

Nicotinate adenine-dinucleotide phosphate

Aarhus, R., Dickey, D.M., Graff, R.M., Gee, K.R., Walseth, T.F., and Lee, H.C. (1996) Activation and inactivation of Ca^{2+} release by $NAADP^+$. *Journal of Biological Chemistry* 271, 8513–8516.

Rutter, G.A. (2003) Calcium signalling: NAADP comes out of the shadows. *Biochemical Journal* 373, e3–e4. And papers cited within.

Confocal microscopy

Rizzuto, R., Brini, M., Murgia, M., and Pozzan, T. (1993) Microdomains with high Ca^{2+} close to $InsP_3$-sensitive channels that are sensed by neighbouring mitochondria. *Science* 262, 744–747.

Williams, D.A. (1993) Mechanims of calcium-release and propagation in cardiac-cells: do studies with confocal microscopy add to our understanding? *Cell Calcium* 14, 724–735.

Web-held images

Many companies, for example those that sell the machinery, and those that sell the probes, have confocal microscopy/Ca^{2+} ion images on their web pages. Many research teams too show their images on university-held web pages, so the use of a search engine with an author's name could easily reveal images from confocal microscopy work. A good place to start would be the site of Molecular Probes. Their URL is **www.probes.com**

8

Reactive oxygen species, reactive nitrogen species, and redox signalling

Compounds such as reactive oxygen species (ROS) and reactive nitrogen species (RNS) are now known to be important as signals in many systems. Early research in this field concentrated on the role of these compounds as destructive agents, targeted against either invading pathogens, or the cellular components themselves. However, more recent research has revealed a role of ROS and RNS in the control of a host of cellular functions. Although the action of many of these molecules is not fully understood, this chapter discusses the current state of our understanding of how they are generated, perceived, and propagate their messages. The main components involved here including those used in phosphorylation (see Chapter 4) and cyclic nucleotides as discussed in Chapter 5.

8.1 Introduction

The cellular production of reactive chemical species, especially those based on the reduced states of molecular oxygen, had been known for several years, but it was probably the realization that endothelium-derived relaxing factor (EDRF) was in fact nitric oxide, and that it had profound physiological effects, that started a new area of research, the role of such chemicals in cell signalling.

There are basically two groups of such chemicals:

- Reactive oxygen species, referred to as ROS or AOS (active oxygen species), which includes the superoxide anion ($O_2^{\bullet-}$) and hydrogen peroxide (H_2O_2).

- Reactive nitrogen species (RNS), which is mainly thought of as the nitric oxide radical, $NO^{\bullet}$. However, this can gain and lose electrons to give the NO^{-} and NO^{+} species too.

At a first glance these are rather surprising groups of molecules to be involved in cell signalling, and on the face of it these compounds would not appear to have the right criteria to be good signalling molecules, and moreover they have the potential capacity to be detrimental to the cells in which they are formed as well as to the cells around them. However, research papers appear regularly citing ever-increasing amounts of experimental evidence for roles for these molecules in the control of cellular functions, including apparent direct effects on gene transcription. Nitric oxide, for example, has been shown to regulate guanylyl cyclase activity, control neurotransmitter release, act as a neurotransmitter, as well as have a role in a bacterial and tumoricidal capacity.

If they are signalling molecules they should be able to exhibit some of the characteristics of signals which were outlined in Chapter 1 (section 1.3). That is, they should be relatively small, move easily from their site of production to their site of action, and have a unique and defined effect. They should also be able to be removed relatively easily and quickly. In fact, these molecules are all characterized by being small inorganic molecules, and as such can diffuse easily to their site of action. Furthermore, neither nitric oxide nor hydrogen peroxide are charged so they can diffuse through membranes, with the potential to move from one cell to another. However, they generally are extremely reactive and have known reactivity towards biological materials, often causing the latter's alteration and loss of function. Molecules such as superoxide and hydrogen peroxide have well-established classical functions in the body, in the destruction of invading organisms as a major part of host defence. In fact, this is not unique to the animal kingdom, with more and more literature now showing a similar role for superoxide and hydrogen peroxide in plants, possibly leading to local cell death and areas of necrosis. However, this characteristic means they are relatively short lived, and almost by default removed once their signal is perceived.

So it can be seen that, even if at face value they seem unlikely to be signalling molecules, if their production is controlled and their presence perceived then they can add to the cells' repertoire of signals. Certainly the amounts of ROS and/or NO present have been shown to correlate to the presence of many stimuli of cells, and to correlate to the responses seen. The following discussion looks at their production, what they might be controlling, and how they might be doing it.

8.2 Nitric oxide

The small gaseous molecule nitric oxide ($NO^{\bullet}$) has been found to be a significant signalling molecule, where its use is not restricted to one particular tissue but where it has functions in various and diverse locations. In fact, originally a factor which caused the relaxation of cells was described, known as endothelium-derived relaxing factor (EDRF), but in 1987 Moncada and colleagues realized that this activity was mediated by the molecule nitric oxide. Furthermore, it was found that some treatments for angina, for example the use of nitroglycerine, are mediated through nitric oxide. The nitroglycerine or other organic nitrates are converted to $NO^{\bullet}$ which causes relaxation of the blood vessels, hence increasing the heart's blood supply. More recently, some functions of brain and other nervous tissue have been shown to be inhibited by compounds which reduce the production of $NO^{\bullet}$. Although not really signalling, macrophages appear to involve the use of $NO^{\bullet}$ in part of their host-defence mechanism. Therefore, the production and roles of nitric oxide have become popular research projects.

Nitric oxide is in fact a free radical, commonly written as $NO^{\bullet}$ (the superscript dot denoting its radical status here). That is, it contains an unpaired electron in its outer electronic orbital. This leads to its increased reactivity as this is an unfavoured electronic state, and one from which a molecule will be keen to escape. Here, the $NO^{\bullet}$ can gain or lose an electron to form NO^{+} or NO^{-} (as discussed above) or can be converted to nitrates or nitrites. Alternatively, it can react with other potential signals such as superoxide to form the compound peroxynitrite, or glutathione to produce S-nitrosoglutathione (GSNO). Peroxynitrite is very reactive, and may be involved in either signalling events or cell death, while GSNO has been suggested to be a form of $NO^{\bullet}$ which may be transported around an organism, perhaps in the organism's vascular system.

As $NO^{\bullet}$ is so reactive, and its half-life is in the order of only 5–10 s, its effects are usually local. However, $NO^{\bullet}$ can readily diffuse across membranes, which means that its signalling action is not restricted to the cell of origin, but neighbouring cells can also feel the effect. The signal is usually turned off by deactivation of $NO^{\bullet}$ where $NO^{\bullet}$ is converted to nitrates and nitrites by oxygen and water.

There are several enzymes which can potentially produce $NO^{\bullet}$, but the most likely is the enzyme nitric oxide synthase (NOS). Here, $NO^{\bullet}$ is formed by the oxidation of L-arginine (Fig. 8.1). The guanidine group of arginine is oxidized in a process which uses five electrons, resulting in the formation of L-citrulline and nitric oxide through an intermediate step in which hydroxy-arginine is formed. The hydroxy-arginine intermediate remains tightly bound to the enzyme and is not released. Usefully for experimental design, the enzyme activity can be inhibited by the addition of L-N^{ω}-substituted arginines such as L-N^{ω}-aminoarginine (L-NAA) or L-N^{ω}-methylarginine (L-NMA). These substituted analogues act competitively with the binding of arginine although long exposure to some of these compounds leads to irreversible inhibition of the enzyme. Therefore, such compounds can be used to assess the role of NOS in $NO^{\bullet}$ production in new species of organisms or under new conditions.

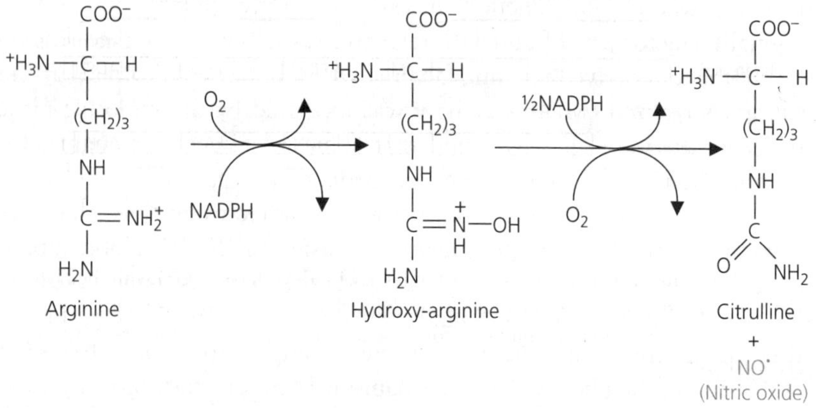

Fig. 8.1 The production of nitric oxide from arginine as catalysed by nitric oxide synthase. NADPH, nicotiamide adenine dinucleotide phosphate (reduced).

Nitric oxide synthase (NOS) is not a single enzyme entity, but rather a family of isoforms. Using the observation that calmodulin was required for activation of the enzyme, Bredt and Snyder first purified the enzyme NOS from brain tissue in 1990. This was denoted bNOS; a protein of approximately 160 kDa. This led to the purification by others of isoforms from macrophages (macNOS) and endothelial cells (eNOS, approximately 133 kDa). Subsequent Southern blot analysis has suggested that mammalian genomes contain three genes which encode NOS isoforms. Two of these are expressed constitutively, an endothelial-type enzyme (eNOS) and a neuronal type (nNOS), the latter of which has an expression pattern that includes non-neuronal tissues. The third gene encodes an inducible form (iNOS) which includes the enzyme seen in macrophages. Interestingly, some forms of the enzyme are seen as dimers *in vivo* and it has been suggested that some forms may exist in an equilibrium between monomeric and dimeric states.

Cloning of the nitric oxide isoenzymes has revealed that they all share a close homology with the enzyme cytochrome P450 reductase (Fig. 8.2): an NADPH-binding site (where NADPH is reduced nicotinamide adenine dinucleotide phosphate) has been identified, as well as areas for flavin adenine dinucleotide (FAD) and flavin mononucleotide (FMN) binding, and comparison of the sequences reveals consensus sequences for the binding of these cofactors. Purification of NOS has shown that each monomer has 1 FAD and 1 FMN although, like other enzymes, the FAD can dissociate slowly and has to be added exogenously to obtain full activity *in vitro*. The similarity to the cytochrome P450 system continues in that NOS also contains a haem prosthetic group—that is, iron protoporphyrin IX—as we saw with guanylyl cyclase in Chapter 5 (section 5.5). Like many haem-containing enzymes NOS can react with and be inhibited by carbon monoxide, giving a characteristic CO-binding absorbance spectrum

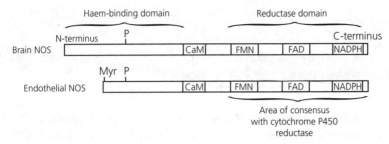

Fig. 8.2 The domain structure of isoforms of nitric oxide synthase (NOS) with the area of homology to cytochrome P450 reductase highlighted. CaM, calmodulin-binding site; FMN, FAD, and NADPH, binding sites for these prosthetic groups/co-factors; Myr, myristoylation site.

suggesting that the haem is attached via a cysteine residue. Such binding of carbon monoxide is characteristic of enzymes which have the capacity to bind to oxygen, as seen with cytochrome oxidase and haemoglobin. It appears that the first step of the catalysis is the binding of the arginine to the haem group, with subsequent oxidation reactions.

In the structure of NOS, between the binding regions for FMN and FAD and that for haem there is a region used for calmodulin binding, which in some cases confers calcium ion control on the enzyme activity. At the N-terminal end of this particular part of the polypeptide is a trypsin-sensitive region. Digestion with trypsin yields two domains, one from the N-terminal end of NOS which can bind to arginine and contains haem, and a second from the other end containing the NADPH- and flavin-binding regions. It has been suggested therefore that the synthase is a bi-domain enzyme, with one domain containing the oxygenase activity with the other domain containing the reductase activity, where the two domains may even be able to function independently of each other. However, the exact stoichiometry of the electron transfer has yet to be fully determined, with electron leakage possible to molecular oxygen which would yield the superoxide ion: also an unstable free radical (see below).

The purification of the brain enzyme showed that tetrahydrobiopterin was tightly bound to NOS and it was thought that it must function in the catalysis as it could take part in an electron transfer role.

Not only was binding of calmodulin to the enzyme NOS crucial to its original purification but it has become apparent that Ca^{2+} is important in the regulation of many isoforms of NOS. The concentrations of calcium involved are in the region to be expected for a calmodulin-activated system, with an EC_{50} of $(2-4) \times 10^{-7}$ M. Neither calcium nor calmodulin appear to affect arginine binding but calmodulin binding seems to regulate the electron transfer activities of the enzyme.

EC_{50}

The EC_{50} value is the concentration of an agonist that elicits a response that is 50% of the maximum response for that agonist.

As mentioned above, whereas some NOS enzymes are produced by the cells constitutively, other forms are inducible, including NOS of macrophages, designated mNOS or iNOS. This is a synthase of approximately 130 kDa and exists as a dimer of two identical subunits, that is, it is a homodimer. It is interesting to note that these isoforms contain calmodulin-binding sites but are in fact unaffected by calmodulin antagonists or Ca^{2+} and indeed calmodulin is tightly bound to these enzymes even in the absence of calcium. Production of new NOS protein molecules of these isoforms—which are inducible by nature—can be stimulated by the presence of, among many other things, interferon-γ (IFN-γ), interleukin-1 (IL-1), and lipopolysaccharide (LPS). Cloning of the transcription start site of the iNOS gene has shown that two distinct regions contain LPS- and interferon-responsive elements. The LPS region lies upstream about 50–200 base pairs (bp) of the transcription start site while region 2, responsive to interferon-γ is about 900–1000 bp upstream from the start site. Other response elements have also been reported, with such data indicating that the levels of iNOS protein in cells is carefully regulated. Furthermore, recently cells other than macrophages have also been shown to contain the inducible form of the enzyme.

As well as control by calcium ions, other control mechanisms for NOS also exist. Consensus sequences for phosphorylation by cAMP-dependent protein kinases (PKAs) have been found in NOS of brain and endothelial cells, although the macrophage form of the enzyme seems to lack them. PKA, protein kinase C (PKC), cGMP-dependent protein kinase, and Ca^{2+}/calmodulin-dependent protein kinase can all phosphorylate the neuronal form of the enzyme, and such phosphorylation results in a decrease in NOS activity.

It appears that it is not only the enzymatic activity of NOS that is regulated by phosphorylation but also the subcellular distribution of the enzyme in endothelial cells. The NOS of these cells is primarily located in the plasma membrane. On addition of bradykinin, a signalling cascade leads to NOS becoming phosphorylated and it is then translocated to the soluble fraction of the cells, albeit in an inactive state. This mechanism would ensure that the NOS enzyme is only active when it is located in a place where the nitric oxide is released from the cell; that is, when it is associated with the plasma membrane. Enzyme location may also be influenced by other factors. For example, the eNOS enzyme is myristoylated at the N-terminal end (see Fig. 8.2), and if this site is removed by site-directed mutagenesis the enzyme alters its location from the membrane to the cytosol. There are also reports that some NOS polypeptides may be palmitoylated, which would have a similar influence on cellular location.

But what does nitric oxide actually do once it is made and released? One of the main cellular targets of NO• is the enzyme guanylyl cyclase, the enzyme responsible for the production of cGMP, itself an important signalling messenger, as discussed in Chapter 5 (section 5.5). It is usually the soluble form of the cyclase which is the target of NO•, but some data have been reported that suggest that the membrane form may too be NO•-regulated to some extent. The activation of soluble guanylyl cyclase is caused by the binding of NO• to the haem group of the enzyme, the presence of iron within the haem group being the key here. Activation of guanylyl cyclase leads to the production of cGMP and a subsequent increase in intracellular cGMP concentration, assuming the cGMP is not removed as quickly by the appropriate phosphodiesterase. cGMP might be responsible for the regulation of serine/threonine kinases, cGMP-dependent protein kinase in particular, or cGMP may regulate the activity of some phosphodiesterases with a resultant modulation of any cAMP response, as described in Chapter 5 (section 5.6). In vascular tissues the activation of guanylyl cyclase by NO• causes muscle relaxation, probably by the activation of a cGMP-dependent protein kinase acting on myosin light chains. One of the well-studied physiological effects in which NO• plays a major role is in fact the control of smooth muscle contraction, and how this controls the flow of blood through the vessels, with probably the most-cited example being that of penile erection. For men with dysfunctional penile responses, the drug Viagra™ is available. It is in fact a misconception that Viagra™ either releases NO• or causes the generation of NO•. What it does is inhibit phosphodiesterase activity, in particular that of phosphodiesterase V. Normally the signalling pathway leads to NO• production, and NO• activates guanylyl cyclase, resulting in elevated cGMP levels, but these are soon removed by phosphodiesterase activity. By inhibiting this phosphodiesterase, the cGMP persists for longer in the cell, and the NO• signal is enhanced. The place in the pathway where Viagra™ acts is shown schematically in Fig. 8.3. Women too are now thought to be able to be helped by Viagra™, with the drug altering the blood supply to the uterus. Interestingly, some people that take Viagra™ have found that they have affected eyesight. This might sound bizarre, until the signalling pathway which relays the message from the light receptors to the brain is studied (see Chapter 10 (section 10.2) and previous discussion in Chapter 5 (section 5.5)). Here, the signalling pathway also involves cGMP, but this time a different isoform of phosphodiesterase is present, type VI, which is obviously affected by Viagra™ in some individuals.

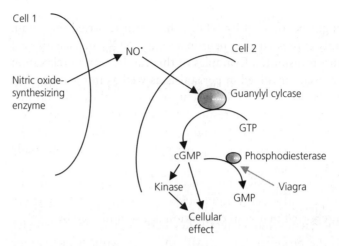

Fig. 8.3 A scheme showing how NO• might fit into a signalling pathway. NO• released from one cell might activate guanylyl cyclase in a neighbouring cell, resulting in an increase in cGMP, leading to a cellular effect. Viagra works by inhibiting the removal of cGMP, and therefore enhances the effects of NO•.

Haem groups

The prosthetic group haem is used commonly in biological systems, in a variety of manners. In haemoglobin, the haem is responsible for oxygen chelation, while in cytochromes it is involved in electron transfer. NO• can bind to many haem-containing proteins, and so alter their function. This might lead to changes in electron transport chain activity in mitochondria for example. NO• binding to haemoglobin is used as an assay for the presence of NO•.

In some enzymes NO• is also able to bind iron that is not associated with haem groups, for example in enzymes containing iron–sulphur complexes, such as NADH-ubiquinone oxidoreductase (where NADH is reduced nicotinamide adenine dinucleotide), otherwise known as complex I of the mitochondrial electron transport chain. The interaction of NO• with such non-haem iron has been implicated in the regulation of translation of some proteins, and even in the alteration of rates of DNA synthesis. Furthermore, the binding of NO• to ferritin can cause the liberation of free iron, which in the presence of oxygen free radicals may lead to lipid peroxidation of membranes through a mechanism that involves the production of hydroxyl radicals (see below). This would potentially be a very detrimental to many of the activities within a cell.

In neuronal tissues NO• is involved many facets of neuronal physiology. NO• has been implicated in the regulation of neurotransmitter release for example, and interestingly both the differentiation and regeneration of neurons also might be affected by NO•. It has been reported that brain NOS is transiently expressed after neuronal injury. NO• can itself also act as a specific neurotransmitter. Examples of this can be seen in relaxation of the smooth muscles involved in peristalsis as well as in the control of penile erections. However, if NO• concentrations rise above the normal very low levels, neurotoxicity can result. NO• can react with another free radical, superoxide, itself a possible product from NOS. This results in the production of a very reactive species, peroxynitrite, which potentially causes the cellular damage.

Besides its role as a cell signalling molecule, like superoxide and its byproducts, NO• can also act in tumoricidal and bactericidal capacities and it has been reported that viral replication is also inhibited by NO•.

Other enzymatic sources of NO•

NOS is not the only source of NO• in biological systems. One enzyme that for many years has been studied for its role in hydrogen peroxide production is xanthine oxidoreductase (XOR; otherwise referred to as xanthine oxidase (XO) or xanthine dehydrogenase). This enzyme can also generate NO•, particularly under anaerobic conditions. Therefore, here we have an enzyme which can generate ROS if oxygen is plentiful, or RNS if there is no oxygen. Both sets of products are powerful signalling molecules, and therefore this enzyme appears to have the ability to switch between signalling pathways, dependent on oxygen levels. Moreover, if oxygen is low but not absent, both ROS and RNS are produced simultaneously, with the subsequent generation of peroxynitrite, a mode of action thought to be involved in antimicrobial activity.

Antimicrobial activity of xanthine oxidoreductase

As XOR can simultaneously produce NO• and ROS, then peroxynitrite can be generated. XOR is found in cells, but also in extracellular fluids, particularly milk, and it has been shown by at least two research groups that milk has antimicrobial activity for which XOR is thought to be the main protagonist.

NO• generation by plants has been known about for a long time, but it was only in 1998 that NO• release was shown to be involved in pathogen-defence responses and the research area opened up. However, the source of NO• in plants has been difficult to determine. A look at the *Arabidopsis* genome database will not reveal the presence of an obvious NOS gene. Although NOS inhibitors in some cases do inhibit both NO• generation and NO•-induced effects, it is clear that there is more than one major source of $NO^{\bullet}$ in plants. The main candidates are nitrate reductase and NOS-like enzymes. Using mutant plants where the nitrate reductase activity is very low, it has been shown for example that this enzyme is key to $NO^{\bullet}$ generation which controls water loss in *Arabidopsis*. Like XOR, nitrate reductase is a molybdenum- and flavin-containing enzyme. The second potential $NO^{\bullet}$-generating enzymes in plants, which are NOS-like, have now been identified, one being a variant of the P protein of the glycine decarboxylase complex. This enzyme is inducible and has been referred to as an iNOS. Clearly the exact roles that such enzymes play in a variety of signalling pathways in plants need to be unravelled.

8.3 Reactive oxygen species: superoxide and hydrogen peroxide

The early discoveries in 1933 by Baldridge and Gerard that phagocytic cells had an increase in oxygen consumption when stimulated has led to a great interest in the molecular events which take place on the activation of these cells. Sbarra and Karnovsky in 1959 overturned the earlier view that the oxygen was used for increased respiration and by 1961 Iyer and his group noted that the oxygen consumption was accompanied by an increase in the hexose monophosphate shunt leading to the production of NADPH. It is now known that the oxygen taken up by phagocytic cells, such as neutrophils, is in fact used in the production of superoxide ions and, as discussed below, many more ROS as a result of further reactions. The production of superoxide, however, involves the direct enzymatic reduction of molecular oxygen by a protein complex situated in the plasma membrane of phagocytic cells: this enzyme complex is known as NADPH oxidase. NADPH oxidase has been best characterized in neutrophils, but is now studied in many cell types.

The importance of the production of ROS in neutrophils is clearly demonstrated by the disease chronic granulomatous disease (CGD). This is a genetic disease which manifests itself as a defect of the NADPH oxidase complex with complete abolition of any superoxide production. People with this disorder suffer from recurrent bacterial and fungal infections and until relatively recently died at an early age.

The primary product from the NADPH oxidase enzyme is the superoxide ion, where the electrons are supplied by intracellular NADPH:

$$2O_2 + NADPH \longrightarrow 2O_2^{\bullet -} + NADP^+ + H^+$$
Superoxide ion

The extra electron supplied to molecular oxygen is in an unpaired state, and hence, like nitric oxide, the superoxide ion is classified as a free radical.

Again, this electronic state is relatively unstable and the superoxide ion is consequently reasonably reactive. For example, it readily undergoes dismutation with the formation of hydrogen peroxide:

> ■ Hydrogen peroxide not itself a free radical but is often grouped in with the oxygen free radicals in discussions.

$$2O_2^{\bullet-} + 2H^+ \longrightarrow H_2O_2 + O_2$$
$$\text{Hydrogen}$$
$$\text{peroxide}$$

Although this dismutation reaction can occur spontaneously, especially at low pH (because it requires the presence of protons to proceed, as shown), it is also catalysed by enzymes called superoxide dismutases (SODs). This enzyme has two main forms, a copper/zinc-containing form which resides in the cytosol of cells and a manganese-containing form which is located in mitochondria.

Both superoxide ions and hydrogen peroxide are reactive towards biological materials, although it is probably the results of a further cascade of reactions which cause the most damage. Superoxide has been shown to cause biological oxidation, especially in hydrophobic environments, while hydrogen peroxide, a relatively weak oxidizing agent, has been found to inactivate some enzymes, usually by the oxidation of thiol groups (see discussion below). In a similar manner to the removal of superoxide by SOD, hydrogen peroxide is catalytically destroyed by either the enzyme catalase or by the glutathione cycle, and hence both superoxide and hydrogen peroxide can be removed quickly once their potential signals are no longer needed: as discussed above, this is a good characteristic of any potential signal.

The importance of the abolition of superoxide is demonstrated by the fact that all aerobic organisms appear to contain at least one isoform of SOD, life in the presence of oxygen being dependent on the destruction of harmful oxygen free radicals. However, the real danger comes when superoxide and hydrogen peroxide react together with formation of hydroxyl radicals:

$$O_2^{\bullet-} + H_2O_2 \longrightarrow OH^{\bullet} + OH^- + O_2$$
$$\text{Hydroxyl}$$
$$\text{radical}$$

This reaction is catalysed by the presence of iron ions (Fe^{2+}) or copper ions, (Cu^{2+}), proceeding via the Haber–Weiss reaction or Fenton reaction.

Hydroxyl radicals are extremely reactive, the free unpaired electron making them very unstable. In fact, it has been mooted that hydroxyl radicals are the most reactive chemicals found in biological systems. Damage can be caused to a cell in the form of oxidation of proteins, oxidation of bases which can lead to DNA-strand breakage, direct attack of deoxyribose moieties, and lipid peroxidation as mentioned above. Usually such reactions proceed via removal of hydrogen atoms from organic molecules, often leading to formation of new radicals and possibly a cascade of further free radical reactions. In the case of lipid peroxidation this can lead to membrane dysfunction.

Other ROS can also be formed as a consequence of the production of superoxide. These include singlet oxygen (O_2^1), which can lead to the destruction of carotenes, haem proteins, and membrane lipids, and also, in phagocytic cells where the haem-containing enzyme myeloperoxidase is present, the production of hypochlorite is seen.

As above when production of $NO^\bullet$ was discussed, it should be emphasized that superoxide can also react with $NO^\bullet$ to produce a very reactive compound, peroxynitrite:

$$NO^\bullet + O_2^{\bullet -} \longrightarrow OONO^-$$
$$\text{Peroxynitrite}$$

Therefore, it is possible that the production of superoxide ions modulates the role of nitric oxide, and the signalling in which it can be involved, albeit with the production of a very reactive compound.

Despite this apparent cascade of dangerous chemicals produced as a consequence of the release of superoxide, both superoxide ions and hydrogen peroxide appear to be released in non-phagocytic cells. Here the function is almost certainly not in host defence. The levels of superoxide are in the order of only 1% of that of neutrophils, while the production is sustained for very long periods of time, possibly constitutively by some cells. Different cell types might use these ROS in different ways but several lines of evidence point to a cell-signalling role in many instances.

It is not only organisms in the animal kingdom that show this activity. For example, there is now great interest in the release of ROS from plants, where they seem to involved in a range of physiological responses, from the defence against pathogen attack and the control of stomatal apertures and hence water loss, to the growth and gravitropic responses of roots.

Evidence for superoxide and hydrogen peroxide acting as a signal

How, if at all, are these reactive, destructive molecules being used as a way of signalling within cells? Studies are often done by adding exogenous

ROS or by the addition of free-radical scavengers and such conditions can promote or reduce the proliferation of cells in culture respectively. Early work by Crawford and colleagues showed that the addition of H_2O_2 or xanthine oxidoreductase/xanthine stimulated the c-*fos* and c-*myc* family of genes and switched on DNA synthesis. Other groups reported that H_2O_2 increased the expression of the genes c-*fos*, c-*jun*, *egr*-1, and JE in other cell lines. There was also an increase in the level of activator protein 1 (AP-1) DNA-binding activity. AP-1 is a transcription factor, a complex composed of *jun* and *fos* gene products. Of significance it was found that the transcription factor nuclear factor κB (NF-κB) was activated by H_2O_2 and it was suggested that ROS were serving as second messengers which directly mediated the release of IκB, the inhibitory subunit, from NF-κB. However, there is still some doubt as to whether this is a direct interaction between H_2O_2 and NF-κB, or whether NF-κB activation is mediated by a signal transduction pathway.

There is little doubt, however, that ROS are acting as signals. Pathways which can lead to altered gene expression, such as those involving MAP kinase and Janus Kinase/ signal transducers and activators of transcription (JAK/STAT) signaling, have been shown to be activated by ROS, while phosphatase activity has been shown to be inhibited. Normally, MAP kinases are rapidly activated in response to several external factors, promoting growth and differentiation. In neutrophils H_2O_2 caused an increase in tyrosine phosphorylation and it was concluded that activation of the MAP kinase was due to stimulation of tyrosine, and maybe threonine, phosphorylation of the kinase mediated by a MAP/ERK kinase, MEK. This activation of the MAP kinase was also concomitant with the inhibition by H_2O_2 of a MAP kinase phosphatase, shutting down the dephosphorylation of MAP kinase.

With the advent of the genome-sequencing projects and the ready availability of gene sequences, global gene-expression studies have been undertaken, showing that ROS both induce and reduce the expression of a plethora of genes. Approximately 3% of the genes expressed have their rates of transcription altered, the 97% not being altered being good evidence that the cells are not just being disrupted and killed.

Reports in the literature also suggest that H_2O_2 can cause activation of both guanylyl cyclase and phospholipase D. An increase in the activity of guanylyl cyclase results in formation of cGMP, as was seen above with NO•, while increased phospholipase D activity would instigate production of lipid-derived signals. Certainly, the production of prostaglandins and thromboxanes has been seen to be stimulated by the presence of hydrogen peroxide in some cell types.

ROS are also implicated in the onset and maintenance of apoptosis programmes (see Chapter 11 for discussion of apoptosis mechanisms), and so there is no doubt that in animals the production and perception of ROS is critical to the functioning of many cells.

Similar signalling by ROS has been noted in plants too. H_2O_2 has been shown to induce cellular-protection genes and was shown to act as a diffusible element that switches on gene expression in adjacent cells, in this case the genes for glutathione S-transferase and glutathione peroxidase, both of which code for products used in the protection of cells. Recently, as with animal cells, global gene-expression studies have been undertaken. In *Arabidopsis* just under 200 genes had their expression increased by H_2O_2, while expression of around 65 was depressed, out of approximately 8000 genes studied (the *Arabidopsis* genome has over 25 000 genes in total).

As well as the activation of gene expression, H_2O_2 has been shown to activate MAP kinases in plants as well. Defined isoforms of MAP kinases have been identified in plants that are activated by hydrogen peroxide and a recent report also shows that a kinase OXI1 is instrumental in H_2O_2 signalling here.

Recent work in yeast has shown that histidine kinases (see Chapter 4; section 4.5) might be involved in hydrogen peroxide signalling into cells, and it is possible that such a mechanism is more widespread than presently appreciated.

The NADPH oxidase complex

One of the most well-characterized sources of superoxide, and hence other ROS, is the enzyme NADPH oxidase. NADPH oxidase is in fact an enzyme complex catalysing the one-electron reduction of molecular oxygen to superoxide. The preferred electron donor is NADPH (K_m approximately 50 mM), and not NADH (K_m approximately 500 mM) and hence the complex was given its name.

NADPH oxidase is a membrane-bound complex consisting of a short electron transport chain, containing only two redox-active groups, a FAD moiety and a cytochrome (Fig. 8.4), although as we shall see both the haem and FAD group reside on the same protein, and the cytochrome is referred to as a flavo-cytochrome.

The cytochrome involved is a b-type cytochrome, located in the plasma membrane and specific granules of neutrophils, and it has been found primarily in the plasma membrane of other cells, such as fibroblasts and mesangial cells from the kidney. The cytochrome is unique in that it has a very low mid-point potential, originally measured as $E_{m7.0} = -245$ mV, considerably lower than that of other eukaryotic cytochromes. This extremely low mid-point potential is however sufficiently low to facilitate the reduction of molecular oxygen to superoxide and it has also been shown that the cytochrome is kinetically competent to act as the direct electron donor to oxygen. Because of this unique characteristic, the unusually low

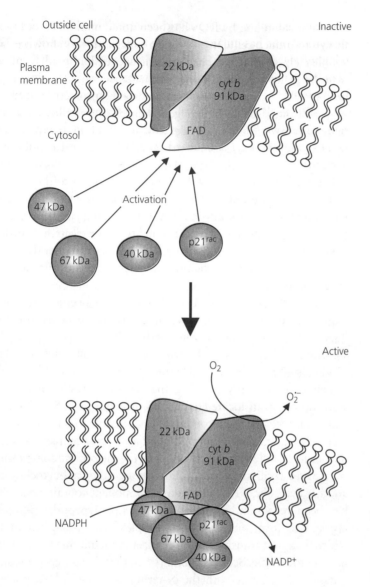

Fig. 8.4 A schematic representation of the NADPH oxidase complex and its activation by translocation of several cytosolic components to the plasma membrane. cyt, cytochrome.

mid-point potential, the cytochrome is often referred to as cytochrome b_{-245} but it is also referred to by the wavelength of its α-band absorbance in the visible spectrum, cytochrome b_{558}. In fact, more recent research has shown that the cytochrome has two haem groups, of different mid-point potentials, and hence the name cytochrome b_{-245} might be misleading.

The cytochrome is also unusual in being a heterodimer, containing a small α-subunit of approximate molecular weight 22 kDa. The small α-subunit has been named p22-*phox*, the *phox* referring to the fact that is was characterized from a phagocytic cell oxidase. The larger β-subunit of the cytochrome has an approximate molecular weight of 76–92 kDa, and is highly glycosylated; it is referred to as gp91-*phox*. The attachment of haem to protein often involves chelation across two histidines, and it is now known that the haem is associated with the larger gp91-*phox* sub-unit, with each haem being held between two α-helices that span the membrane, two histidines being used to hold each haem. To support this topology, gp91-*phox* is believed to contain several membrane-spanning domains.

The FAD group of NADPH oxidase also binds to the gp91-*phox* sub-unit, albeit in a domain that resides on the cytosolic side of the membrane. This domain also contains an NADPH-binding site. Therefore, the large gp91-*phox* contains all that is needed to allow electron transfer from the NADPH in the cytoplasm, through the FAD, subsequently on to the haem and finally to oxygen on the other surface of the membrane, where super-oxide can be released.

Besides the membrane-bound flavo-cytochrome, the oxidase also requires the presence of several cytosolic proteins for full activity (Fig. 8.4). By studying patients with CGD, two polypeptides have been identified as being integral with NADPH oxidase activity. These are of 47 kDa (p47-*phox*) and 67 kDa (p67-*phox*). Both p47-*phox* and p67-*phox* proteins are primarily found in the cytosol of resting neutrophils but on activation are translocated to the plasma membrane. Both have been cloned and sequenced and both contain SH3 domains, which are probably used in their interaction with other polypeptides in the complex. It is thought that the binding of the cytosolic proteins to the membrane complex in some way controls or facilitates electron flow though the flavocytochrome. Recently another cytosolic NADPH oxidase component containing an SH3 domain has been recognized. This is a 40-kDa polypeptide, p40-*phox*, which appears to form an activation complex with p47-*phox* and p67-*phox* which as a unit translocate to the plasma membrane to associate with the flavo-cytochrome. Sequence data reveal that this polypeptide shares a large region of homology with the N-terminus of p47-*phox*, but interestingly no CGD patients appear to have been identified who lack this protein, suggesting that it might have other roles, not just being associated with the NADPH oxidase system.

Other cytosolic proteins required for full activity of NADPH oxidase were found by reconstitution of activity from mixtures of cytosol and membranes. Besides p47-*phox* and p67-*phox* it was found that a hetero-dimeric complex was required. This complex contained a G protein related to Ras, p21^{rac}, and a GDP-dissociation inhibition factor, Rho-GDI. These two proteins undergo a cycle of dissociation and association with the subsequent turnover of GTP (as discussed more fully in Chapter 5, section 5.7).

Chronic granulomatous disease, non-phagocytes, and plants

If the role of ROS as signals in biological systems is to be understood it is necessary to consider what happens in patients who suffer from CGD. Here, the individuals have no NADPH oxidase system functioning, so if ROS are so vital why is signalling in these individuals not compromised?

CDG is a rare inherited disorder which is characterized as a lack of production of oxygen free radicals by neutrophils and the molecular defects have been defined as a lack of NADPH oxidase activity of these cells. This leads to a reduction in host defence with the patient suffering recurrent bacterial and fungal infections, leading to death if untreated.

The most common form of the disease is X-linked, accounting for approximately 50% of patients. This is usually seen as a complete lack of the activity of the gp91-*phox* cytochrome subunit. However, it has been reported that both subunits of the cytochrome are missing in this form of the disease as both subunits are needed for stable incorporation of the cytochrome into the plasma membrane. The gene for gp91-*phox* (CYBB) is located on chromosome Xp21.1. Restriction-fragment length polymorphisms (RFCPs) have been shown to occur in the gp91-*phox* gene with several lesions of the gene implicated in the cause of CGD. Small deletions, nonsense or missense mutations, regulatory region mutations, and splice-site defects have all been identified in CGD patients.

Approximately one-third of CGD patients have defects in p47-*phox*. The gene for p47-*phox* (NCF1) is located on chromosome 7q11.23 and hence the inheritance of defects in this gene is autosomal. A common defect of this gene is a dinucleotide deletion at a GTGT tandem repeat.

The rest of the CGD patients show either a defect in the p22-*phox* gene (CYTA) or a defect in the p67-*phox* gene (NCF2) with approximately 5% having each mutation. The p22-*phox* gene has been located on chromosome 16p24 with CGD being due to deletions, missense mutations, and frameshifts as seen with the gp91-*phox* gene. The p67-*phox* gene (NCF2) has been mapped to chromosome 1q25. Interestingly the polypeptide p40-*phox* was also reduced in amount in patients lacking p67-*phox* although no specific defects of this gene have been reported to be responsible for the occurrence of CGD.

So what of ROS signalling in such individuals? Early work to unravel this problem revealed that NADPH oxidase activity is not confined to phagocytic cells, but exists in cells such as fibroblasts, chondrocytes, and mesangial cells as well. Therefore, there appears to be a conundrum. If the NADPH oxidase has a crucial function in these cells, such as signalling, then how do patients with CGD with a complete lack of activity of NADPH oxidase overcome this difficulty, or are the signals themselves surplus to crucial functioning? Clues first came in 1993 when Meier's group looked at the NADPH oxidase of fibroblasts from CGD patients. It was found that although the neutrophils from an X-linked patient had severely depleted superoxide production when stimulated, the fibroblasts appeared to be unaffected. Further, the cells had a normal content of cytochrome b_{558} although antibodies directed against human neutrophil cytochrome did not recognize a similar cytochrome in fibroblast membranes. They suggested that the low-potential cytochrome b of the NADPH oxidase from neutrophils and that from fibroblasts must be structurally and genetically distinct.

The conundrum was further solved when isoforms of NADPH oxidase were discovered, and in fact there are several forms of the gp91-*phox* protein in humans. Some of these are much longer than those from the original enzyme, containing the original NADPH/FAD/haem domains and also a peroxidase domain. As they contain the oxidase and a peroxidase, these proteins have been referred to as the Duox proteins, from <u>du</u>al <u>ox</u>idase.

At the same time as the new human isoforms were being discovered, research was underway to reveal the source of ROS in plants, and here too a variant of the gp91-*phox* was found. In *Arabidopsis*, for example, several isoforms were found, and interestingly, as with some of the novel human isoforms, these were longer than the original from neutrophils, having an EF-hand towards the N-terminal end. Other species too have been found to have similar proteins, including soybean, tobacco, and pea. However, no Duox variants have been reported in plants.

It can be seen that there are several isoforms of the NADPH oxidase system in many species, from across the animal and plant kingdoms, and no doubt in other species too. Therefore, in CGD, if one isoform is missing, potential signalling roles of others are not affected. The challenge now is to find out how each of these isoforms contribute to any signalling that is taking place, how they are themselves controlled, and what their products are actually doing.

Other sources of superoxide

NADPH oxidase is not the only cellular source of superoxide and other ROS, and in fact electron leakage from most electron transport chains can give up electrons to oxygen with the formation of oxygen radicals. The mitochondrial electron transport chain, for example, can lose electrons from complex I (NADH-ubiquinone oxidoreductase) or complex IV (cytochrome oxidase), and electron leakage can occur in photosynthetic pathways of plants. It has been estimated that the average human doing average exercise will generate 2–3 g of superoxide ions per day! However there are no reported cases of the production of oxygen radicals from this source being used in a constructive way, unless their use during apoptosis is considered. Further, this electron leakage has been implicated in the reduction in the activity of these complexes, and it is thought that damage caused may lead to a further increase in the production of oxygen radicals.

Such a compounding effect is thought to lead to cellular death in some degenerative diseases such as Parkinson's disease and Alzheimer's disease, and in fact in the general ageing process.

A more likely source of oxygen free radicals is from the enzyme xanthine oxidoreductase (XOR). This is a molybdenum- and iron-containing flavoprotein which catalyses the oxidation of hypoxanthine to xanthine and then to uric acid. Molecular oxygen is used as the oxidant and what can be almost considered as a byproduct is hydrogen peroxide. In fact, the addition of hypoxanthine and xanthine oxidoreductase to cultured cells is a good experimental way of introducing hydrogen peroxide to the proximity of cells. However, under low oxygen tension this enzyme can also produce $NO^•$ (see above; section 8.2).

8.4 Redox signalling and molecular mechanisms of hydrogen peroxide signalling

Physiological effects of ROS can be seen and several proteins have been identified that have their activity altered by ROS. However, it is still not clear how ROS are perceived by cells. Although some affects on guanylyl cyclase are seen, this is not the primary way in which ROS are acting. Similarly, some researchers question whether ROS do have a direct effect on some proteins such as NF-κB. Therefore, how is ROS perceived, and how does it affect protein function?

One of the ways in which ROS might be acting is on the redox state of the cell. Cells contain a large concentration of glutathione and other redox-active compounds in the cytoplasm and organelles, and the majority of this glutathione is maintained in a reduced state (GSH). Therefore, if the Nernst equation is used the redox poise of the cytoplasm of a cell can be calculated: it is found in fact to be very negative, around -250 mV. ROS will react with glutathione, oxidizing it (forming GSSG). This has two consequences. First, the amount of available glutathione is lowered, perhaps so that it can no longer partake it certain reactions, and second, the redox state of the cytoplasm becomes more oxidizing. It has been suggested that certain cysteine thiol groups on proteins are maintained in a reduced state, until the redox state of the solution around them alters, allowing them to become oxidized and therefore alter the conformation of the protein on which they reside. This would certainly be an effective mechanism to alter the activity of certain proteins. During apoptosis of animal cells, the redox state of the cytoplasm is said to become approximately 70 mV more oxidizing, certainly enough to affect a thiol group that has an appropriate mid-point potential.

Nernst equation

The Nernst equation can be used to calculate either the proportions of a compound that exist in the oxidized and reduced states in a solution, or can be used to calculate the redox state of a solution.

Alternatively, ROS may act on the proteins directly. Exciting work on the mechanism by which hydrogen peroxide might inhibit the activity of phosphatases has recently been published. While studying the oxidized forms of the enzyme, where it was proposed that a cysteine residue in the active site of the enzyme is oxidized, an interesting cysteine derivative was found, the sulphenyl-amide intermediate (Fig. 8.5). This was formed by a reaction of the sulphur of the cysteine linking to the nitrogen of the serine which was next in the amino acid chain (Fig. 8.6). This species lacked the oxygen and its formation was fully reversible, whereas the irreversible forms, sulphinic acid and sulphonic acid, were only formed at higher hydrogen peroxide concentrations, as observed by peptide analysis using matrix-assisted laser desorption ionization-time-of-flight (MALDI-TOF) mass spectroscopy. Also importantly, the sulphenyl-amide intermediate could react with glutathione to produce glutathionylated derivatives, and hence potentially further signals (see Fig. 8.5).

Irreversibity of higher oxidation states of thiols

Although the formation of the sulphinic acid and sulphonic acid groups was thought to be irreversible, recent work has revealed an enzyme that can catalyse the oxidation of sulphinic acid to sulphenic acid, showing that reversibility is possible, and therefore the formation of such higher oxidation states on thiol groups could be involved in cell signalling processes.

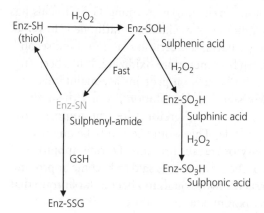

Fig. 8.5 The proposed oxidation states of protein cysteine residues and further derivatives.

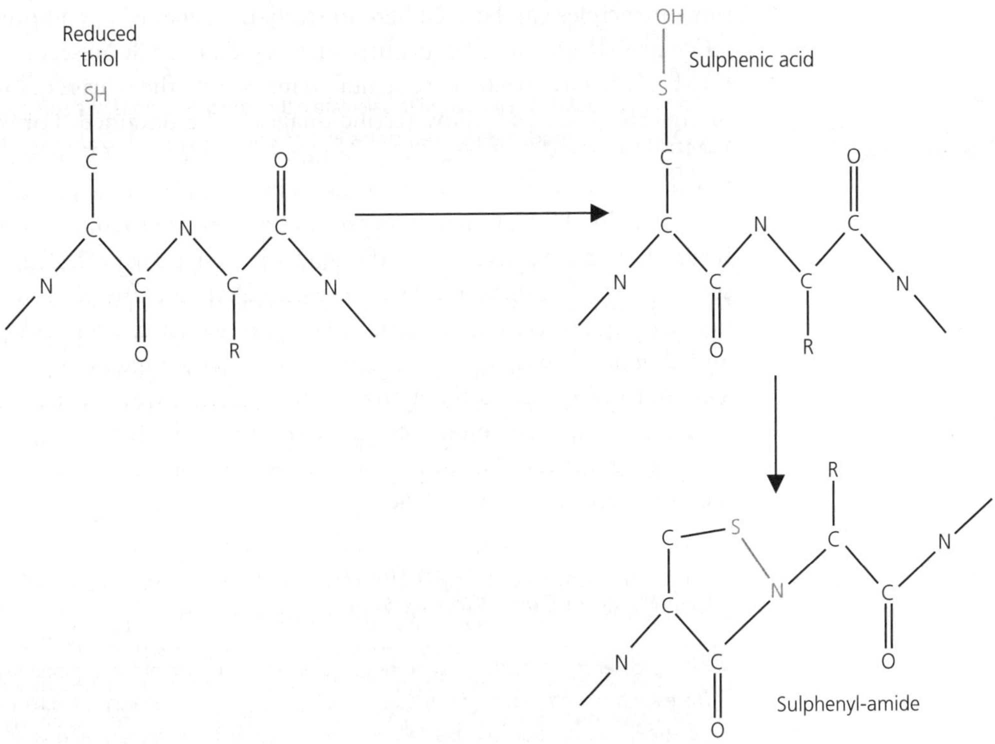

Fig. 8.6 The formation of the sulphenyl-amide derivative.

MALDI-TOF mass spectroscopy

Matrix-assisted laser desorption ionization-time-of-flight mass spectroscopy, along with electrospray ionization mass spectroscopy (ESI-MS), are extremely powerful ways to analyse proteins; either to identify them or study changes that might have taken place in them.

A mechanism such as the formation of the sulphenyl-amide intermediate highlights the fact that hydrogen peroxide and proteins can interact directly, with the alteration of protein conformation and therefore activity of the protein. However, it is unlikely that phosphatases are unique in this, and that they perceive all the hydrogen peroxide signalling needed, and so 'hydrogen peroxide perception proteins' need to be identified.

8.5 Measuring ROS and RNS

In Chapter 7 (section 7.10) the use of fluorescent probes was discussed as a way of finding the concentrations and location of Ca^{2+} ions, but the same principles can be used here to study the generation and presence of ROS and RNS. Specific probes such as diaminofluorescein diacetate (DAF-2DA) have been instrumental in measuring the release of NO• from many cell types, and allow terrific images to be obtained. For example, DAF-2DA has shown that only certain cells release NO• in response to hormones, such as guard cells in leaves in response to the plant hormone abscisic acid (ABA). Often these probes are supplied to the cell in an inactive form, but are converted by the cell to an active form. DAF-2DA is not sensitive to NO•, but once de-esterified to DAF it is. But DAF will not be taken in by the cell. So, to study NO• generation in cells, DAF-2DA is added, it is taken up by the cells, and the cells then metabolize it to DAF. This not only releases the active compound, but traps it inside the cell where it is required. Similar compounds are available to study ROS too, although great care has to be taken as not all compounds used for ROS measurements are very specific.

Delivery of compounds as esters

Often, to deliver compounds or probes to cells, an ester form of the compound is used. The ester forms are often more membrane-permeable, and they are taken up by the cell, de-esterified and trapped in the cell, unable to re-cross the membrane and escape. Once in the cell the de-esterified probe can monitor the signalling molecule that they are sensitive to. The esters themselves are often not sensitive to the signalling molecule, and anyway, any ester not taken up can be simply washed off, so that it is unable to interfere with the data being sought.

To elucidate redox changes in cells green fluorescent protein (GFP) has been engineered, and so it should be possible to get such a protein expressed in the cell, and the redox environment monitored. Often GFP and its derivatives can be monitored with confocal microscopy, allowing quantitative and spatial data to be obtained.

Of course fluorescence is not the only way of measuring ROS, RNS, and redox, and in fact fluorescence is notoriously difficult to quantify. Many other methods also exist for studying the release of these reactive compounds, including the haemoglobin assay for NO•, and luminol luminescence and cytochrome c assays for ROS.

8.6 Carbon monoxide

Recent evidence shows that guanylyl cyclase is not only under control by the gas nitric oxide, and possibly hydrogen peroxide, but that its activation might also be modulated by another gas, carbon monoxide. For example, evidence has been presented to show that this system may be involved in the regulation of insulin secretion as well as in the control of corticotropin-releasing hormone release in the hypothalamus. An isolated form of soluble guanylyl cyclase from bovine lung showed that carbon monoxide could bind to the haem group of the enzyme to form a six coordinate complex.

8.7 SUMMARY

- A group of small inorganic molecules have been discovered that are produced by cells and diffuse to neighbouring cells, or act on the cell that produced them, where they have a role in the control of many cellular functions.

- Here, signalling molecules include nitric oxide, superoxide ions, hydrogen peroxide, and carbon dioxide.

- Nitric oxide is produced in animals primarily by the enzyme nitric oxide synthase (NOS), with arginine as the substrate.

- Cloning of NOS has shown that it contains areas of homology with the enzyme cytochrome P450 reductase. Both contain domains for binding to NADPH, FMN, and FAD, whereas, like cytochrome P450, NOS also contains haem.

- The original role of nitric oxide was reported as a cellular relaxation factor, and it is now known that one of its functions is the control of cGMP production by guanylyl cyclase.

- Other enzymes which can generate $NO^{\bullet}$ include xanthine oxidoreductase and nitrate reductase.

- Oxygen free radicals such as superoxide can be produced by the electron leakage of many electron transport chains, including the mitochondrial complexes.

- Large amounts of superoxide are produced by the enzyme NADPH oxidase, and its role in the killing of invading pathogens is highlighted by its absence in chronic granulomatous disease (CGD).

- Reactive oxygen species, including the non-radical hydrogen peroxide, have been shown to control cellular proliferation and the rates of transcription of cells, perhaps either through a direct action on transcription factors or by the stimulation of MAP kinase-type pathways.

- The molecular mechanisms by which hydrogen peroxide and nitric oxide relay their messages are now being unravelled, but include nitrosylation and oxidation of proteins.

8.8 FURTHER READING

Nicholls, D.G. and Ferguson, S.J. (2002) *Bioenergetics* 3. Academic Press. An example book on redox, now in its 3rd edition.

Nitric oxide

Bredt D.S. and Synder S.H. (1990) Isolation of nitric oxide synthase, a calmodulin-requiring enzyme. *Proceedings of the National Academy of Science, USA* 87, 682–685.

Burnett, A.L., Lowenstein, C.J., Bredt, D.S., Chang, T.S.K., and Snyder, S.H. (1992) Nitric oxide: A physiologic mediator of penile erection. *Science* 257, 401–403.

Chandok, M.R., Ytterberg, A.J., van Wijk, K.J., and Klessig, D.F. (2003) The pathogen-inducible nitric oxide synthase (iNOS) in plants is a variant of the P protein of the glycine decarboxylase complex. *Cell* 113, 469–482.

Desikan, R., Griffiths, R., Hancock, J.T., and Neill, S.J. (2002) A new role for an old enzyme: Nitrate reductase-mediated nitric oxide generation is required for abscisic acid-induced stomatal closure in *Arabidopsis thaliana*. *Proceedings of the National Academy of Science USA* **99**, 16319–16324.

Furchgatt, R.F. (ed.) (1995) Special topic: nitric oxide. *Annual Review of Physiology* **57**, 659–790. A collection of several excellent articles on the production and role of nitric oxide.

Godber, B.L.J., Doel, J.J., Durgan, J., Eisenthal, R., and Harrison, R. (2000) A new route to peroxynitrite: a role for xanthine oxidoreductase. *FEBS Letters* **475**, 93–96.

Guo, F.-Q., Okamoto, M., and Crawford, N.M. (2003) Identification of a plant nitric oxide synthase gene involved in hormonal signalling. *Science* **203**, 100–103.

Hancock, J.T., Salisbury, V., Ovejero-Boglione, M.C., Cherry, R., Hoare, C., Eisenthal, R., and Harrison, R. (2002) Antimicrobial properties of milk: dependence on the presence of xanthine oxidase and nitrite. *Antimicrobial Agents and Chemotherapy* **46**, 3308–3310.

Neill, S.J., Desikan, R., and Hancock, J.T. (2003) Nitric oxide signalling in plants. *New Phytologist* **159**, 11–35.

Palmer, R.M.J., Ferrige, A.G., and Moncada, S. (1987) Nitric oxide release accounts for the biological activity of endothelium derived relaxing factor. *Nature* **327**, 524–526.

Stuehr D.J., Cho H.J., Kwon, N.S., Weise, M.F., and Nathan C.F. (1991) Purification and characterisation of the cytokine induced macrophage nitric oxide synthase, an FAD containing and FMN containing flavoprotein. *Proceedings of the National Academy of Science, USA* **88**, 7773–7777.

Superoxide and hydrogen peroxide

Baldridge, C.W. and Gerard, R.W. (1933) The extra respiration of phagocytosis. *American Journal of Physiology* **103**, 235–236.

Biteau, B., Labarre, J., and Toledano, M.B. (2003) ATP-dependent reduction of cysteine-sulphinic acid by *S. cerevisiae* sulphiredoxin. *Nature* **425**, 980–984.

Cooper, C., Patel, R.P., Brookes, P.S., and Darley-Usmar, V. M. (2002) Nanotransducers in cellular redox signaling: Modification of thiols by reactive oxygen and nitrogen species *Trends in Biochemical Sciences* **27**, 489–492.

Hancock, J.T., Desikan, R., Neill, S.J., and Cross A.R. (2003) New equations for redox and nano signal transduction. *Journal of Theoretical Biology* **226**, 65–68.

Harman, D. (1972) The biologic clock: the mitochondria? *Journal of the American Geriatric Society* **20**, 145–147.

Heyworth, P.G., Cross, A.R., and Curnutte, J.T. (2003) Chronic granulomatous disease. *Current Opinion in Immunology* **15**, 578–584.

Iyer, G., Islam, M.F., and Quastel, J.H. (1961) Biochemical aspects of phagocytosis. *Nature* **192**, 535–542.

Jones, O.T.G. and Hancock, J.T. (2000) NADPH oxidase of neutrophils and other cells. In *Free Radicals in Inflammation* (Winyard, P.G., Blake, D.R., and Evans, Ch.H., eds,) pp. 21–46. Birkhäuser, Basel.

Levine A., Tenhaken, R., Dixon, R., and Lamb, C. (1994) H_2O_2 from the oxidative burst orchestrates the plant hypersensitive disease resistance response. *Cell* **79**, 583–593.

Neill, S.J., Desikan, R., and Hancock, J.T. (2002) Hydrogen peroxide signalling. *Current Opinion in Plant Biology* **5**, 388–395.

Rhee, S.G., Bae, Y.S., Lee, S.-R., and Kwon, J. (2000) Hydrogen peroxide: a key messenger that modulates protein phosphorylation through cysteine oxidation. *Science's Signal Transduction Knowledge Environment*, http://stkesciencemag.org/cgi/content/full/OC_sigtrans; 2000/53/pe1.

Salmeen, A., Anderson, J.N., Myers, M.P., Meng, T.-C., Hinks, J.A., Tonks, N.K., and Barford, D. (2003) Redox regulation of protein tyrosine phosphatase 1B involves a sulphenyl-amide intermediate. *Nature* **423**, 769–773.

Sbarra, A.J. and Karnovsky, M.L. (1959) The molecular basis of phagocytosis. *Journal of Biological Chemistry* **234**, 1355–1362.

Schafer, F.Q. and Buettner G.R. (2001) Redox environment of the cell as viewed through the redox state of the glutathione disulphide/glutathione couple. *Free Radicals in Biology and Medicine* **30**, 1191–1212.

Schreck, R., Rieber, P., and Baeuerle, P.A. (1991) Reactive oxygen intermediates as apparently widely used messengers in the activation of the NF-κB transcription factor and HIV-1. *EMBO Journal* **10**, 2247–2258.

Van Montfort, R.L.M., Congreve, M., Tisi, D., Carr, R., and Jhoti, H. (2003) Oxidation state of the active-site cysteine in protein tyrosine phospatase 1B. *Nature* **423**, 773–777.

Measuring ROS and RNS

Hancock, J.T. and Jones, O.T.G. (1994) Assays of plasma membrane NADPH oxidase. *Methods in Enzymology* 233, 222–229.

Hempel, S.L., Buettner, G.R., O'Malley, Y.Q., Wessels, D.A., and Flaherty, D.M. (1999) Dihydrofluorescein diacetate is superior for detecting intracellular oxidants: comparison with 2′,7′-dichlorodihydrofluorescein diacetate, 5(and 6)-carboxy-2′,7′-dichlorodihydrofluorescein diacetate, and dihydrorhodamine 123. *Free Radicals Biology and Medicine* 27, 146–159.

Østergaard, H., Henriksen, A., Hansen, F.G., and Winther, J.R. (2001) Shedding light on disulphide bond formation: engineering a redox switch in green fluorescent protein. *EMBO Journal* 20, 5853–5862.

Carbon monoxide

Stone, J.R. and Marletta, M.A. (1994) Soluble guanylate cyclase from bovine lung: activation with nitric oxide and carbon monoxide and spectral characterization of the ferrous and ferric states. *Biochemistry* 33, 5636–5640.

Vara, E., Ariasdiaz, J., Garcis, C., and Balibrea, J.L. (1994) Evidence for a cGMP-dependent, both carbon monoxide and nitric oxide mediated, signalling system in the regulation of islet insulin secretion. *Diabetologia* 37, A112.

Insulin and the signal transduction cascades it invokes

9.1 The insulin signalling system

9.2 SUMMARY

9.3 FURTHER READING

The preceding chapters in this book describe the components which may be recruited to transmit signals in biological systems. However, most commonly a variety of such components must come together to form a signalling pathway. Here, in Chapter 9, insulin signalling is used as an example of such a system. As will be discussed, once insulin has been produced, transported, and perceived, a surprisingly diverse signalling transduction mechanism may be used to bring about the final alteration of cellular function. Dysfunction here may cause diabetes, and therefore a discussion of insulin highlights the importance of signalling to normal cellular function.

9.1 The insulin signalling system

In the preceding chapters the separate components of possible signal transduction pathways have been discussed. Often, although a component is found in one pathway, it has profound effects, or in sometimes even more important effects, in another transduction pathways. However, none of these signalling components work in isolation, and therefore it is important to illustrate how they might come together in a pathway. Here, insulin is used as an example of an extremely important hormone which results from, and leads to, complex signalling. It is made in one cell, stored until needed, released when the secretory mechanism is suitably stimulated, travels to its target cell, is recognized by the extracellular surface of that cell and, through a cascade of various messengers, results in many cellular effects. Such a scheme, resulting in the movement of glucose transporters in cells is shown in Fig. 9.1. Some of the effects resulting from insulin binding are listed in Table 9.1.

Some of the insulin-induced effects are seen through modulation of cytosolic enzymes, while other effects include control of gene expression

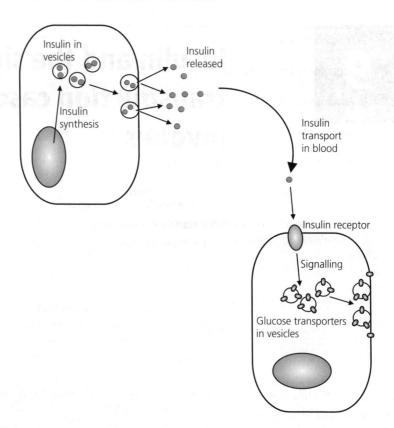

Fig. 9.1 A simplified overview of insulin signalling. Insulin is made and stored in one cell until needed. Under the control of a signal transduction pathway it is released, travels to its site of perception on another cell, and then more signalling controls the movement of glucose-transport proteins to the plasma membrane to alter glucose-uptake rates.

inside the nucleus. In general insulin will promote anabolic processes of a cell while causing a reduction in catabolic processes. For example, insulin will promote the synthesis of glycogen and fatty acids while simultaneously inhibiting their breakdown. An alteration of insulin signalling, either its production, its detection, or subsequent signaling, can lead to the disease diabetes mellitus.

Anabolic and catabolic

Anabolic refers to reactions that require energy, such as the synthesis of fats, while catabolic refers to reactions that transform cellular fuels, such as fats, to usable energy. The sum of anabolism and catabolism may be referred to as metabolism.

Insulin is produced by the cells of the islets of Langerhans. These are cell clusters found in the pancreas. As well as insulin these cells are responsible for the production of another important hormone, glucagon. Glucagon, a single-polypeptide-chain hormone, along with insulin, is

Phosphorylation of IRS1
Activation of PtdIns 3-kinase
Activation of Ras
Phosphorylation of kinases; e.g. MAP kinases, ribosomal S6 kinase
Phosphorylation of phosphatases; e.g. PP1
Phosphorylation of metabolic enzymes; e.g. glycogen
Dephosphorylation of metabolic enzymes; e.g. glycogen synthase/phosphorylase kinase
Translocation of proteins; GLUT-4, insulin receptors
Regulation of gene expression
Modulation of protein synthesis

Table 9.1 A list of some of the cellular effects induced by insulin.

responsible for the regulation of the concentration of glucose in the bloodstream of mammals.

The route for the production of insulin has been studied in a human tumour of the islet cells, where insulin was produced in large amounts. Incorporation of tritiated leucine into the insulin polypeptide allowed analysis of the various steps, and therefore this process is well characterized. Production starts with the synthesis of a single polypeptide chain known as preproinsulin. At the N-terminal end of the preproinsulin molecule is a 19-amino-acid signal sequence, which is relatively hydrophobic and directs the polypeptide to the endoplasmic reticulum. In the lumen of the endoplasmic reticulum the 19 amino acids are removed proteolytically to form proinsulin. Proinsulin is passed subsequently through the Golgi apparatus of the cell and into secretory granules, where a 33-amino-acid stretch of polypeptide is removed from the middle of the chain. When analysed, even from different species, the sequence removed was found to contain a -Lys-Arg- sequence at the C-terminal end with an -Arg-Arg- sequence at the N-terminal end: such conserved amino acids serve as recognition sequences for the protease responsible for the cleavage. Once removed, this leaves two polypeptide chains, an A chain of 21 amino acids and a B chain of 30 amino acids, which are connected by two disulphide bridges. A third disulphide bridge also spans across two cysteine residues in the smaller chain of amino acids, the A chain (Fig. 9.2). The amino acid

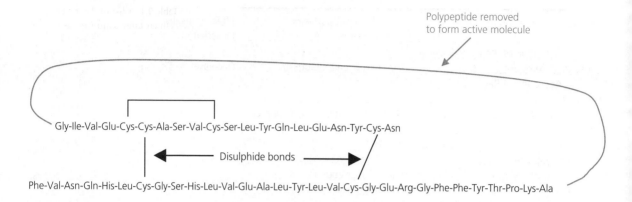

Polypeptide removed
to form active molecule

Gly-Ile-Val-Glu-Cys-Cys-Ala-Ser-Val-Cys-Ser-Leu-Tyr-Gln-Leu-Glu-Asn-Tyr-Cys-Asn

← Disulphide bonds →

Phe-Val-Asn-Gln-His-Leu-Cys-Gly-Ser-His-Leu-Val-Glu-Ala-Leu-Tyr-Leu-Val-Cys-Gly-Glu-Arg-Gly-Phe-Phe-Tyr-Thr-Pro-Lys-Ala

Fig. 9.2 The amino acid sequence of insulin. The disulphide bonds are indicated by the black lines, while the connecting peptide that is removed is shown by the blue line.

sequence of insulin was determined in 1953 by Frederick Sanger who won the Nobel Prize for Chemistry in 1958, while its structure was studied by Dorothy Hodgkin, who won the Nobel Prize for Chemistry in 1964 for her work on vitamin B_{12}.

The three-dimensional structure of insulin, when determined to a resolution of 1.9 Å, showed that as well as the three disulphide bonds, the structure is held together by salt links and hydrogen bonds. The molecule assumes a basically globular structure, with only the N-terminal and C-terminal ends of the B chain reaching out into the solution.

It is worth noting that insulin is not totally unique, and that a second molecule shares great similarity to insulin; that is, insulin-like growth factor (IGF-I). This is a hormone which is similar in both sequence and structure to insulin and is produced in the liver. Its release is controlled by pituitary hormones and the role of IGF-I seems to be in the control of an organism's growth.

Once made, insulin is contained in secretory granules in the cell, until such time as it is needed. Translocation of these secretory granules, and their fusion with the plasma membrane, allowing the release of the insulin into the extracellular fluid, is both under the control of other hormones and in responds to neuronal signals. Hormones arriving at insulin-containing cells will bind to a receptor, and lead to a signal transduction pathway, resulting in insulin release, in the manner shown in Figs 1.2 and 9.1. Kinases such as 5′-AMP-activated protein kinase (AMPK) are thought to be involved here, where inhibition of AMPK in the presence of glucose activates insulin secretion. The resultant level of insulin measurable in the blood is usually very low, only in the order of 10^{-10} M, and it is this concentration which must be perceived by the insulin-responsive cells.

As stated earlier the effects of insulin are fairly numerous. Some of the effects of insulin are seen relatively quickly, while others can take a matter of hours. This suggests a complicated signalling pathway, or set of pathways,

Enzyme	Metabolic pathway involved	Effect of insulin on activity
Phosphorylase kinase	Glycogen metabolism	Decrease
Pyruvate kinase	Glycolysis	Decrease
Lipase	Lipid breakdown	Decrease
Glycogen synthase	Glycogen synthesis	Increase
Pyruvate dehydrogenase	Citric acid cycle	Increase
Acetyl-CoA carboxylase	Fatty acid synthesis	Increase
Hydroxymethylglutaryl-CoA reductase	Cholesterol synthesis	Increase

Table 9.2 Some enzymes that have their activity modulated through the action of insulin.

that are used by insulin, not a straight path leading to a single effect. The immediate effects of insulin include the modulation of the activities of various enzymes in different metabolic pathways, some of which are listed in Table 9.2, and also include an increase in the rate of glucose uptake.

Usually effects such as those as listed in Table 9.2 are stimulated by blood insulin levels in the order of 10^{-9} to 10^{-10} M. Longer exposure of cells to higher concentrations of insulin (approximately 10^{-8} M) will induce protein synthesis of enzymes involved in glycogen synthesis in the liver, enzymes needed in triglycerol synthesis in adipose tissue, and in some cells such as fibroblasts such levels of insulin will stimulate proliferation. However, such high concentrations are generally not physiological, although the measurement of local concentrations of signalling molecules is often not carried out.

Local concentrations

Often experiments are carried out to measure the concentration of hormones or other signals using a group of cells, perhaps from tissue culture or a whole tissue, so an average concentration is recorded, but the actual concentrations present around the surface of a cell, in the immediate locale of a receptor, are not measured. Usually these are extremely hard to measure, so it is possible that the concentrations of signals may be much higher, or lower, than thought.

Clearly, insulin is released by one cell type, and has profound effects elsewhere, in other tissues and cells. Therefore, one of the key events is the

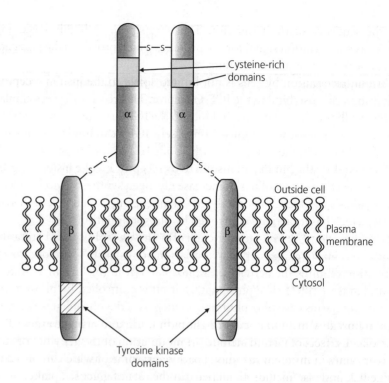

Fig. 9.3 A model of the structure of the insulin receptor.

perception of insulin by cells. Insulin is detected on the target cell mainly by the insulin receptor. This receptor was first purified by Pedro Cuatrecasa. The receptor has a binding affinity for insulin of approximately 10^{-10} M, the same order of magnitude as the concentration of insulin in the blood which needs to be detected. The insulin receptor is a glycoprotein which contains two large α-subunits of 135 kDa and two smaller β-subunits of 95 kDa, and therefore would be described as having an $\alpha_2\beta_2$ subunit structure. The β-subunits are integral to plasma membrane while the α-subunits reside on the outside of the cell membrane and are held to the β-subunits by disulphide bonds as depicted in Fig. 9.3. As was discussed in Chapter 3 (section 3.1), receptors are often found in dimers, especially following ligand binding. Even though the insulin receptor is a tetramer, perhaps it would be interesting to view it as a dimer, where each part of the dimer has two subunits, an α and a β, and therefore the insulin receptor can be thought of as a dimer of dimers. In this way, perhaps, it is not too unusual, and has ligand binding similar to many other receptors.

In fact, even though there are two different subunits in the insulin receptor, the α and β subunits of the receptor are synthesized as a single polypeptide chain of 1382 amino acids, which is subsequently cut up to release the active α and β polypeptides. The precursor peptide has a signal peptide at the N-terminal end which is cleaved off and this leaves the α-subunit, followed by a tetrapeptide with the sequence -Arg-Lys-Arg-Arg-, followed by the sequence of the β-subunit. The tetrapeptide is highly basic and serves as a recognition signal for the protease which processes the polypeptide, releasing the α and β subunits.

Insulin recognition by cells is not limited solely to the insulin receptor, but insulin can also bind to the IGF-I receptor. This receptor is very similar to the insulin receptor but insulin binds to the IGF-I receptor with approximately 100-times-lower affinity. Conversely, IGF-I can bind to the insulin receptor but again the binding is with about 100-times-lower affinity.

So what does the binding of insulin to its receptor invoke inside the cell? One of the effects of insulin is to increase glycogen synthesis by cells. Here, the synthesis of glycogen is catalysed by an enzyme called glycogen synthase, which has its activity controlled by phosphorylation (see Chapter 4, section 4.1, and Fig. 4.2 in particular). In the phosphorylated state the synthase is inactive, and dephosphorylation activates it. Therefore, to get activation requires the activity of a phosphatase, in this instance protein phosphatase 1 (PP1). The phosphatase itself is controlled by phosphorylation, but in this case phosphorylation increases the phosphatase's activity. So how does insulin fit here? Insulin binding to its receptor activates tyrosine kinase activity (discussed in more detail below), which leads to the activation of what has been referred to as insulin-sensitive protein kinase or insulin-stimulated protein kinase (ISPK). This kinase phosphorylates PP1, activating it, and therefore the active PP1 can dephosphorylate glycogen synthase, activating it, and the synthesis of glycogen is increased. As discussed in earlier chapters, futile cycles in metabolism are to be avoided (see section 4.1), and PP1 also has action on phosphorylase to prevent the instant breakdown of the glycogen formed.

As well as the rapid effects on glycogen metabolism, one of the major effects of insulin on cells is the resultant increased capacity of the cell to be able to take up glucose, and so leading to a lowering of blood glucose concentrations, a so-called hypoglycaemic effect. Within approximately 15 min of the application of insulin to adipocytes, their rate of uptake of glucose will have increased between 10 and 20-fold. This short time span would not be sufficient to allow for the synthesis of new protein and hence new glucose transporters. However, analysis has shown that the number of glucose transporters found in the plasma membrane will have greatly increased during this time. Glucose transporters, otherwise known as the GLUT4 proteins, are synthesized and targeted to endosome-like vesicles. They are made as integral membrane proteins, as their role is to transport glucose across the plasma membrane, and as they travel through the cell they remain as integral membrane proteins of vesicles. However, these vesicles remain in the cytoplasm, where the GLUT4 proteins have no chance to function, until the cell is stimulated by insulin. Once stimulated, the vesicles will be transported to the plasma membrane, where the membranes fuse and the GLUT4 transporters become part of the plasma membrane where they can function in the rapid uptake of glucose. On the removal of the insulin signal, the GLUT4 proteins can be endocytosed back into the interior of the cell, and so the uptake of glucose can once again be reduced. Recent work where the transporters have been tagged with a fluorescent marker has allowed this movement of the GLUT4 polypeptides to be visualized using a confocal microscope (see section 1.10 for further discussion of this technique).

So what is the link between the insulin receptor and the vesicles containing the GLUT4, and to investigate the pathway in a logical order, perhaps, how does the insulin receptor relay the message, from the recognition and binding of insulin, to the control of the intracellular mechanisms it affects? The α-chains of the receptor contain cysteine-rich domains and it is on these subunits that insulin binding takes place. The β-subunits on the other hand contain domains which contain tyrosine kinase activity (see Fig. 9.3), and hence the receptors are classed as receptor tyrosine kinases (RTKs). On binding of insulin to the receptor, there will be a defined change in the conformation of the subunits of the receptor, relayed from the binding site on the α-subunits to the β-subunits and subsequently through the membrane to have an influence on events on the intracellular side of the plasma membrane. The first response is that the receptor phosphorylates itself on at least three tyrosine residues; that is, it autophosphorylates (Fig. 9.4).

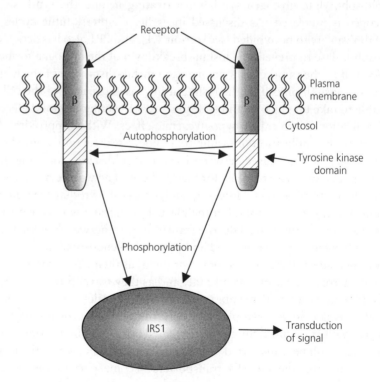

Fig. 9.4 A schematic representation of the phosphorylation events which result from the activation of the insulin receptor.

This has two results. Firstly, autophosphorylation activates the receptor to phosphorylate other cellular proteins. Secondly, even if the insulin is subsequently removed from the receptor, the receptor remains active unless the receptor itself is dephosphoylated by a phosphatase.

However, one of the key events in the signal transduction from the insulin receptor is multi-phosphorylation of a protein known as insulin receptor substrate 1 (IRS1), as illustrated in Fig. 9.4. The IRS proteins are actually a family of at least four proteins (IRS1–IRS4), all having PH domains and phosphotyrosine-binding domains. IRS1 is a protein of 130 kDa, and can be phosphorylated by either the insulin receptor or the IGF-I receptor. IRS1 is highly serine-phosphorylated and also can be highly tyrosine-phosphorylated by these receptors. Six of the sites of tyrosine phosphorylation on this protein have the neighbouring sequence of -Tyr-(Pro)-Met-X-Met- (where X could be a variety of amino acids).

IRS1 acts as a relay protein, and once phosphorylated and therefore activated it can interact and modulate the activity of different pathways. One of the things that IRS1 does is to bind to SH2 domains. As IRS1 is phosphorylated on tyrosine residues, this creates binding sites for the SH2 domains of other proteins (see Chapter 1, section 1.7). One such SH2 domain is contained on the regulatory subunit of PtdIns 3-kinase, causing the kinase's activation. PtdIns 3-kinase is also phosphorylated on a tyrosine by the insulin receptor. As discussed in Chapter 6 (section 6.2), activation of PtdIns 3-kinase will lead to the production in phosphorylated forms of membrane phosphatidylinositol lipids, on the 3 position of the inositol ring. Particularly, $PtdIns(4,5)P_2$ will be converted to $PtdIns(3,4,5)P_3$, which in turn will activate 3-phosphoinositide-dependent kinase (PDK1). This will then be able to phosphorylate and activate protein kinase B (PKB) and perhaps also PKC (see Fig. 9.5). Downstream signalling from these two kinases will lead to translocation of GLUT4 transporter proteins to the plasma membrane through movement of GLUT4-containing vesicles, as shown in Fig. 9.1.

Another substrate of the insulin receptor is the Cbl–CAP complex. Cbl is a proto-oncogene product, and it becomes associated with the insulin receptor aided by the protein CAP, hence the origin of its name, Cbl-associated protein. Cbl is phosphorylated by the activated insulin receptor, and dissociates from the receptor, upon which the Cbl–CAP complex moves to become associated with lipid rafts. Flotillin in the lipid raft associates with the Cbl–CAP proteins, and further signalling is invoked which leads to the translocation of GLUT4-containing vesicles to the plasma membrane (Fig. 9.6). The exact nature of the signalling downstream from the lipid raft has yet to be elucidated.

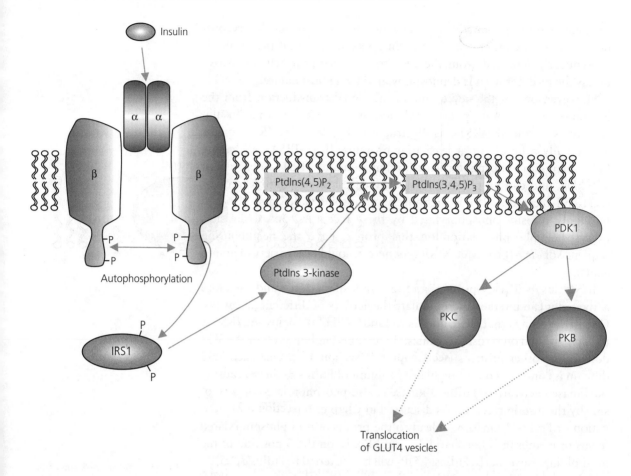

Fig. 9.5 One of the pathways in which insulin signals to control the movement of glucose transporters to the plasma membrane. Here, PtdIns 3-kinase control is the key.

Lipid rafts

Lipid rafts are sub-areas of phospholipid membranes which have been shown to have a structure different from the rest of the membrane. They are like islands of a particular structure, floating in the sea of the membrane. Particular proteins and hence functions have been associated with these rafts, and much interest is now focused on what they actually do, and also how their function might be modulated, perhaps in the form of new drug therapies.

As discussed above, insulin signalling uses a wide range of components, including kinases, phosphatases, adaptor proteins, lipid rafts, and yet-to-be-discovered components, to control the rates of metabolism and the intake of glucose. However, there is another pathway which has been investigated, one very similar to that used by other growth factor-like molecules, and leads to events in the nucleus. This pathway also starts with perception of insulin by its receptor, and also involves the phosphorylation

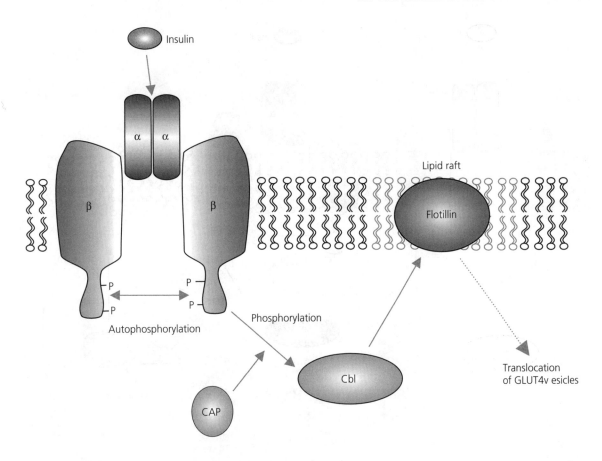

Fig. 9.6 An alternative signalling pathway used by insulin, the Cbl–CAP pathway. CAP, Cbl-associated protein.

of IRS1. Phosphorylation of tyrosine residues in proteins such as IRS1 creates binding sites for SH2 domains, and this is used here. Once phosphorylated, IRS1 interacts with an adaptor protein called GRB2, through the latter's SH2 domains. The GRB2 protein also contains two SH3 domains which can further bind to a guanine nucleotide-releasing protein, in this case Sos (see Chapter 5, section 5.7). Sos can, once activated, catalyse the release of GDP from the monomeric G protein Ras and therefore allow the G protein to take up its active state in which GTP is bound. Ras can then activate the kinase Raf which leads to the phosphorylation and activation of MAP kinase cascades. Such a pathway is illustrated schematically in Fig. 9.7.

Similar pathways, different details

It is perhaps worth noting that the scheme illustrating the insulin-induced pathway leading to gene expression, Fig. 9.7, is similar to that shown in previous chapters, e.g. Fig. 4.8. This highlights the similarities used throughout cell signalling. The principles are often the same, it is the detail which gives specificity.

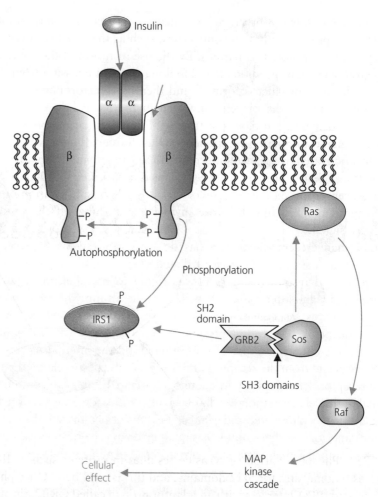

Fig. 9.7 A signalling cascade associated with insulin detection which leads to alterations of gene expression. Binding of insulin to its receptor leads to autophosphorylation of the receptor and phosphorylation of a relay protein, IRS1. The phosphotyrosines formed on IRS1 interact with the SH2 domains of the adaptor GRB2 and in a similar fashion to the EGF receptor pathway this leads to sequential involvement of Sos, Ras, and then a MAP cascade, which transmits the signal into the cell, often resulting in enhanced transcription.

One of the end results in the activation of this MAP kinase cascade is the alteration of gene expression. MAP kinases will, once activated in cytoplasm, translocate to the nucleus, where they have their action, that of phosphorylating the next protein in the pathway, perhaps a transcription factor. In fact, the promoter regions of the genes which have their expression altered by the presence of insulin at the cell surface have also come under close scrutiny. Termed insulin-responsive elements (IREs), such regions have been determined in several genes including those for phosphoenolpyruvate carboxylase (PEPCK), amylase, liver pyruvate kinase, and glyceraldehyde-3-phosphate dehydrogenase (GAPDH). In general they seem to contain two 8-bp elements which are AT-rich, and both of these need to be present, although some genes, such as that for liver pyruvate kinase, do not follow this pattern. By the use of fusion of the luciferase gene to the relevant promoters, and looking for light emission from cells where the gene product is formed, and then by transforming cells with mutated forms of the proteins involved in the MAP kinase pathway, Tavaré and colleagues have illustrated the importance of some of the above signal transduction components in insulin signalling.

Once activated by the insulin receptor, IRS1 can also interact with other proteins, such as a protein called Syp. This is a tyrosine phosphatase and it has been proposed that it is involved in the dephosphorylation of the IRS1 protein itself and therefore involved in turning off the insulin-induced signal.

Like many receptors, once it has done its job, the insulin receptor itself is internalized and down-regulated (see Chapter 3, section 3.4). Radiolabelled insulin (using ^{125}I) was found to be taken into the cell quite rapidly and after 5 min up to 30% of the insulin which had bound to the cell was associated with internal vesicles. Once internalized the environment of the receptor protein will change, there will be an alteration in the conformational of the protein, and the internalized insulin will dissociate from the receptor, and it is generally destroyed by the cell. The receptor, while it is still in the phosphorylated state, can continue to signal to the cell, even though it has been internalized, but normally it will become dephosphorylated and can then be translocated back to the plasma membrane for another round of signalling. Several phosphatases have been identified which may be involved in the dephosphorylation of the insulin receptor. Of course, recycling the receptors is a sensible strategy for the cell, as discussed further in Chapter 3 (section 3.4).

9.2 SUMMARY

- Insulin has varied and profound effects on many aspects of a cell's functioning, ranging from the control of metabolic pathways and the stimulation of protein translocation to the control of gene expression.

- Insulin is composed of two peptide chains, held together by disulphide bonds, hydrogen bonds, and salt bridges.

- Insulin is made by the cells of the islets of Langerhans of the pancreas as a single gene product which is modified by cleavage.

- Once made, insulin is stored in vesicles until needed. On stimulation of the cells, insulin is released and

washed around the body by the blood, until it reaches its target.

- Insulin is detected by cells mainly by the insulin receptor, but can also bind to the receptor for insulin-like growth factor-I (IGF-I).

- The insulin receptor is a tetramer with an $\alpha_2\beta_2$ subunit structure. The α-subunits are on the outside of the cell and are responsible for insulin binding, whereas the β-subunits contain transmembrane domains along with tyrosine kinase domains.

- Binding of insulin to its receptor results in autophosphorylation of the receptor on tyrosine residues, and also results in phosphorylation of other cellular proteins, such as insulin receptor substrate 1 (IRS1).

- Activation of IRS1 leads to activation of other proteins through the interaction of its phospho-tyrosyl residues with the SH2 domains of other polypeptides.

- The PtdIns 3-kinase pathway is recruited by insulin, leading to the formation of PtdIns(3,4,5)P_3, activation of PDK1, activation of PKB/PKC, and translocation of GLUT4 to the plasma membrane.

- The transloction of GLUT4 has also been found to involve the Cbl–CAP complex, and lipid rafts.

- IRS1 also interacts with an adaptor protein such as GRB2 which can lead to activation of MAP kinase cascades and the modulation of gene expression.

- Insulin signalling can be turned off by the internalization of the receptor, which may be recycled back to the plasma membrane. Integral in this is the dephosphorylation of various proteins which are involved in insulin signalling.

9.3 FURTHER READING

Baumann, C.A., Ribon, V., Kanzaki, M., Thurmond, D.C., Mora, S., Shigematsu, S., Bickel, P.E., Pessin, J.E., and Saltiel, A.R. (2000) CAP defines a second signalling pathway required for insulin stimulated glucose transport. *Nature* **407**, 202–207. Also see comment on pp. 147–148.

Blackshear, P.J., Hampt, D.M., App, H., and Rapp, U.R. (1990) Insulin activates the Raf-1 protein kinase. *Journal of Biological Chemistry* **265**, 12131–12134.

Boulton T.G., Nye, S.H., Robbins, D.J., Ip, N.Y., Radziejewska, S.D. Morgenbesser, R.A., DePinho, N., Panayotatos, N., Cobb, M.H., and Yancopoulos, G.D. (1991) ERKs: a family of protein serine/threonine kinases that are activated and tyrosine phosphorylated in response to insulin and NGF. *Cell* **65**, 663–675.

Cuatrecasas, P., Jacobs, S.J., and Avruch, J. (eds) (1990) *Insulin: Handbook of Experimental Pharmacology*, vol. 92. Springer-Verlag, Heidelberg. See chapters by Roth on insulin receptor structure, pp. 169–181, and by Rothenberg, *et al.* On insulin receptor tyrosine kinase, pp. 209–236, in particular.

Da Silva Xavier, G., Leclerc, I., Varadi, A., Tsuboi, T., Moule, S.K., and Rutter, G.A. (2003) Role for AMP-activated protein kinase in glucose-stimulated insulin secretion and preproinsulin gene expression. *Biochemical Journal* **371**, 761–774.

Dent, P., Lavoinne, A., Nakielny, S., Caudwell, F.B., Watt, P., and Cohen, P. (1990) The molecular mechanism by which insulin stimulates glycogen synthesis in mammalian skeletal muscle. *Nature* **348**, 302–308.

Hashimoto, N., Feener, E.P., Zhang, W.-R., and Goldstein, B.J. (1992) Insulin receptor protein-tyrosine phosphatases. *Journal of Biological Chemistry* **267**, 13811–13814.

Kimball, S.R., Vary, T.C., and Jefferson, L.S. (1994) Regulation of protein synthesis by insulin. *Annual Review of Physiology*, **56**, 321–348.

Litherland, G.J., Hajduch, E., and Hundal, H.S. (2001) Intracellular signalling mechanisms regulating glucose transport in insulin-sensitive tissues [Review]. *Molecular Membrane Biology* **18**, 195–204.

Myers, M.G. and White, M.F. (1996) Insulin signal transduction and the IRS proteins. *Annual Review of Pharmacology and Toxicology* **36**, 615–658.

Pearl, L.H. and Barford, D. (2002) Regulation of protein kinases in insulin, growth factor and Wnt signalling. *Current Opinion in Structural Biology* **12**, 761–767.

Pessin, J.E. and Saltiel, A.R. (2000) Signalling pathways in insulin action: molecular targets of insulin resistance. *Journal of Clinical Investigation* **106**, 165–169.

Rutter, G.A., White, M.R.H., and Tavaré, J.M. (1995) Involvement of MAP kinase in insulin signalling revealed by non-invasive imaging of luciferase gene-expression in single living cells. *Current Biology* **5**, 890–899.

Sun, X.J., Rothenberg, P., Kahn, C.R., Backer, J.M., Araki, E. *et al.* (1991) Structure of the insulin receptor substrate IRS-1 defines a unique signal transduction protein. *Nature* **352**, 73–77.

Whitehead, J.P., Clark, S.F., Ursø, B., and James, D. (2000) Signalling through the insulin receptor. *Current Opinion in Cell Biology* **12**, 222–228.

10 Perception of the environment

All cells must be able to perceive elements of their environment, and the same is true for whole organisms. For example, humans need to perceive light to be able to see, and this involves a complex signalling mechanism. Aspects of this signalling are used to illustrate how components described in earlier chapters may come together to form coherent signalling pathways. Other systems used by organisms to perceive their environment are also discussed briefly to highlight the importance and diversity of such systems and to show that most draw on the repertoire of components discussed previously in this book.

10.1 Introduction

One of the essential roles of cell signalling events is to enable the survival of the cell, and there is a need for all cells, whether they compose individual organisms, or are part of a complex organism, to stay alive and prosper, at least until they are no longer needed or have given rise to the next generation. Many cells are in a constant struggle to live, as they have a constant requirement for nutrients, and a need to avoid dangers and toxins. Integral to this is the ability of the cell to respond to environmental changes imposed upon it. Such changes might be shifting of temperatures; for example, a simple organism or plant might have to survive the cold of winter, or the heat of the midday desert sun. In similar manner, a cell might need to respond to an environmental change such as the arrival of chemicals that indicate the proximity of a source of food, or alternatively the proximity of danger, and as such smell becomes a crucial sense. Some signals are released by individual organisms of a species, and sensed by others of the same species. Such signals are pheromones, which are discussed in Chapter 2 (section 2.8), and when they are perceived an appropriate response is needed, perhaps allowing the survival of the individual or mating to prolong the species.

Therefore, whether an organism consists of a single cell or is a multi-cellular structure, it is imperative that cells, and organisms as a whole, can sense their environment. Many organisms have adapted their awareness of their environment to suit their niche in nature, and as such their need to perceive aspects of their environment is somewhat specialized. For example, some animals such as moles and bats have very poor eyesight, not easily detecting and perceiving light, as they live either underground or hunt at night where little light is available, but they do have good auditory perception, while others might have an acute sense of smell or very efficient hearing, such as animals which graze in the open that need to sense the arrival of a predator before it has a chance to get too close.

The perception of environmental signals relies on mechanisms analogous to those discussed in previous chapters, and such machinery is often akin to that used by individual cells within a body. Therefore perception of the environment relies on the use of specific receptors, tuned to the particular environmental factor needed to be detected, and transmission of a signal as a change in the receptor which is recognized in the interior of the cell, with subsequent signalling cascades and cellular response, as discussed in Chapter 3 and subsequent chapters.

Here, a focus on the molecular signalling of the mammalian eye allows a discussion of how many of the components discussed in previous chapters might come together to allow an organism to sense one aspect of its environment; that is, light. Other organisms also have to sense light, and most, if not all, need to have some sense of the array of chemicals around them, sensed by humans as smell and taste, and many need to have auditory senses too.

10.2 Photodetection in the eye

A sense which many of us take for granted is the ability to see. Photodetection requires that the light be perceived, and that a signal is created which the organism can translate, either into a response, as in lower organisms, or an image, as in higher organisms. In humans the eye is a specialized organ that undertakes this role, where the perception of the light is translated into electrical signals that are unscrambled by the brain to create the image. Often, the response is a conscientious act resulting from us seeing the image, a conscientious act which of course is controlled by further cell signalling events. However, it is imperative that the image is continuous. Therefore, it is no good for the cell signalling mechanism involved to 'see' the scene, and send that to the brain, and then stop. It needs to perceive the light, send the signal, and then reverse quickly to the ground state ready for the next photon of light to be perceived. Hence, the signalling needs to be activated rapidly and reversed rapidly, and any proposed mechanism must take this into account. Clearly, such activation and de-activation of signalling components cannot be spontaneous, and there is a time lag in the ability to perceive photons. If a person watches a train pull into the station, the wheels often appear to reverse momentarily as the train slows. This is a consequence of the time lag that the cell signalling events have imposed on the photoreception of the eye. The spokes of the wheel appear to re-align in a way that makes it look as though they have reversed. However, our brain has the ability to translate the signals into a continuous image, which appears to us to be seamless. The screen on a television works in the same way. It does not project a continuous moving image, but rather scans the image on to its surface frame by frame, but we see it as a continuous movie.

At the cellular level, photoreception in humans is the responsibility of two types of specialized cell, the rods and cones, so named because of their shape. These cells form a layer called the retina at the back of the eye on to which the light is focused by the lens. The rods are used in the perception of a wide range of wavelengths of light at low levels whereas the cones function in brighter light and fall into three classes, each with their own wavelength sensitivity. One type of cone senses blue light, one green light, and a third red light. In humans the wavelength maxima for the three classes of cones are 426, 530, and 560 nm, although other species such as fish have slightly different wavelength maxima: 455, 530, and 625 nm. If an animal lacks cones, then their perception of colours is compromised.

The rods and cones are single cells, but ones which are bipolar, having an outer segment and an inner segment (Fig. 10.1). The outer segment is basically an elongated bag containing flattened membranous sacks called discs, which lie one on top of the other perpendicular to the plane of the incident light. A rod may contain up to 1000 of these discs, which measure only about 16 nm thick. The inner segment contains the main body of the cell including the endoplasmic reticular machinery, mitochondria, and a nucleus, but just as importantly it also has a synaptic body, which is used for the transmission of any created signal to the nervous system (Fig. 10.1).

Carrots and the eyes

As discussed in Chapter 2 (section 2.2), during the second world war it was said that eating carrots improved your eyesight in the dark, because carrots are a good source of retinol. However, this was propaganda, as nothing is to be gained if a reasonably healthy diet is maintained.

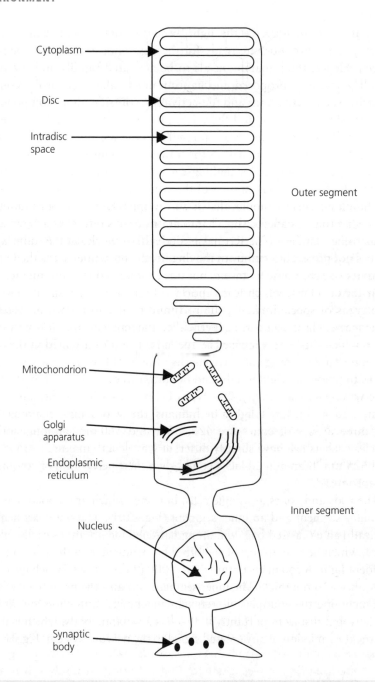

Fig. 10.1 A rod cell: one of the cells involved in photoreception in mammals.

Rhodopsin

Rhodopsin is a protein comprised of a polypeptide called opsin, and a non-protein part (a prosthetic group) called 11-*cis*-retinal. The whole protein, with the polypeptide and prosthetic group together, is called rhodopsin.

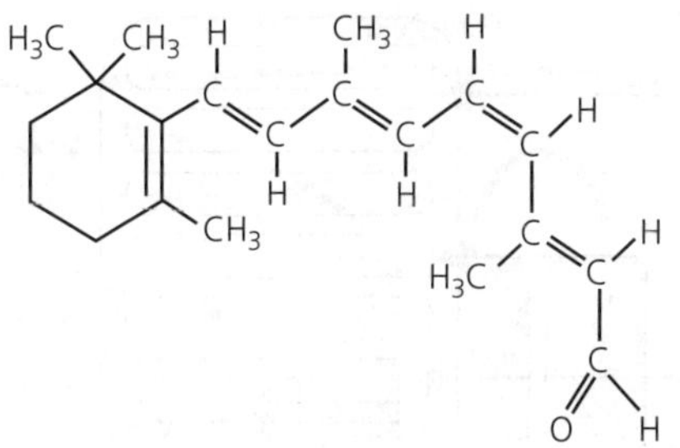

Fig. 10.2 The molecular structure of 11-*cis*-retinal.

The photoreceptive molecule, or chromophore, in rods is rhodopsin. This is a protein, called opsin, that contains 11-*cis*-retinal as a prosthetic group (Fig. 10.2). 11-*cis*-Retinal is derived from vitamin A (all-*trans*-retinol) which is obtained from the diet. In the protein it is attached via a protonated Schiff base linkage to a lysine residue and it is this interaction, along with others, which leads the 11-*cis*-retinal having an absorbance maximum of approximately 500 nm. Interestingly, the rhodopsin also has an extremely large extinction coefficient, which means that it is very efficient at absorbing light, giving the eyes a fantastic light sensitivity.

Bacteriorhodopsin

The rhodopsin-related protein bacteriorhodopsin was one of the first membrane proteins to have its structure solved, and has become the model for seven-spanning proteins, such as the G protein-linked receptor family. Note similarities in the structures shown in Fig. 10.3 and Fig. 3.2.

The opsin polypeptide is like many other membrane proteins in having seven membrane-spanning α-helices (Fig. 10.3). The retinal lies in the centre of the α-helical section and lies in the plane of the membrane. The N-terminal end of the polypeptide, which contains two N-linked oligosaccharides, lies in the intradiscal space while the C-terminal end lies on the cytosolic side of the membrane. It is this region which is important in interactions with other proteins which transmit the signal to the downstream part of the signal transduction cascade, and ultimately to the nervous system. For example, it is here that a G protein binds and is also phosphorylated on serine and threonine residues, leading to a deactivation of the molecule.

On receiving a photon of light the 11-*cis*-retinal photoisomerizes to all-*trans*-retinal, inducing a conformational change in the molecule. This means that the Schiff base linkage to the protein moves by approximately 5 Å. This new molecule, called bathorhodopsin, undergoes several more

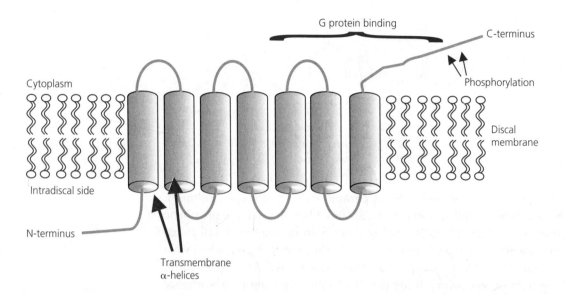

Fig. 10.3 The domain structure of opsin showing its similarity with receptors having seven membrane spans.

conformational changes, each of which can be monitored by the changes in the maximal absorbance wavelength, until metarhodopsin II, otherwise known as photoexcited rhodopsin, is formed in which the Schiff base linkage has become unprotonated. It is the formation of this form which triggers the resultant signalling cascade, resulting from the forced change in conformation of the opsin protein, a change perceived by the G protein to which it is bound.

Rhodopsin is reformed by the hydrolysis of the all-*trans*-retinal from the opsin, its reconversion back to 11-*cis*-retinal which involves all-*trans*-retinol as an intermediate, and then the re-linkage to the opsin.

The same light-sensitive system is also used in the cones, except here the opsin-like molecule has three hydroxyl-group-containing residues which surround the 11-*cis*-retinal and the alteration of its local environment is enough to alter its light-absorbing characteristics and its wavelength maximum (as discussed above).

The C-terminal end of the rhodopsin molecule in the unexcited state is in association with a member of the trimeric G protein family. Here the G protein is transducin, or G_t. Transducin was in fact only the second G protein to be discovered, by both Mark Bitensky and Lubert Stryer. It is composed of an α-subunit of 39 kDa, a β-subunit of 37 kDa, and a γ-subunit of 8.5 kDa. The excited, and therefore activated, rhodopsin acts effectively like a GTP-exchange factor, exchanging the GDP bound to the G protein for a GTP, so causing activation of the G protein. The G protein then dissociates into the $G_t\alpha$ and $G_t\beta\gamma$ subunits. The $G_t\alpha$ subunit interacts with the inhibitory peptide of phosphodiesterase and so reduces the inhibited state of this enzyme, allowing it to catalyse the breakdown of cGMP. This phosphodiesterase normally resides as a complex of two catalytic

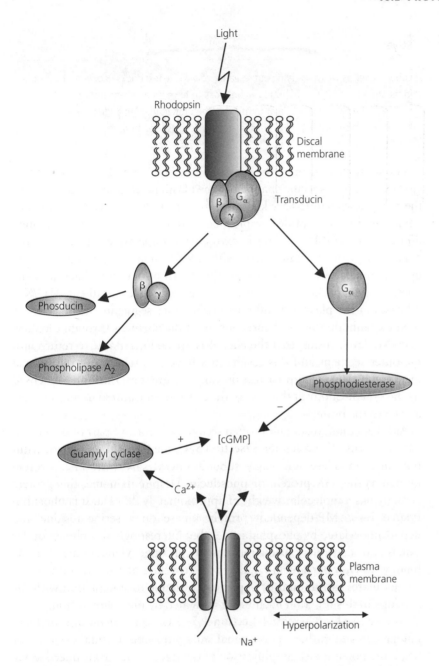

Fig. 10.4 The signal transduction pathway in rod cells showing how light leads to a hyperpolarization of the plasma membrane.

subunits (α and β) which are inhibited by the presence of two inhibitory peptides, the γ subunits (the details of cyclic nucleotide phosphodiesterases (PDEs) are discussed in section 5.6). This inhibitory restraint is removed by the interaction with $G_t\alpha$. The activated phosphodiesterase very rapidly and efficiently hydrolyses cGMP in the cytosol of the cell's outer segment (see Fig. 10.4; also see the section on phosphodiesterases in Chapter 5, section 5.6).

Stryer

Lubert Stryer, as well as carrying out research, also wrote text books. *Biochemistry* is now in its 5th edition and unsurprisingly continues to have an excellent section on the signalling of the eye and the role of transducin, the protein that Stryer discovered.
N.B. *Biochemistry* by Stryer now has Jeremy M. Berg as the first author.

The subsequent drop in the cytosolic concentration of cGMP leads to a closure of cation-specific channels in the plasma membrane of the cells. In the dark these channels are kept open by the binding of cGMP. The channel is a multi-polypeptide complex, each subunit being of approximately 80 kDa. The cGMP binding is extremely co-operative, opening of the channel requiring the interaction with at least three cGMP molecules. The open channels serve as a route for the return of Na^+ ions back into the outer segment of the cells down a large electrochemical gradient. Na^+/K^+-ATPases in the plasma membrane of the inner segments of the cells are used to maintain this gradient. Closure of the channels through the drop in cGMP levels means that the Na^+ ions are no longer able to return into the outer segment and this results in a hyperpolarization of the plasma membrane. This hyperpolarization can be a great as 1 mV and is sensed by the synaptic body at the base of the cell and transmitted to the neurons, and so to the brain.

Although the $G_t\alpha$ subunit is often seen to be the active part of the G protein, as discussed in Chapter 5 (section 5.4), the $G\beta\gamma$ subunit is not without function. Here, the $G_t\beta\gamma$ subunit interacts with another protein known as the 33K protein or phosducin. Despite its name, this protein actually has a molecular weight of approximately 28 kDa. It is phosphorylated by cAMP-dependent protein kinase on a serine residue and dephosphorylated by phosphatase 2A, but interestingly this phosphorylation is light-dependent, being most highly phosphorylated in the dark. By binding to $G\beta\gamma$ phosducin regulates the recycling of the $G_t\alpha$ subunit and so modulates the amount of G protein available for interaction with the rhodopsin, this function itself being regulated by phosphorylation.

As discussed in Chapter 1 (section 1.5), one of the main uses of a signalling cascade, rather than a signal simply turning the final event on or off, is the potential for amplification of the signal as it is transduced along the pathway. The signalling events in the rods of the eye are a very good example of this, as a massive magnification of the signal is seen in this system, each step allowing another degree of amplification. One rhodopsin molecule, once activated, can activate up to 500 transducin G proteins, while a massive amplification is seen in the cleavage of cGMP by the extremely active phosphodiesterase: many thousands of cGMP molecules can be removed by the perception of a single photon.

It was stressed above that the system needs to be rapidly reversed and turned off, and in fact this is achieved by several simultaneous events. Ca^{2+} is normally pumped out of the cell via an exchanger but its re-entry is normally allowed into the outer segments via the cation-specific channels. On closure of these channels following the drop of cGMP the intracellular Ca^{2+} levels decline and this in turn stimulates the synthesis of cGMP by the activation of guanylyl cyclase. Resurrection of the cGMP levels will once again open the cation channels and end the hyperpolarization of the membranes, with the consequent cessation of the neuronal signal to the brain.

However, the rest of the signalling system needs to turn off too. Transducin has, like all heterotrimeric G proteins, an endogenous GTPase activity and thus any bound GTP will be hydrolysed, allowing the reformation of the trimeric inactive G protein, assuming phosducin has not had an influence on the G$\beta\gamma$ levels. The transducin could then be reactivated by rhodopsin, allowing another round of signalling. However, rhodopsin can be turned off. It has been found that rhodopsin may be phosphorylated by rhodopsin kinase, on several serine and threonine residues on the C-terminal end of the polypeptide. Once phosphorylated, the rhodopsin can interact with a polypeptide called arrestin. Interaction with arrestin prevents rhodopsin from interacting with transducin and so prevents a further activation of the cascade. It is interesting to note that a similar mechanism is seen with the β-adrenergic receptors, and in fact that the β-adrenergic receptor kinase can phosphorylate the rhopdopsin molecule in a light-dependent manner, highlighting the structural similarity between these receptor molecules.

Rhodopsin as a receptor

Although rhodopsin has no ligand for binding, it does act just like a receptor, but in this case a receptor for photons. Therefore, the mechanisms of action are just like many of the receptors discussed in Chapter 3.

Although the signalling mechanisms discussed have been mainly elucidated by studies on the rods from the eye, the cones also have an analogous system. However, subtle differences are apparent, such as in the G proteins used. Rods express the G protein subunits α_{t-r}, β_1, and γ_1, while cones express a G protein consisting of a cone-specific α_t, β_3, and a γ-subunit.

To complicate matters even more, phospholipase A_2 activity in rod outer segments is stimulated by light. It has been found that transducin β_γ-subunits increase phospholipase A_2 activity, which is inhibited by transducin α-subunits.

It can be seen therefore that the light perception in the eye serves as a good example of how several signalling components work together to result in a coordinated response (Fig. 10.4). Here, a light receptor (with analogies to a hormone receptor), a G protein, an enzyme, a small-messenger molecule, and an ion channel are all used to transmit the signal, with the cytosolic calcium concentration and another enzymes being involved in the return to the ground state. Other proteins also interact, and other enzymes are involved, giving overall a complex set of signalling events. It is not simply a linear cascade, but rather a complicated interplay involving many actors.

It is not just the absolute intensity of the light to which we are sensitive but rather *changes* in intensity and therefore a biochemical mechanism needs to explain how our eyes can so readily adapt to wide ranges of light intensity, maybe as much as a 10 000-fold difference. Work studying the photoreceptors of amphibians has shown that the activity of nitric oxide synthase and production of nitric oxide has an influence on the metabolism of cGMP, as discussed in Chapter 8 (section 8.2) with other signalling pathways, and therefore such signalling molecules may well be involved in the adaptive nature of the eye. It is, and indeed needs to be, a complex signalling mechanism.

The signalling pathways involved in the light-sensing organs of lower animals differ somewhat from those of vertebrates. In the fruit fly *Drosophila melanogaster* the rhodopsin activates a trimeric G protein of the G_q type, which in turn activates phospholipase C. The phospholipase C catalyses the breakdown of $PtdInsP_2$ in the membrane and therefore produces DAG and $InsP_3$. The $InsP_3$ opens Ca^{2+} channels and allows the release of Ca^{2+} from internal stores, which results in the propagation of the signal. Photo-adaptation and recovery also need the participation of an eye-specific protein kinase C.

10.3 Other environment-perception systems

Plants also need to perceive light and respond to it in order to survive. Leaves on a tree, for example, are optimally aligned to maximize the amount of light which is captured for photosynthesis. If there is no light, plants usually fail to produce chlorophyll, and remain yellow and etiolated, but often grow tall in the search for light. With the restoration of light, the plant produces chlorophyll, a process which itself appears to be light-catalysed, and the plant can once again start photosynthesizing. If a leaf is in the shade, the plant might need to respond by increasing its leaf area compared with that which is exposed to direct sunlight. Similarly, plants are able to detect the time of day, which can be seen, for example, with flowers opening and closing, and also they may sense the time of year. Plant cells have four main ways of monitoring light: phytochromes, UV-B receptors, UV-A receptors, and blue-light receptors. Phytochromes of higher plants, which are responsible for detecting red and far-red light, are coded for by at least five genes, assigned the names *PHYA–PHYE*. Downstream signalling from such light perception will involve many of the signalling components discussed in the previous chapters.

As well as light, it is important that organisms are aware of the presence of chemicals in the environment. Perhaps the occurrence of a chemical suggests the presence of a source of food, the proximity of a partner to mate with, a danger, or perhaps it can be ignored. Whichever response is required, the perception of the chemical is required, and organisms such as prokaryotes have specialized receptors wheres higher animals have specialist tissues and cells dedicated to the job.

Prokaryotes, such as *Escherichia coli*, have a need to detect molecules in the extracellular medium and respond to their presence. Such molecules include attractants such as the sugars galactose and ribose, the amino acids serine, alanine, and glycine, as well as dipeptides. Repellent molecules include acetate, leucine, indole, and metals such as nickel and cobalt. The presence of these molecules is detected by receptors known as methyl-accepting chemotaxis proteins. These proteins form four homologous groups, but they all span the inner bacterial membrane, and are known as Tsr, Tar, Trg, and Tap. Sequence homology between the members of these receptor groups shows that the N-terminal end contains the receptor domain which interacts with the ligand, with a membrane-spanning domain on each side. The C-terminal end contains a domain which propagates the signal into the cell. The C-terminal end is also the site of methylation, which seems to be responsible for the sensitivity of the receptor, enabling the receptors to be receptive to a concentration gradient of the ligand rather than simply acting as an on/off signal, more like a dimmer switch. Methylation is catalysed by a methyltransferase known as CheR while demethylation is catalysed by CheB-phosphate, or CheB-P, which is a methylesterase. The methyl group is supplied from *S*-adenosylmethionine while the targets for transfer are glutamate residues. Different receptor classes are methylated differently, but in general between four and six methylation sites are seen. CheB-P, the methylesterase, removes the methyl group and forms methanol. The CheB protein is controlled by phosphorylation, where CheB is phosphorylated in its most active state; that is, as CheB-P. The non-phosphorylated CheB polypeptide is a target of a kinase CheA.

Many higher animals have specialist chemical-sensing mechansims located in their noses. They are often extremely sensitive to the presence a vast range of chemicals. It is not always clear, however, exactly what it is about the particular chemical which stimulates a certain response.

Molecules which smell the same may have very differing structures. Despite this apparent paradox, in 1991 genes encoding odorant receptors where identified by Buck and Axel, and it is the proteins which they encode that are responsible for the detection of volatile compounds in the air, hence the name odorant receptors. These odorant receptors are found in the neuroepithethial cells of the nose, with each cell only expressing one isoform of the receptors. The human genome has been estimated to contain over 500 genes for odorant receptors, although some of these genes might be non-functional. The proteins the genes encode, that is the receptors, are seven-transmembrane-domain proteins, as found with many plasma membrane receptors (see Chapter 3, section 3.2), and seen with rhodopsin, as discussed above (section 10.2). As discussed with other seven-transmembrane-domain proteins, these odorant receptors are associated with a class of heterotrimeric G proteins, here G_{olf}, which have their influence on adenylyl cyclase, and hence cAMP signalling.

In taste buds too, heterotrimeric G proteins are also important. A G protein analogous to transducin of the rods and cones of the eye has been found here: the taste-specific G protein has been named gustducin. However, not all chemical detection by the taste buds uses the same mechanisms. Four basic tastes are recognizable, each leading to a different cellular response: salty tastes result from an increase in Na^+ movement through Na^+ channels; sour tastes are the result of the blockage of K^+ or Na^+ channels by H^+; sweet and bitter tastes on the other hand are mediated by the presence of G proteins.

Many organisms are responsive to sound, often over a large range of frequencies. Reports that the activity of certain enzymes are correlated to the intensity of sound are fascinating. Studies in the rat have shown that the activity of guanylyl cyclase in the inner ear is inversely related to the volume of sound to which the animals are exposed, suggesting an exciting signal transduction field.

10.4 SUMMARY

- The perception of the environment is as vital for whole organisms as it is for individual cells.

- The response to many environmental factors involves the use of multiple signalling pathways and can be used to illustrate the coordination of signal transduction.

- Many organisms respond to light, either tailoring their growth to optimize light capture, as in plants, or to enabling them to move around and avoid danger, as with animals.

- Plants monitor light by the use of phytochromes, UV-B, UV-A, and blue-light receptors.

- One of the most-specialist, and well-characterized, systems for detecting light is the mammalian eye.

- In the eye, photons of light cause the photoisomerism of retinal, resulting in activation of the protein opsin, which in turn leads to the activation of a trimeric G protein, G_t. The released $G_t\alpha$ causes the activation of a phosphodiesterase which catalyses the rapid breakdown of cGMP. The subsequent drop in cGMP in the cytoplasm leads to the closure of cation-specific channels in the plasma membrane and the formation of an electrochemical potential, propagated to the synaptic region of the cell. From here, electrical signals pass to the brain where the picture is deciphered.

- The signalling system in the eye has to be reversible to allow the perception of subsequent light for a seemingly continuous moving image.

- The control of light perception in the rods and cones of the eyes is a complex interplay of many signalling components, allowing a coordinated response across a wide range of light intensities.

- The presence of chemicals is an environmental factor which needs to be monitored.

- For prokaryotes essential nutrients such as amino acids and sugars need to be sensed.

- Larger animals, such as mammals, can detect air-borne chemicals as smells as well as soluble chemicals as taste.

- The sense of smell uses receptors, linked to heterotrimeric G proteins, having their effect through adenylyl cyclase.

- Taste uses specialized receptor cells, but different types of taste appear to use different signal transduction pathways.

10.5 FURTHER READING

Photodetection in the eye

Benovic, J.L., Mayor, F., Somers, R.L., Caron, M.G., and Lefkowitz, R.J. (1986) Light dependent phosphorylation of rhodopsin by β-adrenergic receptor kinase. *Nature* **321**, 869–872.

Berg, J.M, Tymoczko, J.L., and Stryer, L. (2003) *Biochemistry*, 5th edn. W.H. Freeman & Co, New York. Chapter 32 is a particularly good summary of the signalling of the eye.

Quail, P.H. (1994) Photosensory perception and signal transduction in plants. *Current Opinion in Genetic Development* **4**, 652–661.

Other environment-perception systems

Amsler, C.D. and Matsumma, P. (1995) Chemotaxis signal transduction in *Escherichia coli* and *Salmonella typhimurium*. In *Two-Component Signal Transduction* (Hoch, J.A. and Silhavy, T.J., eds). American Society for Microbiology, Washington, DC.

Buck, L. and Axel, R. (1991) A novel multigene family may encode odorant receptors: a molecular basis for odor recognition. *Cell* **65**, 175–187.

Gillespie, P.G. (1995) Molecular machinery of auditory and vestibular transduction. *Current Opinion in Neurobiology* **5**, 449–455.

Kinnamon, S.C. (1988) Taste transducin: a diversity of mechansims. *Trends in Neurosciences* **11**, 491–496.

McLaughlin, S.K., McKinnon, P.J., and Margolskee, R.F. (1992) Gustducin is a taste-cell-specific G protein closely related to the transducins. *Nature* **357**, 563–569.

Reed, R.R. (2004) After the Holy Grail: establishing a molecular basis for mammalian olfaction. *Cell* **116**, 329–336.

11 Life, death, and apoptosis

Organisms need in many cases to develop from a single cell, and in doing so an anatomy has to be created. This process not only involves the creation of new cells, but also the removal of cells that are not needed. Fully formed organisms too need to remove old and dysfunctional cells, allowing the replacement of damaged cells. It is now recognized that molecular mechanisms are in place that lead to cells committing suicide, a process referred to as apoptosis.

Apoptosis, therefore, is a process which leads to the death of the cell, and as such needs to be very carefully controlled, and it is the signalling involved in the regulation of apoptosis that will be discussed in this chapter. Many of the components involved have not been discussed previously, and therefore several new proteins with confusing names will be introduced, but the principles involved in signalling cascades are the same as discussed before. The processes are initiated with the recognition of extracellular signals or intracellular damage, which leads to cascades of events involving specific interactions, and results in a response, in this case the death of the cell involved.

This chapter highlights that it is not only the survival, adaptation, and normal functioning of the cell that are under the control of cell signalling events, but that ultimately death of the cell may also be brought about by cell signalling cascades. However, the lack of death of cells can be detrimental too, with cancer being a common result.

11.1 Introduction

Discussions in previous chapters have concentrated on the way cells respond to extracellular ligands, which in many cases enable them to adapt and survive. However, cells often need to die, at a precise time and in specific places in an organism. This death of cells results from a precise mechanism and is often referred to as programmed cell death (PCD), and in many cases is termed apoptosis.

Very few of the cells in a multicellular organism are the same age as the body as a whole. If a person says they are 36, few of their cells have been around for 36 years. Cells within an organism are dividing to create new cells continually, and cells are dying continually. Certainly not all cells divide regularly, but many do, and when cells are old and dysfunctional there is a need to get rid of them. It has been estimated that in an average adult human 50–70 billion cells undergo cell death by apoptosis every day. During development the process of apoptosis is also critical. For example, in the nematode *Caenorhabditis elegans* development is reliant on the specific death of 131 cells. In higher organisms too, the formation of organs is dependent on selected cell death as they develop, and studies where apoptosis has been disrupted result in mis-formed and oversized organs.

Other scenarios exist too where there is a need to have the death of a cell within an organism, with the good of the whole organism being the result. For example, if a plant is invaded by a pathogen, one of the responses is the hypersensitive response. Here the plant cells surrounding the invading pathogen die, so depriving the pathogen of nutrients and depriving it of a route to invade the rest of the plant. Dead spots are seen on leaves etc., but the plant as a whole survives to fight another day. The death of cells therefore is essential at times, and has to be controlled carefully. This chapter will give a brief insight into some of the signalling mechanisms involved in the process of apoptosis, but the full details are still to be unravelled by the large research interest in this field.

11.2 An overview of apoptosis

The idea that cells might need to die is not a new one and was originally proposed as long ago as 1842 by Carl Vogt, who studied the development of toads. However, it was not until 1972 that the term apoptosis was actually coined. The word apoptosis is derived from a Greek word which means 'to fall away from', and, among other things, was used to describe the fall of autumn leaves from trees. It is now a widely used term to describe the process of naturally occurring cell death, or cellular suicide. However, it is not always used correctly in the strictest sense, as apoptosis is a particular form of cell death, and a more accurate term in many cases would be programmed cell death, or PCD, meaning that an intracellular mechanism was used to bring about the death of the cell.

Apoptosis, in the true sense, is accompanied by defined morphological changes, which are distinct from the changes seen during death by necrosis. In apoptosis cell shrinkage and membrane ruffling (or blebbing) occur, and the cell disintegrates into small membrane-bound vesicles called apoptotic bodies. Inside the cell chromatin condensation and nuclear fragmentation occur, which are accompanied by breakdown of the DNA into regular-sized fragments, often referred to as DNA laddering (named from the pattern seen on an electrophoretic gel). On the surface of the cell lipids are rearranged in the bilayer of the plasma membrane with the lipid phosphatidylserine becoming exposed to the outside. Therefore, it can be seen that profound changes occur in the cell during this process of apoptosis, and the result is the death of the cell. It needs to be controlled very carefully, as the death of the wrong cells, or death of cells at the wrong time, may be catastrophic to the organism as a whole. Alternatively, the lack of death of the cells can be detrimental too, and over-proliferation of cells without a balancing death of cells can lead to cancer and the growth of tumours.

There are two main cell signalling pathways which have been identified in the control of apoptosis: the intrinsic pathway, or core pathway, which involves the mitochondria, and the extrinsic pathway, which involves cell-surface receptors. These pathways, will be discussed further in sections 11.4 and 11.5 below.

Once cell death has been completed, the remains of the dead cell need to be removed. A multicellular organism cannot survive with a host of dead non-functioning cells as part of its main organs. Neighbouring cells and macrophages phagocytose the dead cells before they are able to release their cytosolic contents. Several genes have been identified that encode proteins involved in the two pathways of phagocytosis which are used here.

Macrophages

Macrophages are derived from monocytes which circulate in the blood of mammals. If macrophages are required, perhaps as a result of tissue injury, monocytes leave the bloodstream and differentiate and develop into macrophages. One of their major roles is to remove unwanted material, either host cells or pathogens.

11.3 Caspases

Apoptosis results in major changes to cell morphology and eventually cell death. The molecular machinery behind these events has to first control such activity, but secondly must bring about the changes seen. One central group of players in this molecular orchestra is the caspases (named from cysteinyl-aspartate-specific proteases). It is the caspases that mediate the cell shrinkage, DNA fragmentation, and other changes that are seen.

Table 11.1 The caspase families. The prefix m denotes that it is of mouse origin, while prefix b denotes bovine origin.

Caspase-1 family	CED-3 family	
	Caspase-3 sub-family	Caspase-2 sub-family
Caspase-1	Caspase-3	Caspase-2
Caspase-4	Caspase-6	Caspase-9
Caspase-5	Caspase-7	
mCaspase-11	Caspase-8	
mCaspase-12	Caspase-10	
bCaspase-13		
Caspase-14		

However, caspases are also involved in other events in cells, besides apoptosis.

Caspases are proteases, proteins that cleave and break down other proteins. In this case they are a class of proteases called the cysteine proteases, and they cleave their target proteins next to an aspartate residue. There are 14 known mammalian caspases, 11 of which have been found in humans. They can be divided into two main families. The first family are those that are related to caspase-1 (otherwise referred to as interleukin-1β-converting enzyme or ICE), and the second family are those that are related to CED-3, the product of a death gene from *C. elegans*. This second family can be further divided into two sub-groups, known as the caspase-2 and caspase-3 sub-families (see Table 11.1). The caspase-1 family are known to be involved in the processing of cytokines (see Chapter 2, section 2.4) while the main role of the caspase-2 and caspase-3 sub-families is in cell death.

As mentioned above, caspases are cysteine proteases, which means that they contain a cysteine residue in their active site, and in fact they all contain the same conserved region around the active cysteine; that is, they all contain the peptide sequence -Glu-Ala-Cys-X-Gly- (where X is any amino acid). Caspase activity is such that they cleave the peptide bond in their target protein on the C-terminal side of an aspartate residue but they don't all have exactly the same target preferences. Three groups of caspases can be distinguished according to their target sequence at which they will cleave. The caspases mainly involved in apoptosis are those in group II, which include caspases-3 and -7, preferring the target sequence -Asp-Glu-X-Asp-, and caspase-2, preferring -Val-Asp-Val-Ala-Asp-, and those in group III, which includes caspases-6, -8, -9, and -10, which cleave at -(Ile/Lev/Val)-Glu-X-Asp-.

The activation of the caspases is itself by proteolytic cleavage, and it has been found that one caspase can cleave and so activate another caspase, leading to activation cascades of caspases. Therefore, inactive caspases are in a non-cleaved form, referred to as the pro-caspases. The first cleavage event is an autocatalytic one, where the caspase cleaves and activates itself. This active caspase can then cleave and activate another caspase. This leads to another way of classifying caspases, either as initiator caspases or effector caspases. The former, initiator caspases, are those that can be at the start of a cascade and undergo auto-cleavage, while the effector caspases are downstream and are activated by other caspases. Initiator caspases include caspases-2 and -9, while effector ones include caspases-3, -6, and -7.

Pro-caspases contain three or four domains, as seen in Fig. 11.1. Firstly there is a pro-domain at the N-terminal end, followed by a large-subunit domain. At the C-terminal end is a small-subunit domain, and between the subunit domains is often a linker region. During activation cleavage takes place between the large-subunit and small-subunit domains, so releasing them as separate proteins. These then interact in a hetero-tetrameric fashion, that is two small subunits and two large subunits come together to form an active enzyme (see Fig. 11.1). The pro-domain may also be cleaved off and released. The active site of the functional enzyme

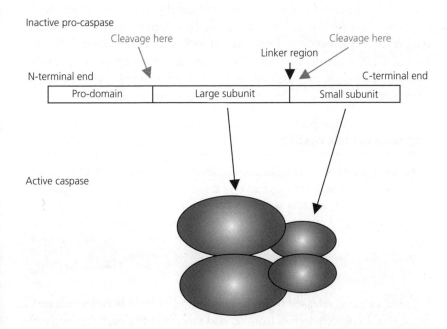

Fig. 11.1 The domain structures of caspases.

is comprised partly from one small subunit and partly from one large subunit, so without protein–protein interaction no activity would be seen.

11.4 The intrinsic pathway

The intrinsic pathway for the control of apoptosis involves the mitochondria of the cell. This organelle is instrumental in normal cellular metabolism as the major site of ATP production, and therefore the main source of usable energy in the cell. However, alongside this important role it has become apparent that mitochondria also have a large part to play in the decision of whether a cell lives or dies.

In response to DNA damage, topoisomerase inhibition, cytoplasmic stress, and many other stimuli, the mitochondria become permeabilized, followed by the release from them into the cytoplasm of many proteins. The permeabilization involves proteins of what is termed the Bcl-2 family. Permeabilization of the mitochondria allows for the release of many proteins but one of the most important is cytochrome c, a 12.5-kDa haem-containing membrane-associated (but also soluble) protein. This protein normally has a redox function, shuttling electrons between Complex III and Complex IV of the electron transport chain. Cytochrome c in mitochondria resides on the outer face of the inner mitochondrial membrane, where it is a peripheral protein associated with lipids. The release of cytochrome c involves its dissociation from these lipids, a process probably facilitated by the presence of free radicals, emanating from the mitochondria themselves. As cytochrome c is such an integral part of the mitochondrial respiratory chain and so used in the the production of ATP for the cell, release of cytochrome c from the mitochondria has potentially two effects. First, it can act as a signalling molecule in the apoptosis pathway, as discussed below, and secondly it can no longer function in the electron transport chain and so ATP production is compromised.

Cytochrome c as a signal

The role of cytochrome c as a redox protein in mitochondria was established many years ago and it could have been claimed that it was a very well-characterized protein. However, its role in apoptosis seems very disparate from its redox role. This illustrates that a wide range of proteins can be involved in signalling, even ones where their roles are thought to be known. This multifunctional role of proteins will become more common in the future, as new roles for old proteins are discovered.

Other proteins released from the mitochondria include endonuclease G, Smac/Diablo (which derives from second mitochondria-derivated activator of caspases/direct IAP (inhibitor of apoptosis protein)-binding protein with low pI), and AIF (apoptosis-inducing factor), another redox protein, but this time a flavin-containing protein.

Members of the Bcl-2 family of proteins may promote apoptosis (pro-apoptotic) or inhibit apoptosis (anti-aopototic). The major pro-apoptoic proteins here include Bax and Bak, which facilitate the release of cytochrome c from the mitochondria. Anti-apopotic proteins include Bcl-2 and Bcl-X_L.

The signalling pathway that leads to apoptosis during the intrinsic route can be outlined as follows, and is shown in Fig. 11.2. The DNA

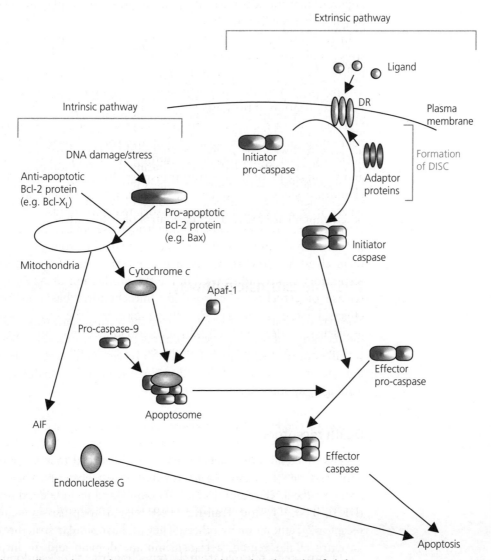

Fig. 11.2 The signalling pathways of apoptosis. Two main pathways have been identified, the intrinsic and extrinsic pathways. DISC, death-inducing signalling complex; DR, death receptor.

damage or cellular stress leads to the activation of pro-apoptotic Bcl-2 proteins, mitochondria become permeable and cytochrome c and other proteins are released. Cytochrome c forms a complex with a protein called Apaf-1 (apoptotic protease-activating factor 1), causing a conformational change in the structure of the Apaf-1 protein which is dependent on the presence of ATP. The Apaf-1 forms an oligomeric complex which then complexes with pro-caspase-9, forming a structure called an apoptosome. Caspase-9 within the apoptosome is cleaved, and activated, and so can cleave further caspases, resulting in the apoptotic effects seen. Other proteins released from the mitochondria also contribute to the morphological changes seen, as endonuclease G may migrate to the nucleus where it causes DNA degradation, and AIF may also lead to nuclear changes including condensation.

The control of apoptosis is critical and Fig. 11.2, and the discussion above, describe a simplified version of events. There are many other proteins that can modulate this pathway. For example, a protein called p53 can neutralize the activity of anti-apoptotic Bcl-2 proteins, while in the cytoplasm there are inhibitor of apoptosis proteins (IAPs). These proteins are inhibitors of capases-3 and -9. During the onset of apoptosis by the intrinsic pathway, the protein Smac/Diablo released from mitochondria inhibits the activity of IAPs, so facilitating the activation of caspases. The intrinsic pathway is therefore a complex interplay between many proteins, some which activate others, and some of which inhibit the activity of others, but all are focused on the control of cellular death.

11.5 The extrinsic pathway

The extrinsic pathway is separate from the intrinsic pathway and involves the activation of protein receptors at the plasma membrane. These are the so-called death receptors (DRs). Death receptors have the role of ligand binding on the outside of the cell, and subsequent intracellular signalling leading to apoptosis.

Death receptors

A class of membrane receptors has been identified that are involved in apoptosis, and these have been termed the death receptors. They are a sub-group of the TNF receptor (TNF-R) family and include death receptor 3 (DR3), TNF-R1, and Trail-R2 (TNF-related apoptosis-inducing ligand receptor 2), among many others. They all have similar structures, having an extracellular region, a transmembrane domain and an intracellular region. The extracellular region is characterized by the presence of up to

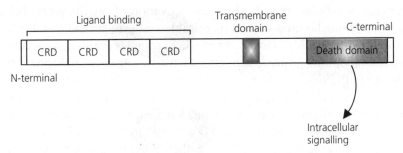

Fig. 11.3 The domain structure of a typical death receptor, in this case DR3. The extracellular region has multiple cysteine-rich domains (CRDs). They all contain a transmembrane domain spanning the plasma membrane, and in the intracellular region is a death domain for protein–protein interaction and signalling. In the immature form they also contain a leader peptide at the N-terminal end for localization to the correct membrane in the cell.

six cysteine-rich domains (CRDs), as can be seen in Fig. 11.3 for the DR3 receptor, which has four such domains. The intracellular region is characterized by the presence of a death domain, and it is this domain that is involved in intracellular signalling.

Ligands of these receptors are trimers, and often are found to be membrane proteins, although soluble forms, created either by cleavage from a membrane protein or from alternative-splicing events, are also known. Binding of the ligand to the receptor leads to trimerization of the receptors, which allows the signalling needed to take place.

Signalling from death receptors

Signalling during the extrinsic pathway starts with ligand binding to the death receptor, and is shown schematically in Fig. 11.2. As discussed above, the ligand are trimers, and ligand binding leads to the trimerization of the receptors themselves. The death domains on the intracellular are protein–protein-interaction domains (similar to those discussed in Chapter 1, section 1.7). Ligand binding, trimerization, and activation of the receptors allow recruitment of adaptor proteins, which bind to the death domains of the receptors. This is a similar mechanism to that seen with the receptor tyrosine kinases in Chapter 4 (section 4.3). The adaptor proteins here also contain death domains, and it is through the interaction of the death domain of the receptor and the death domain of the adaptor that the two proteins can come together. Adaptor proteins involved in apoptosis include FADD (Fas-associated protein with death domain). The adaptor protein FADD also contains a second protein-interaction domain, known as the dead effector domain (DED). These interact with DED domains on initiator caspases, which leads to the formation of a complex of the receptor, the adaptor, and the caspase. This complex is known as

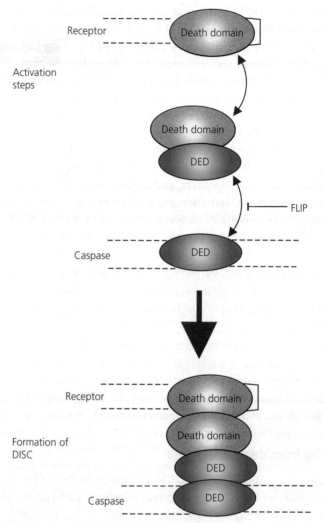

Fig. 11.4 The formation of DISC (death-inducing signalling complex) during the extrinsic pathway. Protein–protein interactions between the death domains of the receptor and the adaptor protein, and the dead effector domains (DEDs) of the adaptor and the caspase, leads to the formation of a complex, termed DISC. This complex results in cleavage and activation of caspases. FLIPs (Flice-like inhibitory proteins) can disrupt caspase recruitment as they also contain DED regions.

DISC (death-inducing signalling complex; Fig. 11.4). The complexes will activate the caspases within them, leading to the activation of the caspase cascade and the resulting apoptosis.

Other adaptor proteins, such as CRADD (caspase and RIP adaptor with death domain) contain a different protein–protein interaction domain for caspase recruitment. CRADD contains a caspase-activation and -recruitment domain (CARD), which allows CRADD to interact with pro-caspase-2.

The process of DISC formation can be modulated by the presence of other proteins which contain a DED region. Such proteins are the two isoforms of FLIP (Flice-like inhibitory proteins), $FLIP_L$ (long isoform, which has two DED domains and a caspase homology domain that is inactive) and $FLIP_S$ (short isoform, which has only two DED domains). Both contain DED regions and can compete with caspases for binding to the DISC, but as they have no caspase activity no active DISC is formed, and therefore apoptosis is not induced.

Modulation of the extrinsic pathway has also been found to occur by the presence of what has been termed decoy receptors. These are receptors that resemble the death receptors, and are able to bind to the same ligands as the death receptors, but they are unable to signal to the rest of the extrinsic pathway. Examples of decoy receptors include DcR2, which has no death domain and therefore cannot recruit the adaptor protein, and DcR1, which is a very truncated version of a death receptor and like DcR2 also has no capacity to signal into the cell. However, the presence of such receptors will compete for ligands and so stop, or reduce, the signalling that takes place through functioning receptors. In this case, DcR1 and DcR2 both bind to Trail and therefore will modulate Trail-induced apoptosis.

Other signalling molecules involved in apoptosis

There are many other proteins involved in the mechanisms of apoptosis, and it is beyond the detail of the discussion in this chapter to include them all. However, there are several non-protein signals that are also involved, notably ROS and RNS such as nitric oxide. It is still unclear what their exact role is, but their production has been linked to the pathways that modulate apoptosis (see **Chapter 8** for discussion on the signalling roles of these molecules).

11.6 SUMMARY

- Apoptosis is a process of cell death controlled by defined cellular pathways.

- Apoptosis is an important process during development as well as being involved in body maintenance during adult life.

- There are two signalling pathways leading to apoptosis: the intrinsic and extrinsic pathways.

- Caspases are important factors which mediate many morphological changes seen during apoptosis.

- Caspases are cysteine proteases; that is, they contain an active cysteine in their active site.

- Caspases cleave other proteins at the C-terminal side of an aspartic acid residue.

- Caspases are themselves activated by protein-cleavage events, existing in the inactive state as pro-caspases.

- Active caspases have a tetrameric structure of two large subunits and two small subunits.

- The intrinsic pathway of apoptosis involves the mitochondria, and the release of proteins from that organelle, including cytochrome c, apoptosis-inducing factor (AIF), and endonuclease G.

- Cytochrome c forms a complex with apoptotic protease-activating factor 1 (Apaf-1) and with caspase-9 to form a structure termed the apoptosome.

- Caspase-9 in apoptosomes becomes active and will cleave and activate further caspases, leading to apoptosis.

- The extrinsic pathway involves death receptors (DRs) at the plasma membrane.

- Death receptors contain multiple cysteine-rich domains and a death domain.

- Through the presence of protein–protein interaction domains, both death domains and dead effector domains (DEDs), the death receptor, adaptor proteins, and initiator caspases form a complex called the death-inducing signalling complex (DISC).

- Formation of DISC leads to caspase cleavage and results in apoptosis.

- DISC formation can be modulated by the presence of Flice-like inhibitory proteins (FLIPs) or decoy receptors.

11.7 FURTHER READING

Antonsson, B. (2004) Mitochondria and the Bcl-2 family proteins in apoptosis signaling pathways. *Molecular and Cellular Biochemistry* **256**, 141–155.

Conradt, B. (2001) Cell engulfment, no sooner ced than done. *Developmental Cell* **1**, 445–447. A paper on the pathways of phagocytosis, which removes cells that have undergone apoptosis.

Cotter, T.G. (ed.) (2003) Programmed cell death. *Essays in Biochemistry*, vol. 39 Portland Press, London. An excellent collection of articles on the topic, written by researchers in the field.

Diedrich, M. (ed.) (2003) Apoptosis, from signalling pathways to therapeutic tools. *Annals of the New York Academy of Sciences* **1010**. A substantial collection of papers on the topic arising from a conference in Luxembourg. It encompasses a vast amount of recent research in 799 pages.

Wang, S.L. and El-Deiry, W.S. (2003) TRAIL and apoptosis induction by TNF-family death receptors. *Oncogene* **22**, 8628–8633.

12

Importance, complexity and the future

12.1 Introduction

12.2 Specificity, subtlety and crosstalk

12.3 Longer-term effects

12.4 The future

12.5 FURTHER READING

On reading this book, or even just dipping into it, it is hoped that the reader will have gained an insight into the mechanisms used by cells to control their activities. However, it is also hoped that many questions may have been raised. In this final chapter, some such questions are discussed briefly, leading to a look to the future of cell signalling research.

12.1 Introduction

This book started with the remark that "Cell signalling has arguably become one of the most important aspects of modern biochemistry and cell biology" and it is hoped that the chapters that have preceded this one have emphasized that point. Organisms, from their very moment of existence, either from a relatively simple cell-division event, or from the instant that the DNA from a sperm enters the egg, rely on functional cell signalling for their very survival. Development, of their limbs, organs, and form, relies on cells signalling to each other and for cells in the neighbourhood to respond appropriately. When fully formed, the obtainment of a food source, the avoidance of predators, and adaptation to environmental stresses, whether that is heat, cold, too much salt, or lack of light, all rely on cells to sense and respond, using cell signalling mechanisms. Therefore, there is no part of modern biology, from environmental sciences to medicine, that is not impacted by the field of cell signalling. Dysfunctional cell signalling, after all, is responsible for a plethora of diseases and conditions.

Cells respond not only to the influences external to the organism, such as pheromones, but also to the internal signals of the organism itself, and the inner workings of both plants and animals are awash with hormones and extracellular signals, such as cytokines. All these have to be received at the right time and by the right cells, and the perception has to be followed by appropriate action, the activation or inhibition of intracellular signal transduction, all of which comes under the umbrella of cell signalling.

12.2 Specificity, subtlety, and crosstalk

It is the uniqueness of the signals used, and the specific perception of the signals that gives the cell the range of responses it needs to survive any given moment or event. Molecules which are structurally very similar can have very different effects, such as seen with cAMP and cGMP, and it is the specific nature of the interactions of these signals that is crucial. Often a cell signalling pathway will start with the specific interaction between a receptor and its ligand, and only then will the cell recognize the presence of the ligand and respond. This perception has to be specific, and the complement of receptors possessed by a cell determines whether the cell is capable of responding to a external signal or not.

It must be remembered, however, that at any given moment in time, a cell is not simply sitting there responding to a single signal. It is not like a telephone operator talking to one caller at a time. It is more like the train operator on the platform when the train doesn't arrive, being surrounded by dozens of people, some asking the same question (Where is my train?), others asking a different but related question (When is the next one due?), and some just shouting nonsense, all in an atmosphere dominated by other unrelated noises. The cell has to respond as efficiently and quickly as possible, and ignore the noise that does not apply to it. By having specific receptors, specific protein interactions, and specific effectors all ready to act, the cell can filter and take action.

However, many conundrums exist. A glance through the chapters above will reveal that the same components exist in many different pathways. The demands on a given component to signal might come from many sources at the same time, with different amplitudes, and perhaps even opposite messages. For example, if both hormone X and hormone Y lead to an increase in the concentration of signal Z inside the cell, when signal Z increases, how does the cell know that it is due to the perception of X or Y? It clearly does, but how? The answer must come from the complexity of the events that are taking place. It is clear from the literature that few signals prompt a single response from cells, and often several events are initiated when a component is activated. Phospholipase C for example generates $InsP_3$ and DAG, both potent signals. Therefore, signalling is not a series of parallel signal transduction pathways as depicted in many books, but rather it is a complex web of events, and it has been argued by some in the literature, although this cannot be the case, that every component can interact with every other component. However, at any moment in time, there are many signals being initiated, reduced, or created, many of which will impact on others. The term crosstalk is often used to indicate that one pathway is interfering with, or at least having an influence on, another. Perhaps a good analogy would be if we consider our perception of music, a signalling event in itself. If one plays a D major cord it sounds harmonic and rather happy. We perhaps respond in a certain way to music in the major keys, they sound joyful and make us pleased. However, change a single note, an F sharp down to an F natural, and all of a sudden we have a minor key, and it sounds very different—it sounds depressing and sad. We respond differently. A little change, a mere semitone among three notes and we have a different response. Perhaps the cells are a same. It is the suite of signals that are responded too, but the singularity, and subtle changes within, can make large differences. Perhaps future computer modelling will reveal the answer.

Some proteins for which the function and role are thought to be well characterized are also now appearing among the suite of signalling molecules available to the cell. For example, cytochrome c has a role as an electron carrier in mitochondria, and it was thought that that is all it did. However, it is now known that under certain circumstances it is released from the mitochondria and acts as a signalling molecule in the process of apoptosis (discussed in Chapter 11, section 11.4). Other proteins with defined functions are also appearing in the list of signalling components. For example, a protein which is instrumental in the metabolic pathway of glycolysis, glyceraldehyde-3-phosphate dehydrogenase, has been shown to be modified by hydrogen peroxide and undergoes a conformational change that modulates the activity of a phospholipase. This multifunctional role of proteins is a theme which will be seen much more commonly as research probes the presence and roles proteins in signalling pathways, and will no doubt contribute to the crosstalk in signalling in cell.

Furthermore, as discussed in Chapter 1 (section 1.5), the cell architecture as we now view it appears to be contrary to the findings that we sometimes see. For example, if the cytoplasm of the cell is a homogeneous fluid entity as described by many, why does calcium only appear in hotspots within it when cells are activated? Why is it that reactive oxygen and nitrogen species also appear in hot-spots? Why don't they diffuse quickly? Why do apparently soluble proteins not interact with each other? What is the cellular architecture that we don't understand? Some receptors which in theory are perceiving extracellular signals appear to exist in the endoplasmic reticulum, but how does the ligand ever get perceived? What is it that we simply don't understand?

Clearly, these and many other questions raised show that we are a long way from a full understanding of cell signalling in any single organism, but future research will no doubt reveal the answers.

12.3 Longer-term effects

Cell signalling often refers in many papers to the here and now, and most of the text above falls into this trap. But we must not neglect the longer term. A hint to what might be happening in the long term may come from the studies using microarrays. In such studies, it is often found that in response to the perception of a extracellular ligand there is a profound change in gene expression. In these papers, the authors usually try to classify the types of genes which are expressed, according to the type of protein they are thought to encode. It is no surprise to find that a significant proportion of such genes encode products that are involved in signalling, such as receptors, kinases, G proteins etc. Therefore, the perception of the ligand has not just led to the immediate response, but has led to the alteration of the complement of signalling components within the cell, and therefore has altered the way the cell might respond a second time. Such events no doubt enable a cell to adapt to their new environment, or allow specialist cells to adapt for the good of the whole organism. Sometimes the genes which have their expression raised will even encode the proteins that produced the original signal, perhaps a way of the cells amplifying and emphasizing the original message, making it last longer with a more profound effect, until signalled to do otherwise by the perception of another different ligand.

But it does not always stop there. A careful look at the type of signalling proteins encoded by the genes for which expression is raised, or lowered, in microarray studies often reveals that a significant number are transcription factors, which themselves will alter the expression of yet more genes. It is an assumption without the studies to back it up, but it is almost certain that these transcription factors will alter the expression of more components of signal transduction pathways. Therefore, a ligand will put in place a cascade of events, leading to gene expression, leading to production of new proteins, which lead to altered gene expression, which will lead to altered signalling, some of which might involve even more transcription factors, and so the cascade might continue. Therefore, the long-term ramifications of signalling need to be considered further in many cases. Only further external and internal influences will modify such on-going events. Clearly therefore, a cell's functioning and activity are controlled by signalling which has taken place in the recent past, as well as the signalling taking place in the present, all of which will be influenced by the signalling of the future. It is complex, dynamic, and fluid system, which we are only just starting to understand.

Mechanisms used in cell signalling also exist that are not discussed in detail above. Mechanical stresses and pressures have been shown to have profound effects on cells, for example in auditory perception. Pressure pulses appear to be sensed by plants too, and others have used mechanical shock to study gene expression in cells from both plants and animals.

12.4 The future

The above discussion highlights the complexity of signalling and perhaps paints a negative view of the future, but recent technological advances means that studies in cell signalling, along with all aspects of molecular biology, are at an exciting point. Microarrays, and similar technologies, now allow the study of expression of all the genes in a genome at the same time, while genome-sequencing studies are expanding the availability of such arrays, across a wide range of species, from prokaryotes to humans. These sequencing studies also allow for the comparison of protein function across species, so protein domains with new functions will no doubt come to light. Proteomics, with mass spectroscopy, is also opening the door to holistic studies of cells and promises to yield a wealth of information.

RNA interference, and the on-going study of diseases and mutants, allows signalling to be studied where one or more components have been removed, or at least had their activity ablated. Therefore, the ramifications of the lack of a function to the cell as a whole, and to the web of signalling involved, can be studied. At the moment, often such studies are hard to interpret, as the lack of a protein might do very little, suggesting that cells have degeneracy in their signalling, and removal of one signal allows another simply to take over. Or, removal of one component has a profound effect, and authors of papers claim that they have found the key component, only for other researchers to remove another protein and claim that their protein is in fact the key one. How can they both be right? Perhaps they are, and it is the interplay and subtlety of signalling that is important.

There is no doubt that individual studies are making a great advances into our understanding of cell signalling, but a full understanding is some way off. Assimilation of data, with authors taking a holistic view of the workings of the cell, will no doubt reveal how signalling truly works and allow manipulation of signalling events which will lead to the creation of better crops and the curing of many diseases.

12.5 FURTHER READING

Chapman, K.E. and Higgins, S.J. (eds) (2001) Regulation of gene expression. *Essays in Biochemistry*, vol. 37. Portland Press, London. A collection of articles on control of gene expression and including a chapter on signal transduction pathways.

Latchman, D.S. (2003) *Eukaryotic Transcription Factors*, 4th edn. Academic Press. The topic of transcription factors has not been covered in detail above, so excellent texts such as this are recommended.

Pawson, T. (2004) Specificity in signal transduction: from phosphotyrosine-SH2 domain interaction to complex cellular systems. *Cell* **116**, 191–203.

Wan, X., Steudle, E., and Hartung, W. (2004) Gating of water channels (aquaporins) in cortical cells of young corn roots by mechanical stimuli (pressure pulses): effects of ABA and of $HgCl_2$. *Journal of Experimental Botany* **55**, 411–422.

■ INDEX